THERMODYNAMICS OF POLYMER SOLUTIONS

MMI PRESS POLYMER MONOGRAPH SERIES
Edited by Hans-Georg Elias

This series will present international accounts of research and developments in the specialized areas of macromolecules.

Volume 1 Thermodynamics of Polymer Solutions
By Michio Kurata
Translated from the Japanese by Hiroshi Fujita

Additional volumes in preparation:

Volume 2 ^{13}C-NMR Spectroscopy: A Working Manual with Exercises
By E. Breitmeier and G. Bauer
Translated from the German by Bruce K. Cassels

Volume 3 Polymer Analytics
By M. Hoffmann, H. Kromer, and R. Kuhn
Translated from the German by Horst G. Stahlberg

Volume 4 Practical Macromolecular Organic Chemistry
By D. Braun, H. Cherdron, and W. Kern
Translated from the German by Kenneth J. Ivin

ISSN 0275-7265

The publisher will accept continuation orders for this series, which may be cancelled at any time and which provide for automatic billing and shipping of each title in the series upon publication. Please write for details.

THERMODYNAMICS OF POLYMER SOLUTIONS

Michio Kurata
Institute for Chemical Research, Kyoto University
Uji, Japan

Translated by

Hiroshi Fujita
Department of Macromolecular Science, Osaka University
Toyonaka, Japan

harwood academic publishers
chur • london • new york

Published under license by:

Harwood Academic Publishers GmbH
Poststrasse 22
CH-7000 Chur, Switzerland

Editorial Office for the United Kingdom:
61 Grays Inn Road
London WC1X 8TL

Editorial Office for the United States of America:
Post Office Box 786
Cooper Station
New York, New York 10276

Library of Congress Cataloging in Publication Data

Kurata, Michio, 1925-
Thermodynamics of polymer solutions.

(MMI Press polymer monograph series, ISSN 0275-7265 ; v. 1)
Translation of: Kōbunshi yōeki ron.
Includes bibliographical references and index.
1. Polymers and polymerization—Thermal properties.
2. Solution (Chemistry) I. Title. II. Series.
QD381.9.T54K8713 1982 547.7. 82-9381
ISBN 3-7186-0023-4 AACR2

ISBN 3-7186 0023-4 ISSN 0275-7265

Printed in the United States of America

Contents

Preface to the English Edition

This book is devoted to a discussion of the thermodynamic theories of osmotic equilibrium, phase separation, concentration fluctuations, and sedimentation equilibrium in polymer solutions. An especially detailed exposition is made of two relatively new developments, the phase equilibrium in polydisperse polymer solutions and the light scattering and sedimentation equilibrium in mixed-solvent systems. The correct determination of polymer molecular weights requires caution both in the choice of experimental conditions, such as temperature and solvent, and in the analysis of observed results. I hope that this book will prove to be useful to those who are interested in molecular weight determination.

When Professor Fujita told me of his interest in preparing the English edition of my Japanese book, the original of this book, I was pleased to give him permission for undertaking the work, since I had appreciated his capability through friendly relations over the past three decades. I would like to acknowledge with many thanks his effort in the translation as well as his advice in the revision of the original Japanese test for the English edition. I am also indebted to Dr. R. Kikuchi and Dr. K. Šolc, who kindly permitted me to quote their unpublished papers concerning the etymology of the word "spinodal" and the phase equilibrium in quasiternary systems, respectively, and to Professor E. F. Casassa, who made many valuable suggestions and comments which were so useful for improving the content of my book.

Michio Kurata

Translator's Note

Few current textbooks on polymer solution thermodynamics are comparable in quality and coverage to the classic volume written by Tompa more than two decades ago. The unavailability of a comprehensive treatment has been a great inconvenience for anyone seeking to study more recent developments in the theory of the equilibrium properties of polymer solutions. Thus when Michio Kurata's monograph "Kobunshi Yoeki Ron" appeared in 1975 (Asakura Publishing Company, Tokyo), I felt that access to a work of such excellence ought not to be limited to the Japanese polymer community. This English edition is the fruit of that conviction. It is a translation of the first three chapters of Professor Kurata's book. The original contains a fourth chapter, on the statistics of isolated macromolecules, which is omitted here since comparable treatments are readily available elsewhere. Thus, the content of the present volume is fully consistent with the title, "Thermodynamics of Polymer Solutions." Errors which had crept into the Japanese text have been corrected, and some sections have been revised to achieve a conceptually clearer presentation.

I owe a very special debt to Michio Kurata. Over the years he has been a constant friend and an exemplary scientific colleague. I am grateful beyond measure for his encouragement and advice throughout the preparation of this text. More directly, and most generously, he aided by rewriting parts of the Japanese text for my use in preparing the English version.

Professor Edward F. Casassa of Carnegie-Mellon University gave unstintingly of his time and effort in polishing the English of my first draft. His critical reading of the entire manuscript also led to many suggestions for improving the presentation. His contributions are reflected on nearly every page of the final text. I acknowledge his patience and friendship with gratitude.

Hiroshi Fujita

Thermodynamics of Solutions

CHAPTER 1

Thermodynamics of Solutions

1.1. Partial Molar Quantities and Composition Variables

1.1.1. Molar and Partial Molar Quantities

Suppose a solution is made of a sample of polystyrene and benzene. One might consider it a two-component system. This is not correct. Any synthetic polymer is a mixture of molecules of different molecular weights. Hence it is itself a multicomponent system and its solutions are multicomponent systems as well. An extensively purified protein preparation might be considered a single thermodynamic component, but we seldom study two-component solutions consisting of a purified protein and pure water. Usually, the solvent is itself an aqueous solution of a simple salt and/or a buffer, and thus the solution is a system of at least three components. For these reasons it is appropriate to begin our study of the thermodynamics of polymer solutions with a general discussion of multicomponent systems.

We distinguish component substances making up a multicomponent solution by indices i ($i = 0, 1, \ldots$) and denote the amount of component i by n_i. The total amount n of an $(r + 1)$-component solution is then represented by

$$n = \sum_{i=0}^{r} n_i \tag{1.1.1}$$

We assign component 0 to the **principal solvent**, which is the liquid component in a single-solvent solution and which is the liquid component present in the largest relative amount in a mixed-solvent solution.

The SI† basic unit for the amount of a substance is the **mole**, designated here by the symbol mol. One mole of any substance contains as many elementary entities as there are atoms in 0.012 kilogram of carbon-12. When the mole is used, one must specify the elementary entities, which may be atoms, molecules,

†International System of Units

ions, electrons, other particles, or specified groups of such particles. The number of elementary entities per mole is called the **Avogadro constant** and denoted here by N_A. Its latest value is

$$N_A = 6.0220 \times 10^{23} \text{ mol}^{-1} \tag{1.1.2}$$

According to the International System of Units (SI units), the amount of substance, like length, mass, time, electric current, and thermodynamic temperature, is treated as a basic quantity having an independent dimension. Thus N_A is not a dimensionless or pure number but a constant of dimensions mol^{-1}. For this reason we do not call it Avogadro's number.

If we let the amount of each component in a solution be increased by a factor λ at fixed temperature T and pressure p, the volume V of the solution is increased by λ. This fact is expressed mathematically by

$$V(T, p, \lambda n_0, \lambda n_1, \ldots, \lambda n_r) = \lambda V(T, p, n_0, n_1, \ldots, n_r) \tag{1.1.3}$$

i.e., V at constant T and p is a linear homogeneous function of $r + 1$ variables n_0, $n_1, \ldots, n_r$. Then, according to Euler's theorem (Appendix A.2), we find that

$$V(T, p, n_0, n_1, \ldots, n_r) = \sum_{i=0}^{r} n_i V_i \tag{1.1.4}$$

where V_i is defined by

$$V_i \equiv (\partial V/\partial n_i)_{T,p,n_j} \tag{1.1.5}$$

and is called the **partial molar volume** of component i. The subscript n_j here implies that the amounts n_j of all components other than component i are held fixed. It is convenient to introduce the **molar volume** V_m, defined by

$$V_m \equiv V/n \tag{1.1.6}$$

where, as mentioned above, n is the total amount of a given solution. In terms of the **mole fractions** x_i $(i = 0, 1, \ldots, r)$

$$x_i \equiv n_i/n = n_i/\Sigma_j n_j, \quad \sum_{i=0}^{r} x_i = 1 \tag{1.1.7}$$

Eq. (1.1.6) combined with Eq. (1.1.4) may be written

$$V_m = \Sigma_i x_i V_i \tag{1.1.8}$$

In what follows, as in this equation, the range of summation is omitted unless it is not self-evident. Of the $r + 1$ mole fractions only r, say $x_1, x_2, \ldots, x_r$, can be varied independently.

Quantities (or properties) treated in thermodynamics are classified into two groups: **extensive** and **intensive**. Extensive quantities (like volume) are proportional to the amount of matter constituting a system of given composition. Intensive quantities (like temperature and pressure) are independent of the amount of a system. If Y stands for an extensive quantity, the corresponding molar Y_m and partial molar Y_i quantities are intensive [see Eq. (A.21); equations in the Appendix are quoted with the letter A]. For any extensive quantity Y, equations corresponding to Eqs. (1.1.4), (1.1.5), and (1.1.8) are

$$Y = \Sigma_i n_i Y_i \tag{1.1.9}$$

$$Y_i \equiv (\partial Y/\partial n_i)_{T,p,n_j} \tag{1.1.10}$$

$$Y_m = \Sigma_i x_i Y_i \tag{1.1.11}$$

It can be shown that the partial molar quantities $Y_i (i = 0, 1, \ldots, r)$ are not independent. Since Y is a function of T, p, and $n_i (i = 0, 1, \ldots, r)$, its total derivative dY is expressed by

$$dY = \left(\frac{\partial Y}{\partial T}\right)_{p,n} dT + \left(\frac{\partial Y}{\partial p}\right)_{T,n} dp + \sum_{i=0}^{r} Y_i dn_i \tag{1.1.12}$$

where the subscript n indicates that the moles of all $r + 1$ components are fixed. On the other hand, total differentiation of Eq. (1.1.9) gives

$$dY = \Sigma_i (n_i dY_i + Y_i dn_i) \tag{1.1.13}$$

Equating this to Eq. (1.1.12) gives

$$\left(\frac{\partial Y}{\partial T}\right)_{p,n} dT + \left(\frac{\partial Y}{\partial p}\right)_{T,n} dp - \sum_{i=0}^{r} n_i dY_i = 0 \tag{1.1.14}$$

At constant temperature and pressure, this reduces to

$$\Sigma_i n_i (dY_i)_{T,p} = 0 \tag{1.1.15}$$

which, upon division by n, gives

$$\sum_{i=0}^{r} x_i (dY_i)_{T,p} = 0 \tag{1.1.16}$$

As has been noted above, Y_i is an intensive quantity so that at constant T and p, it depends only on the composition of the solution, thus on $x_1, x_2, \ldots, x_r$. Hence $(dY_i)_{T,p}$ can be written

$$(dY_i)_{T,p} = \sum_{j=1}^{r} (\partial Y_i/\partial x_j)_{T,p,x_k} dx_j, \quad (k \neq 0) \tag{1.1.17}$$

Putting this into Eq. (1.1.16) and exchanging the order of the summations give

$$\sum_{j=1}^{r}\left[\sum_{i=0}^{r} x_i(\partial Y_i/\partial x_j)_{T,p,x_k}\right] dx_j = 0, \quad (k \neq 0) \tag{1.1.18}$$

Since the dx_j ($j = 1, 2, \ldots, r$) are independent of one another, it follows from Eq. (1.1.18) that

$$\sum_{i=0}^{r} x_i(\partial Y_i/\partial x_j)_{T,p,x_k} = 0, \quad (j = 1, 2, \ldots, r;\ k \neq 0) \tag{1.1.19}$$

Equations (1.1.14) through (1.1.16), along with Eq. (1.1.19), play a very important role in the thermodynamics of multicomponent systems. Each shows a relation among the Y_i of the components making up a solution. When Y is the Gibbs free energy, these equations are called **Gibbs-Duhem relations**.

Equation (1.1.11) allows the computation of Y_m when the Y_i ($i = 0, 1, \ldots, r$) are given. The converse equation, which allows Y_i to be expressed in terms of Y_m, can be derived in the following way. Choose $x_1, x_2, \ldots, x_r$ as the independent composition variables and rewrite Eq. (1.1.11) as

$$Y_m = (1 - \sum_{i=1}^{r} x_i) Y_0 + \sum_{i=1}^{r} x_i Y_i \tag{1.1.20}$$

Differentiation with respect to x_j and introduction of Eq. (1.1.19) yield

$$(\partial Y_m/\partial x_j)_{T,p,x_k} = Y_j - Y_0, \quad (j = 1, 2, \ldots, r;\ k \neq 0) \tag{1.1.21}$$

The $r + 1$ relations in Eqs. (1.1.20) and (1.1.21) may be regarded as a set of simultaneous equations for $r + 1$ unknowns $Y_0, Y_1, \ldots, Y_r$ and solution of this set of equations yields the desired formulas:

$$Y_0 = Y_m - \sum_{i=1}^{r} x_i(\partial Y_m/\partial x_i)_{T,p,x_k} \tag{1.1.22a}$$

$$Y_j = Y_m + (\partial Y_m/\partial x_j)_{T,p,x_k} - \sum_{i=1}^{r} x_i(\partial Y_m/\partial x_i)_{T,p,x_k}, \quad (j = 1, 2, \ldots, r;\ k \neq 0) \tag{1.1.22b}$$

For a two-component system these reduce to

$$Y_0 = Y_m - x_1(\partial Y_m/\partial x_1)_{T,p} \tag{1.1.23a}$$

$$Y_1 = Y_m + (1 - x_1)(\partial Y_m/\partial x_1)_{T,p} \tag{1.1.23b}$$

Suppose that the curve A_0PA_1 in Fig. 1.1 represents a given Y_m versus x_1 relation. The compositions $x_1 = 0$ and $x_1 = 1$ correspond, respectively, to the pure components 0 and 1. Thus points A_0 and A_1 represent $Y_0{}^o$ and $Y_1{}^o$, the molar quantities of the pure liquid components. We draw a straight line tangent to the curve at point P and denote its points of intersection with the vertical lines at $x_1 = 0$ and $x_1 = 1$ by B_0 and B_1, respectively. The slope of the tangent equals $\partial Y_m/\partial x_1$. Hence the segments C_0B_0 and C_1B_1 in the figure represent the second terms on the right-hand side of Eqs. (1.1.23a) and (1.1.23b), and points B_0 and B_1 give Y_0 and Y_1, respectively. The physical meaning of point D_1 will be explained in § 1.1.4.

A triangular coordinate system is used to represent graphically the composition of a three-component solution. As shown in Fig. 1.2, through a point P inside an equilateral triangle with sides of unit length, we draw three straight lines, each parallel to one of the sides, and let their lengths between P and the sides 12, 20, and 01 correspond to the mole fractions x_0, x_1, and x_2, respectively. This is permissible because the sum of these lengths is unity. By defining the triangular coordinates x_0, x_1, and x_2 in this way we can represent any composition of a three-component solution by a point inside the triangle. In particular, the apices of the triangle, 0, 1, and 2, represent the pure components 0, 1, and 2, respectively, and each side of the triangle corresponds to one of the three two-component mixtures.

The relation between Y_m and the composition of a three-component solution is represented graphically by a surface above the composition triangle, as

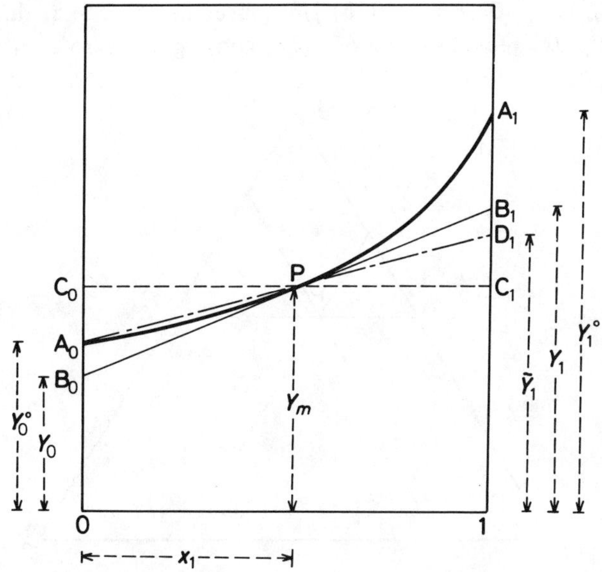

Fig. 1.1. Illustration of molar Y (Y_m), partial molar Y (Y_0 and Y_1) and apparent molar Y ($\tilde{Y}_1$) in a two-component solution.

illustrated in Fig. 1.3, if Y_m is plotted perpendicular to the plane of the composition triangle. A plane is tangent to this surface at point P and its intersections with the vertical lines erected at the apices 0, 1, and 2 are denoted by B_0, B_1, and B_2, respectively. Then the heights of B_0, B_1, and B_2 above the composition triangle equal the values of Y_0, Y_1, and Y_2 for the composition corresponding to point P.

Now, returning to the problem concerning solution volume, we designate by $\Delta_m V$ the change in volume which occurs when $r + 1$ pure components, each having a molar volume V_i^o, are mixed at fixed temperature and pressure. The quantity $\Delta_m V$ is called the **volume of mixing.** The volume V of the solution is expressed by

$$v = \Sigma_i n_i V_i \tag{1.1.24}$$

When $\Delta_m V = 0$, then $V_i = V_i^o$. For some solutions, $\Delta_m V$ may become negative. A two-component solution with negative $\Delta_m V$ gives a V_m versus x_1 curve which, as illustrated in Fig. 1.4, is convex downward. In such a case, if $\Delta_m V$ is sufficiently large, V_1 may become negative in some range of composition (Fig. 1.4). According to Eq. (1.1.19), we have

$$(\partial V_1/\partial x_1)_{T,p} = -[(1 - x_1)/x_1](\partial V_0/\partial x_1)_{T,p} \tag{1.1.25}$$

Hence if $\partial V_0/\partial x_1 \neq 0$ as $x_1 \to 0$, $\partial V_1/\partial x_1$ becomes infinite at this limit. Conversely, if $\partial V_1/\partial x_1$ remains finite, $\partial V_0/\partial x_1$ converges to zero as $x_1 \to 0$. Figures

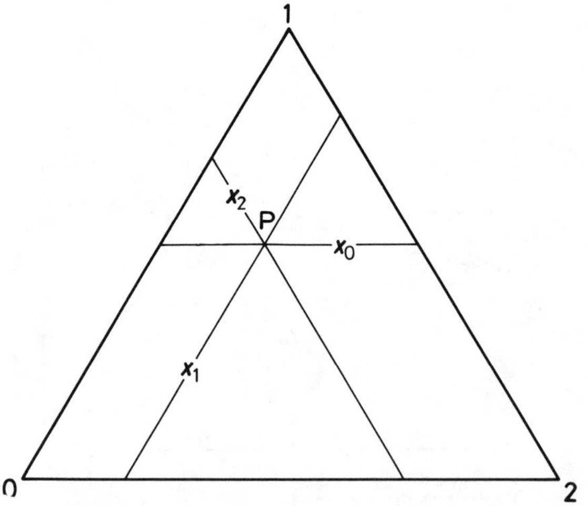

Fig. 1.2. The composition triangle for a three-component system. x_0, x_1, and x_2 represent the mole fractions of components 0, 1, and 2 corresponding to point P.

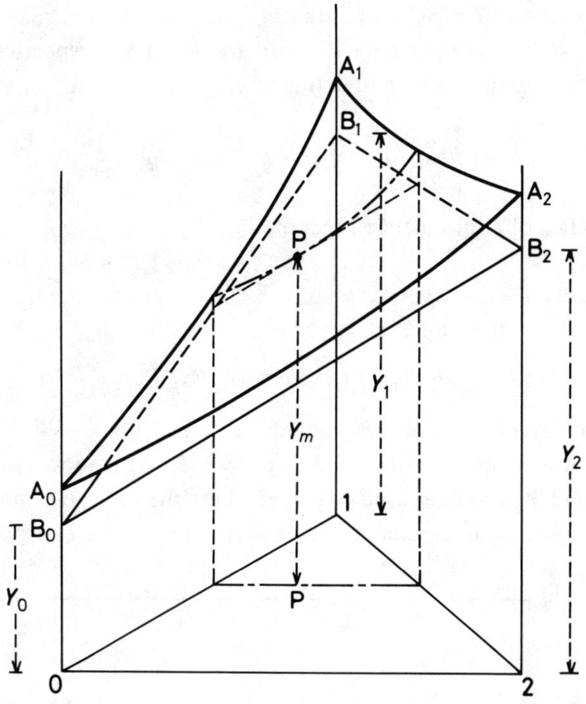

Fig. 1.3. Illustration of molar Y (Y_m) and partial molar Y (Y_0, Y_1, and Y_2) in a three-component solution.

1.5(a) and (b) illustrate these limiting properties of V_0 and V_1. The property observed in volume measurements is always of type (a). Thus for **dilute solutions** in which $x_1 \ll 1$ we may assume that

$$V_0 \simeq V_0{}^o \qquad (1.1.26)$$

1.1.2. Composition Variables, Molar Mass, and Molecular Weight

The composition of a solution can be expressed in terms of various sets of variables other than the mole fractions. Typical composition variables widely used in solution theory are summarized below, together with the relations between different sets of variables.

a. Mole Fraction x_i. The defining equation for x_i has been given in Eq. (1.1.7). The composition of an $(r + 1)$-component solution is completely determined by r mole fractions $x_1, x_2, \ldots, x_r$, which will usually be designated by a single letter x in the subsequent discussion.

b. Mass Fraction or Weight Fraction w_i. We let the total mass of a solution and the mass of component i be denoted by q and q_i, respectively. The mass fraction w_i of component i is then defined by

$$w_i \equiv q_i/q = q_i/\sum_{j=0}^{r} q_j, \quad \sum_{i=0}^{r} w_i = 1 \tag{1.1.27}$$

If the **molar mass** of component i is denoted by M_i then

$$q_i = n_i M_i \tag{1.1.28}$$

Hence we obtain the relations

$$w_i = x_i M_i/\Sigma_j x_j M_j, \quad x_i = (w_i/M_i)/\Sigma_j (w_j/M_j) \tag{1.1.29}$$

The weight fraction w_i' of component i is given by $w_i' = W_i/W$, with W and W_i being the weight of the solution and the weight of component i, respectively. When W and W_i are measured at places where the acceleration due to gravity is different, w_i' does not equal w_i. However, this case is exceptional, and no

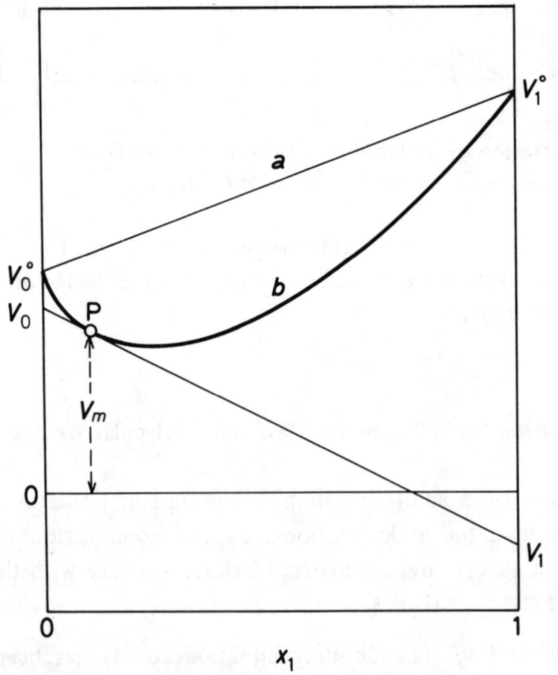

Fig. 1.4. Molar volume V_m and partial molar volumes V_0 and V_1 for a two-component solution. The graph illustrates the case in which V_1 becomes negative in some range of composition x_1.

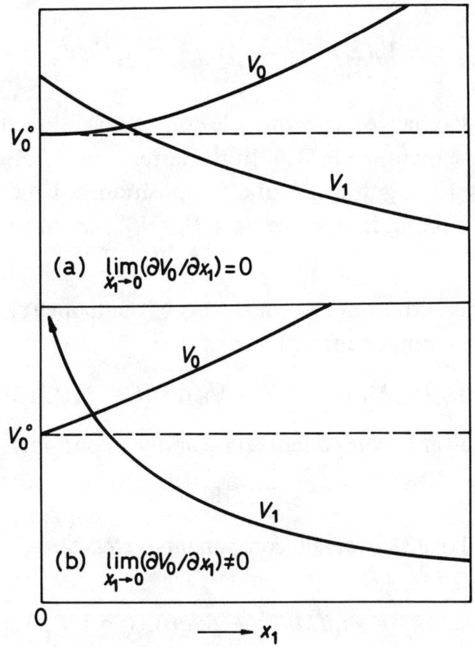

Fig. 1.5. Partial molar volumes V_0 and V_1 in dilute binary solutions.

further distinction between mass fraction and weight fraction will be made in the subsequent discussion.

c. **Volume Fraction ϕ_i.** The volume fraction ϕ_i of component i is here defined by

$$\phi_i \equiv x_i V_i^o / \sum_{j=0}^{r} x_j V_j^o = w_i v_i^o / \sum_{j=0}^{r} w_j v_j^o, \quad \sum_{i=0}^{r} \phi_i = 1 \tag{1.1.30}$$

where v_i^o denotes the **specific volume** of pure liquid component i, i.e.,

$$v_i^o \equiv V_i^o / M_i \tag{1.1.31}$$

Here, "specific" means "per unit mass." The composition of a solution, expressed in terms of mole fractions or weight fractions, is independent of pressure and temperature. However, since V_i^o and v_i^o depend on these factors and, in addition, the dependence varies with the kind of substance, it is meaningless to express the composition in terms of volume fractions without specifying the chosen V_i^o or v_i^o.

Some authors define volume fractions ϕ_i' in terms of the **partial** molar volumes V_i:

$$\phi_i' \equiv n_i V_i / \sum_{j=0}^{r} n_j V_j = n_i V_i / V = x_i V_i / V_m \tag{1.1.32}$$

However, in general, V_i changes with the solution composition and in some cases becomes negative, as shown in Fig. 1.4. In the latter case, ϕ_i' also becomes negative and loses the physical significance of a composition variable. For a system in which the volume of mixing is zero we have $V_i = V_i^o$, so that ϕ_i' coincides with ϕ_i.

d. Molality m_i. This is the amount (in moles) of component i ($i = 1, 2, \ldots, r$) relative to unit mass of component 0. Thus

$$m_i \equiv n_i/(n_0 M_0) = x_i/(x_0 M_0), \quad (i = 1, 2, \ldots, r) \tag{1.1.33}$$

The molality of the solvent (component 0) is simply a constant:

$$m_0 \equiv 1/M_0 \tag{1.1.34}$$

Summation of Eq. (1.1.33) over all components gives $\sum_{j=0}^{r} m_j = 1/x_0 M_0$. Hence

$$x_i = m_i / \sum_{j=0}^{r} m_j = m_i / (M_0^{-1} + \sum_{j=1}^{r} m_j), \quad (i = 1, 2, \ldots, r) \tag{1.1.35}$$

where Eq. (1.1.34) has been used.

e. Volume Molality or Molar Concentration c_i. This is defined as the amount of component i contained in unit volume of solution. Thus

$$c_i \equiv n_i/V = x_i/V_m = m_i/\nu_M, \quad (i = 0, 1, \ldots, r) \tag{1.1.36}$$

where

$$\nu_M \equiv V/(n_0 M_0) \tag{1.1.37}$$

Physically, ν_M is the volume of solution containing unit mass of the principal solvent (component 0). Among $r + 1$ variables c_i we have the relation

$$\sum_{i=0}^{r} c_i V_i = 1 \tag{1.1.38}$$

Therefore, if we choose r independent variables $c_1, c_2, \ldots, c_r$, the concentration c_0 becomes a dependent variable.

In statistical mechanics, the number of molecules η_i in unit volume of solution is frequently used as a composition variable. This is related to c_i by

$$\eta_i \equiv N_A c_i = N_i/V, \quad \left[\sum_{i=0}^{r} \eta_i V_i = N_A \text{ by Eq. (1.1.38)} \right] \tag{1.1.39}$$

where N_i is the total number of molecules of component i in the solution with volume V. The quantity η_i is called the **molecular concentration** or **molecular number density**.

f. Mass Concentration C_i. The mass of component i contained in unit volume of solution is C_i, i.e.,

$$C_i \equiv n_i M_i/V = x_i M_i/V_m = m_i M_i/v_M = c_i M_i, \quad (i = 0, 1, \ldots, r) \quad (1.1.40)$$

We have the relations

$$\sum_{i=0}^{r} C_i = \Sigma_i n_i M_i/V = \rho \quad (1.1.41a)$$

$$\sum_{i=0}^{r} C_i V_i/M_i = \Sigma_i C_i v_i = 1 \quad (1.1.41b)$$

where ρ is the density of the solution, and v_i is the partial specific volume of component i, i.e.,

$$v_i \equiv V_i/M_i \quad (1.1.42)$$

As has been mentioned above in Fig. 1.5, we have $V_m \simeq V_0 \simeq V_0{}^o$ for dilute solutions, in which $x_1 \ll 1, x_2 \ll 1, \ldots, x_r \ll 1$. Hence for such solutions we have

$$\begin{aligned} x_i &\simeq (M_0/M_i) w_i \simeq (V_0{}^o/V_i{}^o)\phi_i \simeq M_0 m_i \simeq V_0{}^o c_i \\ &= (V_0{}^o/M_i) C_i \end{aligned} \quad (1.1.43)$$

g. On SI Units. The term molarity has been widely used in chemistry for the number of moles of a solute contained in one liter of solution. According to the basic ideas on which the SI units were established, this statement is not logically precise. To illustrate: if we measure the length L of a rod and obtain 1.5 m or 150 cm, we might record the result as

$$L = 1.5 \text{ m} = 150 \text{ cm} \quad (1.1.44)$$

Our understanding in doing this is that L is a symbol for a physical quantity called the length of the rod, not for the number of meters or centimeters. If we were to assign a symbol for such a number, we would logically have to use different symbols for a physical quantity depending on the units used to measure it. This gives rise to a complication. It is incorrect to call L the number of meters. In the same sense, it is irrelevant to call the stoichiometric amount n of a substance the number of moles. The definition that c_i is the amount of component i contained in unit volume of solution is acceptable, whereas the above-mentioned definition of molarity is not acceptable because it refers to a specific unit "one liter." The old convention $\ell = 1.000028\ \mathrm{dm}^3$ was banned a decade ago. The symbol ℓ is now only a special symbol for dm^3, and the symbol $\mathrm{m}\ell$ for cm^3. The term molarity sounds similar to the term molality. For these reasons the SI would ban it.

A symbol of a physical quantity represents the product of a pure number and a unit. Thus it is meaningful to rewrite Eq. (1.1.44) as $L/\mathrm{m} = 1.5$ or $L/\mathrm{cm} = 150$, where the solidus notation, as usual, denotes a fraction. It is common practice to tabulate or plot data in the form of pure numbers. Column headings or coordinate legends on graphs should therefore be dimensionless like L/m or L/cm. They should not be written L, $L(\mathrm{m})$, $L(\mathrm{cm})$; for example, $L(\mathrm{m})$ implies length multiplied by meters. The argument of a mathematical function, such as x in $\log x$, should be dimensionless. Thus it is wrong to write the chemical potential μ^g of an ideal gas as

$$\mu^g(T,p) = \mu^o(T) + RT\,\ell n\, p \tag{1.1.45}$$

where T is the absolute temperature, p is the pressure, and R is the gas constant, i.e.,

$$R = 8.3143\ \mathrm{J\,K^{-1}\,mol^{-1}} \tag{1.1.46}$$

Here J is the SI unit for energy, i.e., the Joule, which is related to the SI base units, m, kg, and s, and to the SI unit of force, N (the Newton), by $\mathrm{J} = \mathrm{Nm} = \mathrm{kg\ m^2\ s^{-2}}$. Equation (1.1.45) is incorrect because the argument p in the logarithm is not dimensionless. The correct expression for μ^g should be

$$\mu^g(T,p) = \mu^\theta(T) + RT\,\ell n\,(p/p^\theta) \tag{1.1.47}$$

where p^θ is a standard value of pressure, conventionally chosen to be 1 atmosphere ($1.01325 \times 10^5\ \mathrm{N\,m^{-2}}$), and μ^θ is the standard chemical potential at T. Reference [1] should be consulted for details on the use of SI units.

We should not confuse molar mass M with **molecular weight** M_r. A computation will serve to illustrate the distinction. According to van't Hoff's law, the osmotic pressure π of a solution is given by

$$\pi = RTC/M \tag{1.1.48}$$

where C is the mass concentration of the solute, M is the molar mass of the solute, R is the gas constant, and T is the absolute temperature. Suppose that osmotic pressure and density measurements on a cyclohexane-polystyrene solution at 34°C give $\pi = 0.75$ cm of solution at $C = 0.01\ \mathrm{g\ cm^{-3}}$ and $\rho = 0.78\ \mathrm{g\ cm^{-3}}$ Then, with the gravitational constant $g = 9.81\ \mathrm{m\ s^{-2}}$

$$\frac{\pi}{C} = \frac{(0.75 \times 10^{-2}) \times (0.78 \times 10^{3}) \times (9.81)\,(\mathrm{m})\,(\mathrm{kg\,m^{-3}})\,(\mathrm{ms^{-2}})}{10\,(\mathrm{kg\,m^{-3}})}$$
$$= 5.73\ \mathrm{m^2\,s^{-2}} = 5.73\ \mathrm{J\,kg^{-1}} \qquad (1.1.49)$$

Substituting this value into Eq. (1.1.48), we get

$$M = \frac{8.31 \times 307}{5.73}\ \mathrm{kg\,mol^{-1}}$$
$$= 445\ \mathrm{kg\,mol^{-1}} = 445000\ \mathrm{g\,mol^{-1}} \qquad (1.1.50)$$

On the other hand, the quantity M_r is the molecular mass M/N_A relative to one-twelfth of the atomic mass of the ^{12}C nuclear species, i.e., $0.001\ \mathrm{kg\,mol^{-1}}/N_A$. Hence, referring to Eq. (1.1.50)

$$M_r = \frac{M/N_A}{(0.001/N_A)\ \mathrm{kg\,mol^{-1}}} = \frac{1000M}{\mathrm{kg\,mol^{-1}}} = 445000 \qquad (1.1.51)$$

Here $M/\mathrm{kg\,mol^{-1}}$ denotes the numerical value for M expressed in $\mathrm{kg\,mol^{-1}}$. In general

$$M_r = 1000\ M/\mathrm{kg\,mol^{-1}} = M/\mathrm{g\,mol^{-1}} \qquad (1.1.52)$$

Unlike M, M_r is a dimensionless number.

1.1.3. Partial Specific Quantities

Equation (1.1.19) shows that the partial molar quantities Y_i of the components making up a solution are interdependent. Similar relations can be derived for compositions expressed in variables other than the mole fractions x_i. For example, when the weight fractions $w_1, w_2, \ldots, w_r$ are used as the independent composition variables, we obtain in-place of Eq. (1.1.17)

$$(dY_i)_{T,p} = \sum_{j=1}^{r} (\partial Y_i/\partial w_j)_{T,p,w_k}\, dw_j, \quad (k \neq 0) \qquad (1.1.53)$$

which, together with Eq. (1.1.29), is substituted into Eq. (1.1.16) to give

$$\sum_{i=0}^{r} (w_i/M_i)\,(\partial Y_i/\partial w_j)_{T,p,w_k} = 0, \quad (j = 1, 2, \ldots, r;\ k \neq 0) \qquad (1.1.54)$$

Differentiation of Y with respect to q_i, the mass of component i, at fixed T, p, and masses of all components other than component i yields

$$y_i \equiv (\partial Y/\partial q_i)_{T,p,q_j} = (1/M_i)\,(\partial Y/\partial n_i)_{T,p,n_j} = Y_i/M_i \qquad (1.1.55)$$

If this is inserted in Eq. (1.1.54), we obtain

$$\sum_{i=0}^{r} w_i (\partial y_i/\partial w_j)_{T,p,w_k} = 0, \quad (j = 1,2,\ldots,r;\ k \neq 0) \tag{1.1.56}$$

The quantity y_i is called the **partial specific** Y of component i. The partial specific volume v_i defined in Eq. (1.1.42) is an example of this kind of thermodynamic quantity.

The following relations can be derived in a similar way:

$$\sum_{i=0}^{r} (\phi_i/V_i^o)(\partial Y_i/\partial \phi_j)_{T,p,\phi_k} = 0, (j = 1,2,\ldots,r;\ k \neq 0) \tag{1.1.57}$$

$$\sum_{i=0}^{r} m_i(\partial Y_i/\partial m_j)_{T,p,m_k} = 0, (j = 1,2,\ldots,r;\ k \neq 0) \tag{1.1.58}$$

$$\sum_{i=0}^{r} c_i(\partial Y_i/\partial c_j)_{T,p,c_k} = 0, (j = 1,2,\ldots,r;\ k \neq 0) \tag{1.1.59}$$

$$\sum_{i=0}^{r} (C_i/M_i)(\partial Y_i/\partial C_j)_{T,p,C_k} = \sum_{i=0}^{r} C_i(\partial y_i/\partial C_j)_{T,p,C_k} = 0, \quad (j = 1,2,\ldots,r;\ k \neq 0) \tag{1.1.60}$$

When Y is the Gibbs free energy, these relations are all called Gibbs-Duhem equations.

1.1.4. Apparent Molar and Specific Quantities

In regard to Fig. 1.1 we referred to the graphical method of intercepts for determining Y_0 and Y_1 in a two-component solution from experimental data giving Y_m as a function of x_1. This method, though elegant, is not recommended in practice, because even a slight inaccuracy in drawing the tangent may lead to gross errors in both Y_0 and Y_1. As is shown in Fig. 1.1, the line connecting points A_0 and P is extended to the right and allowed to intersect the vertical axis for the pure component 1 at D_1. The ordinate $\widetilde{Y}_1$ of this point is given by

$$\widetilde{Y}_1 = Y_0^o + \frac{Y_m - Y_0^o}{x_1} = \frac{Y_m - (1 - x_1)Y_0^o}{x_1} = \frac{Y - n_0 Y_0^o}{n_1} \tag{1.1.61}$$

where $Y - n_0 Y_0^o$ is the increase in Y produced by the addition of amount n_1 of solute (component 1) to pure solvent, and $\widetilde{Y}_1$ is the corresponding increment per mole of solute. We call $\widetilde{Y}_1$ the **apparent molar** Y. Equation (1.1.61) may be rewritten $Y = n_0 Y_0^o + n_1 \tilde{Y}_1$, so that we obtain at once

$$\begin{aligned} Y_1 &= (\partial Y/\partial n_1)_{T,p,n_0} = \tilde{Y}_1 + n_1 (\partial \tilde{Y}_1/\partial n_1)_{T,p,n_0} \\ &= \tilde{Y}_1 + m_1 (\partial \tilde{Y}_1/\partial m_1)_{T,p} \end{aligned} \tag{1.1.62}$$

It has been shown [2] that, though it also requires the graphical determination of a derivative, this equation affords a more accurate evaluation of Y_1 than does the method of intercepts.

In the case of volume, Y_m and hence $\tilde{Y}_1$ can be measured directly. For $Y = V$ we obtain, by dividing Eqs. (1.1.61) and (1.1.62) by the molar mass M_1,

$$\tilde{v}_1 = (1/M_1)(V - n_0 V_0^o)/n_1 = (V - q_0 v_0^o)/q_1 \tag{1.1.63}$$

$$v_1 = \tilde{v}_1 + q_1(\partial \tilde{v}_1/\partial q_1)_{T,p,q_0} \tag{1.1.64}$$

where q_0 and q_1 are the masses of components 0 and 1, respectively. In terms of the density ρ of the solution and ρ_0 of the solvent, Eq. (1.1.63) may be written [3]

$$\tilde{v}_1 = \frac{1}{q_1}\left(\frac{q_0 + q_1}{\rho} - \frac{q_0}{\rho_0}\right) = \frac{1}{\rho}\left(1 - \frac{q_0 \Delta\rho}{q_1 \rho_0}\right) \tag{1.1.65}$$

with

$$\Delta\rho \equiv \rho - \rho_0 \tag{1.1.66}$$

The density increment $\Delta\rho$ may be measured as a function of q_1 by successive addition of the solute to a fixed mass q_0 of solvent. The data obtained may be analyzed by Eqs. (1.1.64) and (1.1.65) to evaluate v_1.

For an $(r + 1)$-component solution the apparent specific volume $\tilde{v}_i$ of component i is

$$\tilde{v}_i = (V - V')/q_i = (1/\rho)\,[1 - (q'\Delta\rho/q_i\rho')] \tag{1.1.67}$$

with

$$\Delta\rho \equiv \rho - \rho' \tag{1.1.68}$$

Here V', q', and ρ' denote, respectively, the volume, mass, and density of the r-component solution consisting of all components of the $(r + 1)$-component solution except component i. The partial specific volume v_i of component i can be calculated from

$$v_i = \tilde{v}_i + q_i(\partial \tilde{v}_i/\partial q_i)_{T,p,q'} \tag{1.1.69}$$

Finally, we derive the expansion for v_i in powers of m_i. Dividing Eq. (1.1.4) by $n_0 M_0$ and referring to Eqs. (1.1.33), (1.1.37), and (1.1.42), we obtain

$$v_M = v_0 + \sum_{i=1}^{r} m_i V_i \tag{1.1.70}$$

Equation (1.1.5) gives

$$V_i = (\partial V/\partial n_i)_{T,p,n_j} = (\partial v_M/\partial m_i)_{T,p,m_j}, (i = 1, 2, \ldots, r) \tag{1.1.71}$$

Expansion of the right-hand side in powers of m_j yields, at constant T and p,

$$\begin{aligned} v_i &= V_i/M_i \\ &= \frac{1}{M_i}\left[\left(\frac{\partial v_M}{\partial m_i}\right)^o_{m_k} + \sum_{j=1}^{r}\left(\frac{\partial^2 v_M}{\partial m_i \partial m_j}\right)^o m_j + \ldots\right], \\ &(i = 1, 2, \ldots, r) \end{aligned} \tag{1.1.72}$$

If this expansion, together with

$$v_M = v_M{}^o + \sum_{i=1}^{r}\left(\frac{\partial v_M}{\partial m_i}\right)^o m_i + \frac{1}{2!}\sum_{i=1}^{r}\sum_{j=1}^{r}\left(\frac{\partial^2 v_M}{\partial m_i \partial m_j}\right)^o m_i m_j + \ldots \tag{1.1.73}$$

is substituted into Eq. (1.1.70), then

$$v_0 = v_M{}^o - \frac{1}{2}\sum_{i=1}^{r}\sum_{j=1}^{r}\left(\frac{\partial^2 v_M}{\partial m_i \partial m_j}\right)^o m_i m_j + \ldots \tag{1.1.74}$$

where the superscript o indicates infinite dilution. Equation (1.1.74) shows that $v_M{}^o = v_0{}^o$, as expected, and also that the concentration expansion of v_0 contains no term of first order of m_i, in agreement with what was mentioned in relation to Fig. 1.5. Thus we may set $v_0 \simeq v_0{}^o$ for dilute solutions. The relations given above find applications in § 1.4.4, 2.3.5, etc.

1.2. The Laws of Thermodynamics

1.2.1. The First and Second Laws

First, we consider a **closed** system, i.e., one which cannot exchange matter with its surroundings. We denote by $d'q$ an infinitesimal amount of heat absorbed by the system from the surroundings and by $-d'w$ an infinitesimal work done on the system by the surroundings. The **first law of thermodynamics** may then be expressed by

$$dU = d'q - d'w \tag{1.2.1a}$$

where dU denotes the change in the **internal energy** U of the system brought about by the heat and work changes $d'q$ and $d'w$. The symbol d' is used here for the differentiation operator to indicate that $d'q$ and $d'w$ are inexact differentials, i.e., heat and work changes are not unique functions of a change in state but rather depend on the path of the change. On the other hand, dU is an exact differential, the total difference of a differentiable function of state U.

If the system is subjected to no external force save pressure p_0 normal to the boundary surface, only compression work is possible, i.e., $-d'w = -p_0 dV$, where V is the volume of the system. In this case, Eq. (1.2.1a) becomes

$$dU = d'q - p_0 dV \tag{1.2.1b}$$

In particular, if the pressure distribution inside the system is uniform and the compression work is done quasistatically (so that mechanical equilibrium is maintained within the system) the pressure p of the system can be equated to the external pressure p_0. Then the above equation can be written

$$dU = d'q - p dV \tag{1.2.1c}$$

The **second law of thermodynamics** can be put in the form

$$dS \geqslant d'q/T_0 \tag{1.2.2a}$$

where T_0 is the uniform temperature of the surroundings of a system and S is the **entropy**, an extensive quantity of state of the system. If the temperature inside the system is uniform and heat exchange between the system and the surroundings proceeds quasistatically, T_0 may be equated to the system temperature T, and Eq. (1.2.2a) is written

$$dS \geqslant d'q/T \tag{1.2.2b}$$

The equality in Eq. (1.2.2b) can be taken as defining the entropy function and the absolute temperature. The latter can be shown to be identical with the temperature scale defined by the ideal gas. A system in which pressure and temperature are uniform is not necessarily homogeneous in every respect. For example, a multicomponent solution with uniform distributions of pressure and temperature may be inhomogeneous in composition. The equality sign in Eq. (1.2.2b) holds if the change in state is quasistatic, i.e., reversible.

Any spontaneous change occurring in nature is irreversible, and obeys, together with the generally valid relation Eq. (1.2.1b), the inequality in Eq. (1.2.2a). Eliminating $d'q$ from these two equations, we obtain

$$dU - T_0 dS + p_0 dV < 0 \tag{1.2.3}$$

Now we suppose that a system is displaced from a certain initial state to a neighboring state in such a way that the process involves changes δU, δS, δV in the independent variables in Eq. (1.2.3). If we find that

$$\delta U - T_0 \delta S + p_0 \delta V \geqslant 0 \tag{1.2.4}$$

in this change, the process is a **virtual** one, and the system cannot pass spontaneously to the specified neighboring state, because Eq. (1.2.4) contradicts Eq. (1.2.3). Accordingly, if Eq. (1.2.4) holds for all virtual changes to neighboring states, the system must remain in the initial state. In other words, the initial state of the system is an equilibrium state. Thus Eq. (1.2.4) is the criterion for equilibrium of a closed system. It applies regardless of whether the system is homogeneous with respect to temperature and pressure as well as to composition.

Equation (1.2.4) is modified when the system is subject to constraints, as described below.

a. System With U and V Fixed (Isolated System). Equation (1.2.4) for this system becomes

$$(\delta S)_{U, V} \leqslant 0 \tag{1.2.5}$$

Hence a system at constant U and V is at equilibrium when the entropy S either decreases or remains constant for all virtual displacements. This means that the equilibrium state is reached when S is a maximum at given U and V. In this case, if U and V are changed, the system undergoes a spontaneous change toward a new equilibrium state characterized by a maximum S for the modified values of U and V.

b. System With S and V Fixed. The criterion for equilibrium of a system with constant S and V corresponds to the state which gives a minimum U for specified values of S and V:

$$(\delta U)_{S, V} \geqslant 0 \tag{1.2.6}$$

c. System With V Constant and T Uniform and Constant. Since T_0 may be equated to T, the criterion for equilibrium is given by

$$(\delta U)_{T, V} - T(\delta S)_{T, V} = [\delta(U - TS)]_{T, V} \geqslant 0 \tag{1.2.7}$$

In terms of the **Helmholtz free energy** A defined by

$$A \equiv U - TS \tag{1.2.8}$$

Eq. (1.2.7) is written

$$(\delta A)_{T, V} \geqslant 0 \tag{1.2.9}$$

This indicates that equilibrium of a system at specified T and V is realized when A is a minimum.

d. System With Both T and p Uniform and Constant. Since T_0 and p_0 may be equated to T and p, respectively, Eq. (1.2.4) is written

$$(\delta U)_{T,p} - T(\delta S)_{T,p} + p(\delta V)_{T,p} = [\delta(U - TS + pV)]_{T,p} \geqslant 0 \quad (1.2.10)$$

The **Gibbs free energy** G is defined by

$$G \equiv U - TS + pV \quad (1.2.11)$$

Hence Eq. (1.2.10) is equivalent to

$$(\delta G)_{T,p} \geqslant 0 \quad (1.2.12)$$

which means that G at equilibrium assumes a minimum value obtainable for the specified T and p.

Next, we shall find relations among thermodynamic quantities of a system in different equilibrium states. First, we consider $U(S, V)$ and find dU when S and V are allowed to change by dS and dV, respectively. For this purpose we must use a reversible path, so that Eqs. (1.2.1c) and (1.2.2b) with the equality sign may be combined to give

$$dU = TdS - pdV \quad (1.2.13)$$

Another extensive function of state, the **enthalpy** H of a system, is defined by

$$H \equiv U + pV \quad (1.2.14)$$

The total derivative of this function is combined with Eq. (1.2.13) to give

$$dH = TdS + Vdp \quad (1.2.15)$$

Similarly, we obtain for dA and dG

$$dA = -SdT - pdV \quad (1.2.16)$$

$$dG = -SdT + Vdp \quad (1.2.17)$$

Since these relations, like Eq. (1.2.13), are exact linear differential forms, it follows that

$$T = (\partial U/\partial S)_V = (\partial H/\partial S)_p \quad (1.2.18)$$

$$p = -(\partial U/\partial V)_S = -(\partial A/\partial V)_T \quad (1.2.19)$$

$$S = -(\partial A/\partial T)_V = -(\partial G/\partial T)_p \quad (1.2.20)$$

$$V = (\partial H/\partial p)_S = (\partial G/\partial p)_T \quad (1.2.21)$$

The quantities appearing in these relations all refer to equilibrium states of a closed system. From these relations we can derive various useful relations by partial differentiation. Among others, the following are of frequent use:

$$U = A + TS = A - T(\partial A/\partial T)_V = -T^2 [\partial(A/T)/\partial T]_V \qquad (1.2.22)$$

$$(\partial S/\partial V)_T = (\partial p/\partial T)_V \qquad (1.2.23)$$

$$(\partial U/\partial V)_T = -p + T(\partial p/\partial T)_V \qquad (1.2.24)$$

$$C_V \equiv (\partial U/\partial T)_V = -T(\partial^2 A/\partial T^2)_V = T(\partial S/\partial T)_V \qquad (1.2.25)$$

Here C_V is called the **heat capacity at constant volume**. In these relations, T and V are chosen as independent state variables. The corresponding relations for T and p as the independent variables are as follows:

$$H = G + TS = G - T(\partial G/\partial T)_p = -T^2 [\partial (G/T)/\partial T]_p \qquad (1.2.26)$$

$$(\partial S/\partial p)_T = -(\partial V/\partial T)_p = -\alpha V, \quad \alpha \equiv (1/V)(\partial V/\partial T)_p \qquad (1.2.27)$$

$$(\partial H/\partial p)_T = V - T(\partial V/\partial T)_p \qquad (1.2.28)$$

$$C_p \equiv (\partial H/\partial T)_p = T(\partial S/\partial T)_p \qquad (1.2.29)$$

$$\kappa \equiv -(1/V)(\partial V/\partial p)_T \qquad (1.2.30)$$

The quantities α, C_p, and κ are called the **isobaric thermal expansivity**, the **heat capacity at constant pressure**, and the **isothermal compressibility**, respectively.

In actual measurements, conditions other than constant volume or constant pressure are often employed. Letting the variables held fixed be denoted by σ, we can express the heat capacity C_σ at constant σ by

$$\begin{aligned} C_\sigma &= T\left(\frac{\partial S}{\partial T}\right)_\sigma = T\left[\left(\frac{\partial S}{\partial T}\right)_V + \left(\frac{\partial S}{\partial V}\right)_T \left(\frac{\partial V}{\partial T}\right)_\sigma\right] \\ &= C_V + T\left(\frac{\partial p}{\partial T}\right)_V \left(\frac{\partial V}{\partial T}\right)_\sigma \\ &= C_V - T\frac{(\partial V/\partial T)_p}{(\partial V/\partial p)_T}\left(\frac{\partial V}{\partial T}\right)_\sigma \end{aligned} \qquad (1.2.31)$$

where Eqs. (A.11) and (A.12) have been applied. For the case $\sigma = p$, Eq. (1.2.31) gives a very useful relation between C_p and C_V.

1.2.2. Chemical Potential

A system that can exchange matter as well as heat and work with its surroundings is called **open**. Suppose that it consists of $r + 1$ components, which do not react chemically. By applying Eq. (1.2.6), the criterion for equilibrium of this system may be written

$$(\delta U)_{S, V, n_0, n_1, \ldots, n_r} \geqslant 0 \tag{1.2.32}$$

since a system with fixed n_i ($i = 0, 1, \ldots, r$) is closed. Hence for any set of S, V, and $n_0, n_1, \ldots, n_r$ we find an equilibrium value of $U(S, V, n_0, n_1, \ldots, n_r)$. The total derivative of this U is represented by

$$\begin{aligned} dU &= \left(\frac{\partial U}{\partial S}\right)_{V,n} dS + \left(\frac{\partial U}{\partial V}\right)_{S,n} dV + \sum_{i=0}^{r} \left(\frac{\partial U}{\partial n_i}\right)_{S,V,n_j} dn_i \\ &= TdS - pdV + \sum_{i=0}^{r} \mu_i dn_i \end{aligned} \tag{1.2.33}$$

with

$$\mu_i \equiv (\partial U/\partial n_i)_{S,V,n_j} \tag{1.2.34}$$

Equations (1.2.18) and (1.2.19) may be applied to $(\partial U/\partial S)_{V,n}$ and $(\partial U/\partial V)_{S,n}$, respectively; since all n_i are fixed in these partial derivatives, the theory for closed systems can be utilized as it stands. The quantity μ_i defined by Eq. (1.2.34) is called the **chemical potential** of component i.

Transformation of Eq. (1.2.33) by use of the defining equations (1.2.14), (1.2.8), and (1.2.11) for enthalpy H, Helmholtz free energy A, and Gibbs free energy G, respectively, yields

$$dH = TdS + Vdp + \Sigma_i \mu_i dn_i \tag{1.2.35}$$

$$dA = -SdT - pdV + \Sigma_i \mu_i dn_i \tag{1.2.36}$$

$$dG = -SdT + Vdp + \Sigma_i \mu_i dn_i \tag{1.2.37}$$

Therefore, μ_i can also be expressed in the following three ways:

$$\mu_i = (\partial H/\partial n_i)_{S,p,n_j} = (\partial A/\partial n_i)_{T,V,n_j} = (\partial G/\partial n_i)_{T,p,n_j} \tag{1.2.38}$$

The last expression indicates that μ_i is the partial molar Gibbs free energy of component i. The general relations for partial molar quantities, Eqs. (1.1.9), (1.1.11), (1.1.14), and (1.1.19), yield, when Y is taken to be G,

$$G = \sum_{i=0}^{r} n_i \mu_i \tag{1.2.39}$$

$$G_m = \sum_{i=0}^{r} x_i \mu_i \tag{1.2.40}$$

$$-S_m dT + V_m dp - \sum_{i=0}^{r} x_i d\mu_i = 0 \tag{1.2.41}$$

$$\sum_{i=0}^{r} x_i (\partial\mu_i/\partial x_j)_{T,p,x_k} = 0, \quad (j = 1, 2, \ldots, r;\ k \neq 0) \tag{1.2.42}$$

As we have already noted, Eqs. (1.2.41) and (1.2.42) are the **Gibbs-Duhem relations.**

Equations (1.2.18) through (1.2.30), valid for closed systems, also hold for open systems as they stand, with the understanding that all n_i are fixed in the partial differentiation. Thus if A is given as a function of T, V, and $n_0, n_1, \ldots, n_r$, use of Eq. (1.2.19) and appropriate subsequent equations allows all thermodynamic quantities of the system to be computed. For this reason the quantity A is called the **characteristic** or **generating function** or the **thermodynamic potential** for the set of state variables T, V, and n_i $(i = 0, 1, \ldots, r)$. The corresponding function for the set of state variables T, p, and n_i $(i = 0, 1, \ldots, r)$ is the Gibbs free energy G.

If Eqs. (1.2.20), (1.2.21), and (1.2.26) involving G are differentiated with respect to n_i and the order of differentiation is exchanged, we obtain

$$S_i = -(\partial\mu_i/\partial T)_{p,x} \tag{1.2.43}$$

$$V_i = (\partial\mu_i/\partial p)_{T,x} \tag{1.2.44}$$

$$H_i = -T^2\, [\partial\,(\mu_i/T)/\partial T]_{p,x} \tag{1.2.45}$$

Here the subscript n has been replaced by x because μ_i is an intensive quantity and so depends on the composition of the system.

1.3. Phase Equilibrium

1.3.1. Equilibrium Criteria for Diffusible Components

If intensive quantities such as temperature, pressure, and density vary continuously with position in a system, we may say that the system consists of a single **phase**. Thus a solution subject to a field of centrifugal forces, in which there is set up a continuous variation of composition with radial distance, is regarded as a one-phase system. But a system in which a liquid solution is in contact with its vapor at a free surface consists of two phases. Here the question is: what conditions have to be satisfied in order that more than one phase may coexist at equilibrium in a multicomponent system? To answer it we take two $(r + 1)$-component systems, α and β, each at equilibrium, and place them in contact. Heat, work, and matter are allowed to pass freely through the surface of contact, but it is assumed that the entire system is separated from its surroundings by a membrane impermeable to matter. Thus the entire system is closed and consists of two phases, each open to the other.

The conditions for equilibrium of this system can be derived from the general equilibrium criterion Eq. (1.2.6). They read

$$T^\alpha = T^\beta = T \tag{1.3.1}$$

$$p^\alpha = p^\beta = p \tag{1.3.2}$$

$$\mu_i^\alpha = \mu_i^\beta = \mu_i, \quad (i = 0, 1, \ldots, r) \tag{1.3.3}$$

which indicate that for the combined system $\alpha + \beta$ to be in equilibrium, the temperature, pressure, and chemical potential of each component must assume the same values T, p, and μ_i in phases α and β. Since we shall be particularly concerned with Eq. (1.3.3), which determines the equilibrium for diffusive transport of matter, we present only the derivation of Eq. (1.3.3), omitting that of Eqs. (1.3.1) and (1.3.2), which refer to thermal and mechanical equilibria.

The criterion for equilibrium of the closed system $\alpha + \beta$ at constant temperature and pressure is given by Eq. (1.2.12). This may be written

$$(\delta G)_{T,p} = (\delta G^\alpha)_{T,p} + (\delta G^\beta)_{T,p} \geqslant 0 \tag{1.3.4}$$

where G^α and G^β denote the Gibbs free energies of the phases. The free energy G is a function only of T and p, because the combined system $\alpha + \beta$ is closed. However, since phase α (or phase β) is an open system, G^α (or G^β) is a function of T, p, and n_i^α (or n_i^β). We define n_i by

$$n_i = n_i^\alpha + n_i^\beta \tag{1.3.5}$$

The virtual changes δn_i^α and δn_i^β are subjected to the restriction

$$\delta n_i^\alpha + \delta n_i^\beta = \delta n_i = 0 \tag{1.3.6}$$

By application of Eq. (1.2.37) we can write $(\delta G^\alpha)_{T,p}$ and $(\delta G^\beta)_{T,p}$ as

$$(\delta G^\alpha)_{T,p} = \sum_{i=0}^{r} \mu_i^\alpha \delta n_i^\alpha \tag{1.3.7a}$$

$$(\delta G^\beta)_{T,p} = \sum_{i=0}^{r} \mu_i^\beta \delta n_i^\beta \tag{1.3.7b}$$

Therefore, Eq. (1.3.4) gives

$$(\delta G)_{T,p} = \sum_{i=0}^{r} (\mu_i^\alpha - \mu_i^\beta)\, \delta n_i^\alpha \geqslant 0 \tag{1.3.8}$$

where Eq. (1.3.6) has been used. The condition in Eq. (1.3.8) must hold for all possible virtual displacements of n_i^α. For example, if $\delta n_0{}^\alpha = a > 0$ and $\delta n_i{}^\alpha = 0$ $(i \neq 0)$, it is required that $\mu_0{}^\alpha \geqslant \mu_0{}^\beta$. If $\delta n_0{}^\alpha = -a < 0$ and $\delta n_i{}^\alpha = 0\ (i \neq 0)$, $\mu_0{}^\alpha$

must be equal to or smaller than $\mu_0{}^{\beta}$. Hence it is necessary that $\mu_0{}^{\alpha} = \mu_0{}^{\beta}$ for Eq. (1.3.8) to hold for both positive and negative $\delta n_0{}^{\alpha}$. From the similar argument for all other components we ultimately arrive at Eq. (1.3.3) as the necessary condition for equilibrium with respect to diffusion of matter. Conversely, if Eq. (1.3.3) is put in Eq. (1.3.8), we obtain

$$(\delta G)_{T,p} = 0 \qquad (1.3.9)$$

which indicates that Eq. (1.3.3) is not only necessary but also sufficient for the equilibrium of diffusible matter.

Equations (1.3.1) through (1.3.3) can also be applied to equilibrium between different parts of a one-phase system. It follows that the equilibrium conditions for a system subjected to no external force are that the temperature, pressure, and chemical potential of each component be uniform throughout the entire system.

1.3.2. Stability of Equilibrium with Respect to Diffusion

We understand from Eq. (1.3.9) that when a closed system at fixed temperature and pressure is at equilibrium with respect to diffusion of matter, the Gibbs free energy of the system assumes an extremal value. In order to know whether this value is a maximum or minimum, it is necessary to examine the sign of the second-order variation of G for virtual displacements of $n_i{}^{\alpha}$.

Thus if

$$(\delta^2 G)_{T,p} > 0 \qquad (1.3.10)$$

the extremal G in question corresponds to a minimum, and the equilibrium is stable. On the contrary, if $(\delta^2 G)_{T,p} < 0$, the equilibrium is unstable. If

$$(\delta^2 G)_{T,p} = 0 \qquad (1.3.11)$$

the equilibrium is said to be neutral, and its stability depends on the sign of still higher-order variations of G.

The variation $(\delta^3 G)_{T,p}$ cannot be always positive for bidirectional variations of $n_i{}^{\alpha}$. Therefore, when Eq. (1.3.11) is obeyed, a stable equilibrium state is obtained if

$$(\delta^3 G)_{T,p} = 0 \quad \text{and} \quad (\delta^4 G)_{T,p} > 0 \qquad (1.3.12)$$

As will be shown in the next section, the condition (1.3.11) corresponds to spinodals, and this condition plus the first of the conditions (1.3.12) to critical points.

Equation (1.3.10) or (1.3.12) for stable equilibrium is concerned only with the stability for very small displacements of matter in the system. It might be possible that G changes to a lower minimum when there is an exchange of larger amounts of matter between different parts of a system. When this is possible, an equilibrium state corresponding to a local minimum in G for small displacements of matter is called **metastable** to distinguish it from the **truly stable** equilibrium state which is realized at the smallest G.

1.3.3. Phase Equilibrium in Binary Solutions

First, we discuss the stability of a binary solution which is homogeneous in temperature, pressure, and composition. In such a solution, the chemical potential of each component is constant everywhere. According to the criteria given in § 1.3.1, the solution is in a certain equilibrium state. We seek the condition for this to be a stable state.

We assume that the solution as a whole is closed and consider in it a region α which is separated from the rest, the region β, by a thin membrane that is completely deformable, diathermal, and permeable to both components 0 and 1. We denote the amount of component i ($i = 0, 1$) in the region ν ($\nu = \alpha, \beta$) by n_i^ν and put $n^\nu = n_0{}^\nu + n_1{}^\nu$. Though not mandatory, it is convenient for mathematical simplicity to let

$$n^\alpha \ll n^\beta \tag{1.3.13}$$

The mole fraction of component 1 is uniform throughout the solution:

$$n_0{}^\nu = (1 - x_1)n^\nu, \quad n_1{}^\nu = x_1 n^\nu \tag{1.3.14}$$

The criterion for stability of the solution, Eq. (1.3.10), gives

$$(\delta^2 G)_{T,p,n_0,n_1} = (\delta^2 G^\alpha)_{T,p} + (\delta^2 G^\beta)_{T,p} > 0 \tag{1.3.15}$$

where G^ν is the Gibbs free energy of region ν and $n_i = n_i^\alpha + n_i^\beta$ ($i = 0, 1$).

We suppose that small amounts of components 0 and 1, $\delta n_0{}^\alpha$ and $\delta n_1{}^\alpha$, are transferred virtually from region β to region α at the given T and p. Then $(\delta^2 G^\nu)_{T,p}$, the second-order variation of G in region ν, can be represented by

$$(\delta^2 G^\nu)_{T,p} = \frac{1}{2} \sum_{i=0}^{1} \sum_{j=0}^{1} \bar{\mu}_{ij}{}^\nu \delta n_i{}^\nu \delta n_j{}^\nu, \quad (\nu = \alpha, \beta) \tag{1.3.16}$$

where

$$\bar{\mu}_{ij}{}^\nu \equiv (\partial^2 G/\partial n_i \partial n_j)_{T,p}^\nu = (\partial \mu_i/\partial n_j)_{T,p,n_k}^\nu = \bar{\mu}_{ji}{}^\nu \tag{1.3.17}$$

Introduction of Eq. (1.3.16) into Eq. (1.3.15) and use of the constraint that $\delta n_i^\alpha + \delta n_i^\beta = \delta n_i = 0$ give

$$(\delta^2 G)_{T,p,n_0,n_1} = \frac{1}{2} \sum_{i=0}^{1} \sum_{j=0}^{1} (\bar{\mu}_{ij}^{\alpha} + \bar{\mu}_{ij}^{\beta})\, \delta n_i^\alpha \delta n_j^\alpha > 0 \tag{1.3.18}$$

The derivative $\bar{\mu}_{ij}^{\nu}$ ($\nu = \alpha, \beta$) should be evaluated at the equilibrium of matter. Thus

$$\bar{\mu}_{11}^{\nu} = \left(\frac{\partial \mu_1}{\partial x_1}\right) \left[\frac{\partial}{\partial n_1}\left(\frac{n_1}{n_0 + n_1}\right)\right]_{n_0}^{\nu} = \frac{1 - x_1}{n^\nu} \left(\frac{\partial \mu_1}{\partial x_1}\right) \tag{1.3.19}$$

where the superscript ν has been omitted from the quantities which are to take the same values for regions α and β at equilibrium. With the condition from Eq. (1.3.13) we find that Eq. (1.3.19) gives

$$\bar{\mu}_{11}^{\beta} \ll \bar{\mu}_{11}^{\alpha} \tag{1.3.20a}$$

and, in a similar way,

$$\bar{\mu}_{00}^{\beta} \ll \bar{\mu}_{00}^{\alpha} \tag{1.3.20b}$$

$$\bar{\mu}_{01}^{\beta} = \bar{\mu}_{10}^{\beta} \ll \bar{\mu}_{01}^{\alpha} = \bar{\mu}_{10}^{\alpha} \tag{1.3.20c}$$

Therefore, Eq. (1.3.18) can be simplified to

$$(\delta^2 G)_{T,p,n_0,n_1} = \frac{1}{2} \sum_{i=0}^{1} \sum_{j=0}^{1} \bar{\mu}_{ij}^{\alpha} \delta n_i^\alpha \delta n_j^\alpha > 0 \tag{1.3.21}$$

One might think that the desired criterion for stable equilibrium of a homogeneous binary solution at fixed temperature and pressure could be obtained by finding mathematical conditions which make this quadratic form of δn_i^α always positive. However, the matter is not quite as simple as this.

To explain, consider the case in which δn_0^α and δn_1^α are chosen so as to satisfy the relation

$$\delta n_0^\alpha / \delta n_1^\alpha = (1 - x_1)/x_1 \tag{1.3.22}$$

In this case, no composition variation takes place in regions α and β. In other words, the thermodynamic state of the entire solution remains unchanged before and after the virtual displacement of matter; what happens is an infinitesimal movement of the membrane separating region α from region β. It is then entirely meaningless to speak of $(\delta^2 G)_{T,p}$, because the only meaningful $(\delta^2 G)_{T,p}$ corresponds to virtual displacements that induce a change in the state of a given system. Thus Eq. (1.3.21) should be restricted to pairs of variations δn_0^α and δn_1^α which are not related by Eq. (1.3.22). However, for any given δn_1^α there is always a $\delta n_0{}^\alpha$ which satisfies this relation, provided that $\delta n_0{}^\alpha$ and $\delta n_1{}^\alpha$ are allowed to vary independently. We can be freed from such a difficulty only in the case where $n_0{}^\alpha$ is fixed; then, any change in $n_1{}^\alpha$ gives rise to a change in the composition of region α and hence in the state of the entire solution. A moment's thought suffices to establish that this restriction on $n_0{}^\alpha$ is a natural consequence of the condition that the solution must be homogeneous, i.e., the equilibrium compositions of regions α and β are the same. If the two regions represent phases with different compositions separated by a well-defined boundary surface, no such restriction is needed, and both $n_0{}^\alpha$ and $n_1{}^\alpha$ can be changed independently.

To keep $n_0{}^\alpha$ constant means to define region α as a portion of the given solution enclosed by a membrane impermeable to component 0. One might worry that, with such a semipermeable membrane, a pressure difference (osmotic pressure) would appear between the regions α and β, and thus the condition of uniform pressure would be violated. This concern is unnecessary, however, because the membrane is assumed to be perfectly deformable. In fact, throughout the above discussion, we have tacitly assumed that the membrane enclosing region α stretches and shrinks so as to keep the pressure of the system uniform when matter is transported through it. If the membrane is rigid, i.e., region α is defined as having a fixed volume, virtual displacements of matter are indeed accompanied by the generation of a pressure difference between the two regions, and the constraint that pressure is uniform throughout the solution is violated. In this case, Eq. (1.3.21) is no longer applicable.

Now, with $n_0{}^\alpha$ fixed, Eq. (1.3.21) simplifies to

$$(\delta^2 G)_{T,p,n_0,n_1} = \frac{1}{2}\,\bar{\mu}_{11}{}^\alpha(\delta n_1{}^\alpha)^2 = \frac{1-x_1}{2n^\alpha}\left(\frac{\partial \mu_1}{\partial x_1}\right)_{T,p}(\delta n_1{}^\alpha)^2 > 0 \qquad (1.3.23)$$

which indicates that of necessity

$$(\partial\mu_1/\partial x_1)_{T,p} > 0 \quad \text{or} \quad (\partial\mu_0/\partial x_1)_{T,p} < 0 \qquad (1.3.24)$$

if a one-phase binary solution at constant T and p is to be thermodynamically stable. Either condition follows from the other by virtue of the Gibbs-Duhem relation.

Referring to Fig. 1.6, we see that a system with a μ_1 versus x_1 dependence like curve a is stable over the entire range of composition x_1, since $(\partial\mu_1/\partial x_1)_{T,p} > 0$ for all x_1 from 0 to 1. In the case characterized by curve b, the system is unstable between points N′ and N″, and thus cannot exist as a homogeneous single phase over the entire range of composition. The compositions at points N′ and N″ are determined by the condition

$$(\partial\mu_1/\partial x_1)_{T,p} = 0 \qquad (1.3.25)$$

When the temperature is varied at a fixed pressure, the relation between μ_1 and x_1 may undergo a change from a shape like curve b to one like curve a. In the course of this change the function ought to pass through a stage represented by curve c at a certain intermediate temperature T_c. Here it has an inflection point C at which the curve is horizontal. At point C we have

$$(\partial\mu_1/\partial x_1)_{T,p} = 0, \quad (\partial^2\mu_1/\partial x_1{}^2)_{T,p} = 0, \quad (\partial^3\mu_1/\partial x_1{}^3)_{T,p} > 0 \tag{1.3.26}$$

The last inequality follows from Eq. (1.3.12). The first two conditions leave only one freely assignable variable out of the three state variables T, p, and x_1 characterizing a binary solution. Thus point C is determined uniquely for a given pressure. It is called the **critical solution point** or simply the **critical point.**

Since

$$\begin{aligned} \bar{\mu}_{11}{}^{\alpha} &= (\partial\mu_1/\partial n_1)^{\alpha}_{T,p,n_0} = (1/n_0{}^{\alpha}M_0)(\partial\mu_1/\partial m_1)_{T,p} \\ &\equiv (1/n_0{}^{\alpha}M_0)\mu_{11} \end{aligned} \tag{1.3.27}$$

where m_1 is the molality of component 1, the condition for stable equilibrium, Eq. (1.3.23), can be written

$$\mu_{11} > 0 \tag{1.3.28}$$

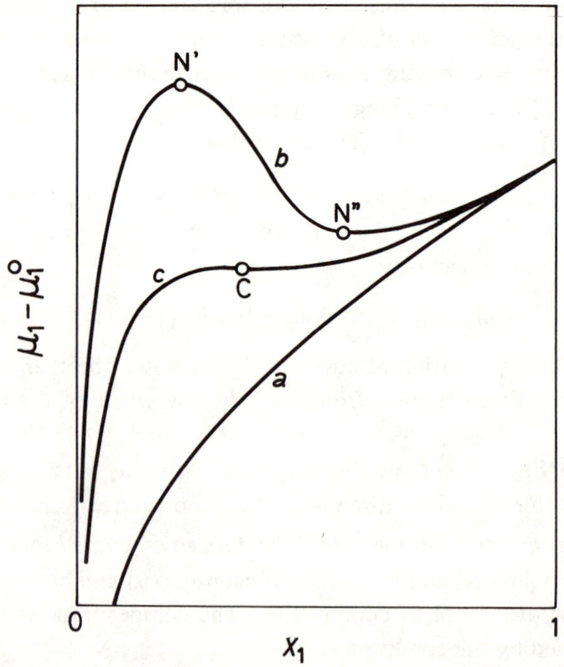

Fig. 1.6. Typical μ_1 versus x_1 relations in two-component solutions.

Further, Eqs. (1.3.25) and (1.3.26) are transformed to

$$\mu_{11} = 0 \tag{1.3.29}$$

and

$$(\partial\mu_1/\partial m_1)_{T,p} = 0,\ (\partial^2\mu_1/\partial m_1{}^2)_{T,p} = 0,\ (\partial^3\mu_1/\partial m_1{}^3)_{T,p} > 0 \tag{1.3.30}$$

respectively. The molality m_1 is often more convenient than the mole fraction x_1 for treating binary solutions at fixed temperature and pressure.

We can discuss the same problem as above on a plot of G_m as a function of x_1. From Eq. (1.1.23b) it follows that

$$\mu_1 = G_m + (1 - x_1)(\partial G_m/\partial x_1)_{T,p} \tag{1.3.31}$$

Putting this in Eqs. (1.3.24) and (1.3.26), we obtain for stable equilibrium

$$(\partial^2 G_m/\partial x_1{}^2)_{T,p} \geqslant 0 \tag{1.3.32}$$

and for the critical point

$$(\partial^2 G_m/\partial x_1{}^2)_{T,p} = 0,\ (\partial^3 G_m/\partial x_1{}^3)_{T,p} = 0,\ (\partial^4 G_m/\partial x_1{}^4)_{T,p} > 0 \tag{1.3.33}$$

As is shown in Fig. 1.7, part of the G_m curve for an unstable system becomes convex upward (i.e., $\partial^2 G_m/\partial x_1{}^2 < 0$). We may draw a tangent common to two points on such a curve. The intersections of this tangent with the vertical lines at $x_1 = 0$ and $x_1 = 1$ give the chemical potentials of components 0 and 1, respectively, at the compositions corresponding to both points P′ and P″, at which the common tangent touches the G_m curve. Thus, denoting the values of x_1 at P′ and P″ by $x_1{}'$ and $x_1{}''$, respectively, we have

$$\mu_0(x_1{}') = \mu_0(x_1{}'') \tag{1.3.34a}$$

$$\mu_1(x_1{}') = \mu_1(x_1{}'') \tag{1.3.34b}$$

Consequently, the two solutions having the compositions $x_1{}'$ and $x_1{}''$ coexist maintaining thermodynamic equilibrium.

Now, suppose that a system which initially had a composition x_1 specified by point P in Fig. 1.7 has separated into two phases characterized by points P′ and P″. The law of mass conservation gives

$$n = n' + n'' \tag{1.3.35a}$$

$$x_1 n = x_1{}'n' + x_1{}''n'' \tag{1.3.35b}$$

where n is the total amount of the initial solution, and n' and n'' are the amounts of the separated phases. These two equations may be solved for n' and n'' to give

$$\frac{n'}{n} = \frac{x_1'' - x_1}{x_1'' - x_1'}, \quad \frac{n''}{n} = \frac{x_1 - x_1'}{x_1'' - x_1'} \tag{1.3.36}$$

Hence n'/n'' equals the ratio of the lengths of line segments $\overline{\mathrm{QQ''}}/\overline{\mathrm{QQ'}}$. This relation is called the **lever rule**.

Letting the molar Gibbs free energy of the solution after phase separation be denoted by $G_m{}^*$, we have

$$\begin{aligned} G_m{}^* &= (n'/n)G_m{}' + (n''/n)G_m{}'' \\ &= G_m{}' + [(x_1 - x_1{}')/(x_1{}'' - x_1{}')]\,(G_m{}'' - G_m{}') \end{aligned} \tag{1.3.37}$$

This $G_m{}^*$, which corresponds geometrically to point P* in Fig. 1.7, is smaller than G_m at point P. Hence, though it is stable in the sense that it satisfies Eq. (1.3.32), the one-phase system specified by point P can assume a lower G_m value, $G_m{}^*$, when it undergoes phase separation. It may thus be concluded that a solution in states in either the region between points P′ and N′ or the region between points N″ and P″ is only **metastable**. Such a solution attains the truly stable state only after phase separation.

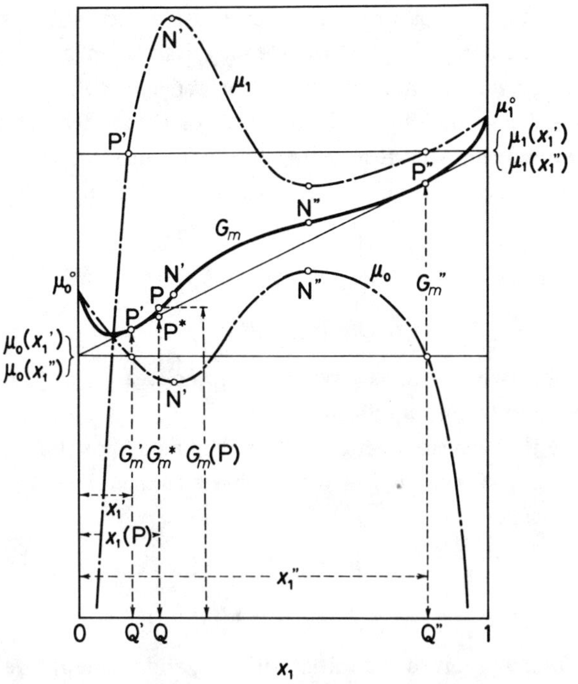

Fig. 1.7. Molar Gibbs free energy (G_m) and chemical potentials (μ_0 and μ_1) as functions of composition x_1 in a two-component solution which undergoes separation into two phases.

A curve of μ_0 versus μ_1 can be constructed from the composition dependences of μ_0 and μ_1 shown in Fig. 1.7. It is illustrated by the dot-dash line in Fig. 1.8, which has one node or binode and two cusps or spinodes [4]. The binode corresponds to points P′ and P″ on the G_m versus x_1 curve, while the spinodes correspond to the inflection points N′ and N″ on the same curve. The other dot-dash lines in Fig. 1.8 indicate μ_0 versus μ_1 relations at different temperatures. The solid line which joins the binodes at different temperatures maps the locus of coexisting phases on the (μ_0, μ_1) plane. The dashed lines which connect the corresponding spinodes represent the locus of boundary points of the unstable region at different temperatures. These curves meet and terminate at a point C on the μ_0 versus μ_1 curve for $T = T_c$. When they are projected on the T versus x_1 plane, the solid and dashed curves shown in Fig. 1.9 are obtained. The solid curve is called the **binodal** and the dashed curve the **spinodal**. The relevance of these names is clear from their correspondence with the characteristic points in the (μ_0, μ_1) plane. In this connection, we note that many textbooks on chemical thermodynamics go directly from the G_m versus x_1 diagram (Fig. 1.7) to the T versus x_1 diagram (Fig. 1.9), skipping an explanation of the characteristic behavior of μ_0 versus μ_1 relations (Fig. 1.8). This procedure leaves the reader unsatisfied as to why projections on the (T, x_1) plane of the coexisting phase points and the inflection points on the (G_m, x_1) plane are termed binodals and spinodals.

Point C in Fig. 1.9 marks the common vertex of the binodal and spinodal. This vertex, at which the spinodal is tangent to the binodal, defines the critical point of a binary liquid system: the critical temperature T_c and the critical

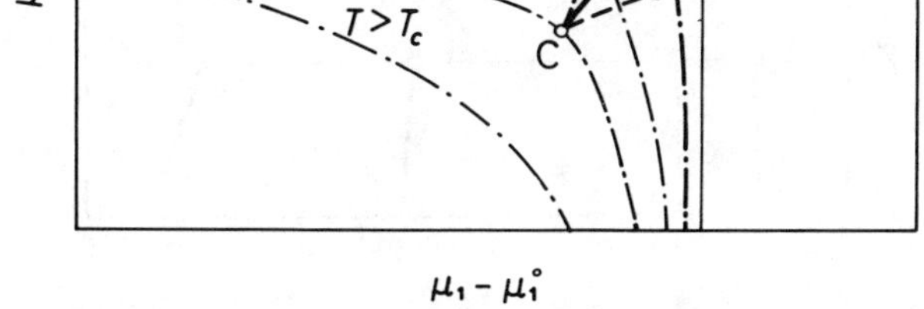

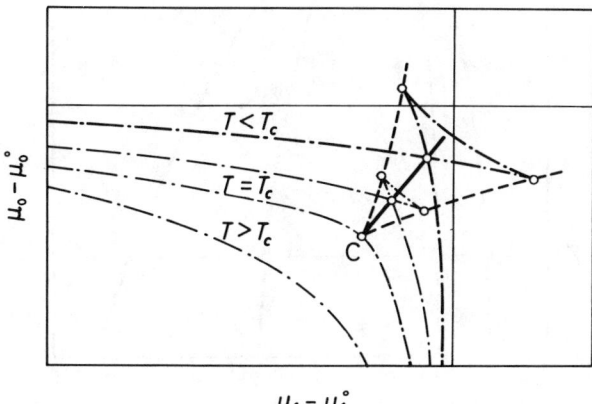

Fig. 1.8. Relations between μ_0 and μ_1 in a two-component solution at temperatures above and below the critical temperature T_c.

composition x_{1c}. If the ordinate in Fig. 1.9 is taken in the direction of increasing temperature, the system characterized by this phase diagram undergoes phase separation at temperatures below T_c. In this case, point C is called the **upper critical point**. In some systems, phase separation occurs at temperatures above T_c; the diagram is then inverted and point C is called the **lower critical point**. In other systems, the binodal describes a closed curve and has both upper and lower critical points, and in still others it consists of a pair of bell-shaped and inverted bell-shaped curves separated by a range of temperature. In this last case, the two-phase system at low temperature becomes a single phase at higher temperature and again undergoes phase separation as the temperature is raised further.

1.3.4. Phase Equilibrium in Multicomponent Solutions

In a way similar to that used in deriving Eq. (1.3.21), we can show that the condition for stable equilibrium of a ternary solution in a single phase is

$$(\delta^2 G)_{T,p,n} = \frac{1}{2} \sum_{i=1}^{2} \sum_{j=1}^{2} \bar{\mu}_{ij}^{\alpha} \delta n_i^{\alpha} \delta n_j^{\alpha} > 0 \qquad (1.3.38)$$

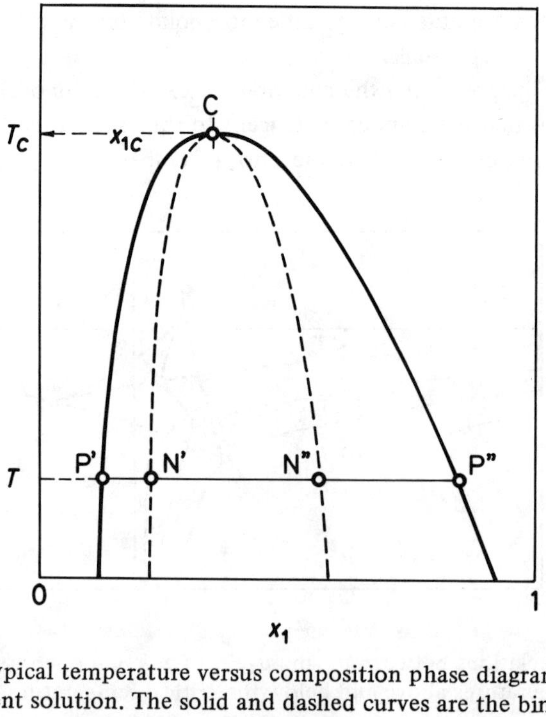

Fig. 1.9. Typical temperature versus composition phase diagram for a two-component solution. The solid and dashed curves are the binodal and spinodal, respectively.

As before, region α is taken to contain a fixed amount of component 0. Necessary and sufficient conditions for the above quadratic form in δn_i^α to be positive definitive are

$$\bar{\mu}_{11}{}^\alpha > 0 \tag{1.3.39}$$

and

$$\begin{vmatrix} \bar{\mu}_{11}{}^\alpha & \bar{\mu}_{12}{}^\alpha \\ \bar{\mu}_{21}{}^\alpha & \bar{\mu}_{22}{}^\alpha \end{vmatrix} = \bar{\mu}_{11}{}^\alpha \bar{\mu}_{22}{}^\alpha - (\bar{\mu}_{12}{}^\alpha)^2 > 0 \tag{1.3.40}$$

where the relation $\bar{\mu}_{12} = \bar{\mu}_{21}$ has been used. The condition $\bar{\mu}_{22}{}^\alpha > 0$ follows from these two inequalities.

Now we have the relation

$$\bar{\mu}_{ij}{}^\alpha = (n_0{}^\alpha M_0)^{-1} \mu_{ij} \tag{1.3.41}$$

where

$$\mu_{ij} \equiv (\partial \mu_i / \partial m_j)_{T,p,m_k} = \mu_{ji} \tag{1.3.42}$$

Hence Eqs. (1.3.39) and (1.3.40) are rewritten

$$\mu_{11} > 0, \quad \mu_{11}\mu_{22} - (\mu_{12})^2 > 0 \tag{1.3.43}$$

Although $(\partial \mu_i / \partial x_j)_{x_k} \neq (\partial \mu_j / \partial x_i)_{x_k}$, we have the symmetry relation $\mu_{ij} = \mu_{ji}$, which implies that it is convenient to use molalities m_i as composition variables in discussing the stability of equilibrium of multicomponent solutions in terms of μ_i.

According to the Gibbs phase rule, the state of a single-phase ternary solution can be described by four intensive variables. If T, p, m_1, and m_2 are taken as these variables, states of the system at fixed T and p can be represented on the m_1, m_2 plane. Each of the three curves corresponding to the conditions $\mu_{11} = 0$, $\mu_{22} = 0$, and $\mu_{11}\mu_{22} - (\mu_{12})^2 = 0$ represents the boundary of an unstable region on this plane, but only one has physical meaning as the spinodal of the solution at the specified T and p. For $\mu_{11} = 0$ or $\mu_{22} = 0$ we have $\mu_{11}\mu_{22} - (\mu_{12})^2 < 0$. This means that the curves corresponding to $\mu_{11} = 0$ and $\mu_{22} = 0$ lie inside the unstable region bounded by the curve corresponding to $\mu_{11}\mu_{22} - (\mu_{12})^2 = 0$. It follows, therefore, that the spinodal is given by

$$\mu_{11}\mu_{22} - (\mu_{12})^2 = 0 \tag{1.3.44}$$

and that, in general, $\mu_{11} > 0$ and $\mu_{22} > 0$ on this curve. Since, according to Eq. (A.13),

$$(\partial \mu_2 / \partial m_2)_{\mu_1} = [\mu_{11}\mu_{22} - (\mu_{12})^2] / \mu_{11} \tag{1.3.45}$$

Eq. (1.3.44) is equivalent to

$$(\partial\mu_2/\partial m_2)_{T,p,\mu_1} = 0 \tag{1.3.46}$$

If μ_2, at fixed T, p, and μ_1, is plotted against m_2, we obtain a curve like that illustrated in Fig. 1.6. The maximum and minimum of such a curve, if they exist, are cross-sections of the spinodal, since Eq. (1.3.46) is satisfied at these points. Thus by an argument similar to that for binary solutions we find

$$(\partial^2\mu_2/\partial m_2{}^2)_{T,p,\mu_1} = 0 \tag{1.3.47}$$

and Eq. (1.3.46) as the conditions that determine the critical point for a ternary solution at fixed T and p. The state which satisfies these conditions is represented, for each T and p, by a point on the m_1, m_2 plane.

We now express the various relations derived above in terms of the molar Gibbs free energy G_m and the mole fractions x_1 and x_2. First, using Eq. (1.1.35), we get

$$\begin{aligned}\mu_{ij} &= \sum_{k=1}^{2} (\partial\mu_i/\partial x_k)_{x_\ell} (\partial x_k/\partial m_j)_{m_\ell} \\ &= (m_0 + m_1 + m_2)^{-1}\left[(\partial\mu_i/\partial x_j)_{x_\ell} - \sum_{k=1}^{2} x_k (\partial\mu_i/\partial x_k)_{x_\ell}\right]\end{aligned} \tag{1.3.48}$$

Next, it follows from Eq. (1.1.22b) that

$$\mu_i = G_m + G_i - \sum_{k=1}^{2} x_k G_k \tag{1.3.49}$$

where G_i is defined by

$$G_i \equiv (\partial G_m/\partial x_i)_{T,p,x_\ell} \tag{1.3.50}$$

Differentiation of Eq. (1.3.49) with respect to x_j gives

$$(\partial\mu_i/\partial x_j)_{x_\ell} = G_{ij} - \sum_{k=1}^{2} x_k G_{kj} \tag{1.3.51}$$

where

$$G_{ij} \equiv (\partial^2 G_m/\partial x_i \partial x_j)_{T,p,x_\ell} \tag{1.3.52}$$

With Eq. (1.3.51), we can write Eq. (1.3.48) as

$$\mu_{ij} = (\Sigma_k m_k)^{-1}\left(G_{ij} - \sum_{k=1}^{2} x_k G_{kj} - \sum_{\ell=1}^{2} x_\ell G_{i\ell} + \sum_{k=1}^{2}\sum_{\ell=1}^{2} x_k x_\ell G_{k\ell}\right) \tag{1.3.53}$$

In anticipation of an extension to solutions containing more than three components, we rewrite Eq. (1.3.53) in compact form, using matrices defined by

$$\underset{\sim}{\mu} \equiv \begin{bmatrix} \mu_{11} & \mu_{12} \\ \mu_{21} & \mu_{22} \end{bmatrix} \qquad \underset{\sim}{G} \equiv \begin{bmatrix} G_{11} & G_{12} \\ G_{21} & G_{22} \end{bmatrix}$$
$$\underset{\sim}{x} \equiv \begin{bmatrix} x_1 & x_1 \\ x_2 & x_2 \end{bmatrix} \qquad \underset{\sim}{E} \equiv \begin{bmatrix} 1 & 0 \\ 0 & 1 \end{bmatrix} \tag{1.3.54}$$

The resulting form is

$$\begin{aligned} \underset{\sim}{\mu} &= (\Sigma_k m_k)^{-1} (\underset{\sim}{G} - \underset{\sim}{x}^t \underset{\sim}{G} - \underset{\sim}{G}\underset{\sim}{x} + \underset{\sim}{x}^t \underset{\sim}{G}\underset{\sim}{x}) \\ &= (\Sigma_k m_k)^{-1} (\underset{\sim}{E} - \underset{\sim}{x}^t) \underset{\sim}{G} (\underset{\sim}{E} - \underset{\sim}{x}) \end{aligned} \tag{1.3.55}$$

where $\underset{\sim}{x}^t$ is the transpose of $\underset{\sim}{x}$. Thus, denoting the determinants for $\underset{\sim}{\mu}$ and $\underset{\sim}{G}$ by $|\mu|$ and $|G|$, respectively, we obtain

$$|\mu| = (\Sigma_k m_k)^{-2} (1 - x_1 - x_2)^2 |G| \tag{1.3.56}$$

Equation (1.3.44) for the spinodal is therefore rewritten as

$$|G| = G_{11}G_{22} - (G_{12})^2 = 0 \tag{1.3.57}$$

We also get

$$(\partial G_1/\partial x_1)_{T,p,G_2} = 0 \tag{1.3.58}$$

corresponding to Eq. (1.3.46), and

$$\begin{aligned} \left(\frac{\partial^2 G_1}{\partial x_1{}^2}\right)_{T,p,G_2} &= \left[\frac{\partial}{\partial x_1}\left(\frac{|G|}{G_{22}}\right)\right]_{G_2} = \frac{1}{G_{22}}\left(\frac{\partial |G|}{\partial x_1}\right)_{G_2} \\ &= \frac{1}{G_{22}}\left[\left(\frac{\partial |G|}{\partial x_1}\right)_{x_2} - \left(\frac{\partial |G|}{\partial x_2}\right)_{x_1}\left(\frac{G_{21}}{G_{22}}\right)\right] \\ &= \frac{1}{(G_{22})^2}\left[\left(\frac{\partial |G|}{\partial x_1}\right)_{x_2} G_{22} - \left(\frac{\partial |G|}{\partial x_2}\right)_{x_1} G_{21}\right] = 0 \end{aligned} \tag{1.3.59}$$

corresponding to Eq. (1.3.47). In deriving the latter equation, we have used the condition $|G| = 0$. Equation (1.3.59) may be rearranged in the form

$$\begin{vmatrix} (\partial |G|/\partial x_1)_{x_2} & (\partial |G|/\partial x_2)_{x_1} \\ G_{21} & G_{22} \end{vmatrix} = 0 \tag{1.3.60}$$

Combination of Eq. (1.3.58) and (1.3.60) allows us to determine the critical point.

Figure 1.10 shows schematically a $G_m(x_1, x_2)$ surface above the composition triangle in a case where separation into two phases takes place. The thick dashed line on the G_m surface represents the spinodal, and point C indicates the critical point. By constructing a plane $B_0B_1B_2$ tangent to the G_m surface in the region of the upward indentation, we can obtain two points of contact P′ and P″. Thus we get the relations:

$$\mu_0(T, p, x_1', x_2') = \mu_0(T, p, x_1'', x_2'') \tag{1.3.61a}$$

$$\mu_1(T, p, x_1', x_2') = \mu_1(T, p, x_1'', x_2'') \tag{1.3.61b}$$

$$\mu_2(T, p, x_1', x_2') = \mu_2(T, p, x_1'', x_2'') \tag{1.3.61c}$$

where (x_1', x_2') and (x_1'', x_2'') are the compositions corresponding to points P′ and P″, respectively.

Now, assuming that a ternary solution of composition (x_1, x_2) is separated

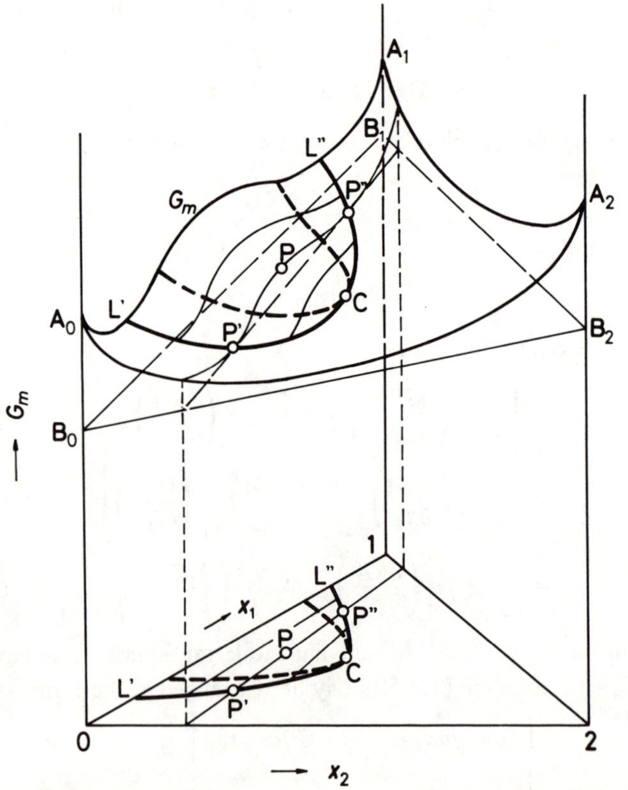

Fig. 1.10. The molar Gibbs free energy surface above the composition triangle for a three-component solution which undergoes separation into two phases.

into two phases characterized by points P′ and P″, we have by the law of mass conservation

$$n = n' + n'' \tag{1.3.62a}$$

$$x_1 n = x_1' n' + x_1'' n'' \tag{1.3.62b}$$

$$x_2 n = x_2' n' + x_2'' n'' \tag{1.3.62c}$$

where n, n', and n'' denote the amounts of the initial solution, the P′ phase, and the P″ phase, respectively. From the first and second equations and from the first and third equations we derive, respectively,

$$\frac{n''}{n} = \frac{x_1 - x_1'}{x_1'' - x_1'} \quad \text{and} \quad \frac{n''}{n} = \frac{x_2 - x_2'}{x_2'' - x_2'} \tag{1.3.63}$$

(the corresponding expressions for n'/n are omitted). Equating these two expressions, we obtain

$$(x_2 - x_2')/(x_1 - x_1') = (x_2'' - x_2')/(x_1'' - x_1') \tag{1.3.64}$$

Figure 1.11 shows points P (x_1, x_2), P′ (x_1', x_2'), and P″ (x_1'', x_2'') projected on the composition triangle. In considering the location of P relative to P′ and P″, we note that $n'' < n$. Equation (1.3.64) requires the two dashed lines in the figure to have the same inclination relative to side $\overline{02}$. Thus points P, P′, and P″ must be colinear. It can also be shown that $n'/n'' = \overline{PP''}/\overline{PP'}$, which is the **lever**

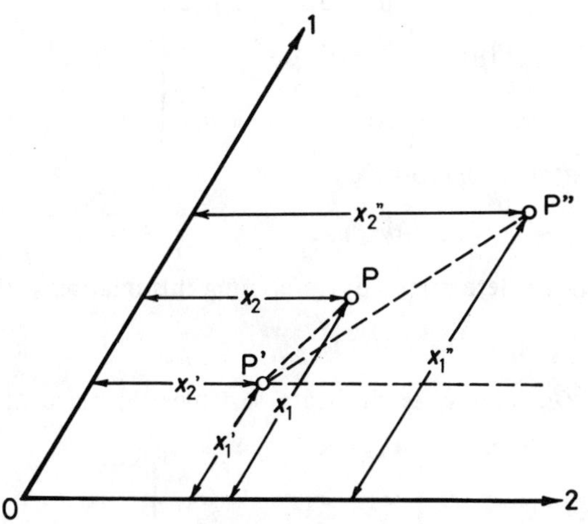

Fig. 1.11. Diagram explaining the lever rule in a three-component system. Points P″, P, and P′ are colinear.

rule for ternary solutions. The line connecting points P′ and P″ is called a **tie line.**

By letting the tangent plane $B_0B_1B_2$ be displaced continuously over the two-phase region of the G_m surface, while maintaining the double contact, we obtain the thick solid curve L′P′CP″L″ in Fig. 1.10 as the locus of points P′ and P″. This curve is the binodal of the ternary solution. Points P′ and P″ merge at the critical point C, which is equivalent to a **plait point** in geometry. The binodal touches the spinodal tangentially at point C. The system is unstable in the region bounded by the spinodal, metastable in the region between the spinodal and the binodal, and absolutely stable in the region outside the binodal.

The details of two-phase equilibrium at specified T, p, x_1, and x_2 can be found from Eqs. (1.3.61a,b,c) and (1.3.64) for x_1', x_2'; x_1'',x_2''. This set of simultaneous equations can in principle be solved, and the solutions, as functions of x_1 and x_2, will yield the binodal at the specified T and p. The resultant binodal does not always assume as simple a shape as that illustrated in Fig. 1.10. This is also the case with the spinodal determined as a solution of Eq. (1.3.57) and we may find more than one critical point from Eqs. (1.3.57) and (1.3.60).

It is not difficult to extend the above treatments of ternary solutions to the general (r + 1)-component solution. We simply replace the upper bound 2 for the double sum in Eq. (1.3.38) by r. Since Eq. (1.3.41) is valid also for the general system, the condition for stable equilibrium is given by $|\mu| > 0$, where $|\mu|$ is the determinant of order r with elements μ_{ij}. In place of Eq. (1.3.44), we have for the spinodal

$$|\mu| = \begin{vmatrix} \mu_{11} & \mu_{12} & \cdots & \mu_{1r} \\ \mu_{21} & \mu_{22} & \cdots & \mu_{2r} \\ \cdots & \cdots & \cdots & \cdots \\ \mu_{r1} & \mu_{r2} & \cdots & \mu_{rr} \end{vmatrix} = 0 \qquad (1.3.65)$$

which may be written in the form

$$(\partial\mu_1/\partial m_1)_{T,p,\mu_2,\ldots,\mu_r} = 0 \qquad (1.3.66)$$

The critical points are determined by combining this relation with

$$(\partial^2\mu_1/\partial m_1^2)_{T,p,\mu_2,\ldots,\mu_r} = 0 \qquad (1.3.67)$$

If we choose G_m and x_i as the state variables, in place of μ_i and m_i, the spinodal is determined by

$$|G| = \begin{vmatrix} G_{11} & G_{12} & \cdots & G_{1r} \\ G_{21} & G_{22} & \cdots & G_{2r} \\ \cdots & \cdots & \cdots & \cdots \\ G_{r1} & G_{r2} & \cdots & G_{rr} \end{vmatrix} = 0 \qquad (1.3.68)$$

and this may be combined with

$$\begin{vmatrix} \partial|G|/\partial x_1 & \partial|G|/\partial x_2 & \dots & \partial|G|/\partial x_r \\ G_{21} & G_{22} & \dots & G_{2r} \\ \dots & \dots & \dots & \dots \\ G_{r1} & G_{r2} & \dots & G_{rr} \end{vmatrix} = 0 \qquad (1.3.69)$$

to calculate the critical point. In Eq. (1.3.69), the partial derivative $\partial|G|/\partial x_i$ is evaluated with mole fractions other than x_i fixed. The theory of general multi-component systems described above is due originally to Gibbs [5].

If we wish to use weight fractions w_i as the composition variables, we may replace G_m in the above equations for the spinodal and the critical points by G_w, the Gibbs free energy per unit mass of solution $[G_w \equiv G/(q_0 + q_1 + \cdots + q_r)]$. If volume fractions ϕ_i defined by Eq. (1.1.30) are chosen as the composition variables, then G_ϕ is defined by

$$G_\phi \equiv G/\Sigma_i n_i V_i^o \qquad (1.3.70)$$

and $G_i^{(\phi)}$ and $G_{ij}^{(\phi)}$, respectively, by

$$G_i^{(\phi)} \equiv (\partial G_\phi/\partial \phi_i)_{T,p,\phi_j}, \quad G_{ij}^{(\phi)} \equiv (\partial^2 G_\phi/\partial \phi_i \partial \phi_j)_{T,p,\phi_k} \qquad (1.3.71)$$

Then Eqs. (1.3.68) and (1.3.69) hold, with substitution of G_{ij} by $G_{ij}^{(\phi)}$.

1.4. Activity and Activity Coefficients

1.4.1. Ideal Solutions

We imagine a process at constant temperature and pressure in which $r + 1$ pure liquids, in an amount (in moles) of n_i of the i-th liquid, are mixed to form a homogeneous solution. The changes in volume, enthalpy, entropy, and Gibbs free energy in this process are expressed as follows:

$$\Delta_m V \equiv V - \sum_{i=0}^{r} n_i V_i^o = \sum_{i=0}^{r} n_i \Delta_m V_i \qquad (1.4.1)$$

$$\Delta_m H \equiv H - \sum_{i=0}^{r} n_i H_i^o = \sum_{i=0}^{r} n_i \Delta_m H_i \qquad (1.4.2)$$

$$\Delta_m S \equiv S - \sum_{i=0}^{r} n_i S_i^o = \sum_{i=0}^{r} n_i \Delta_m S_i \qquad (1.4.3)$$

$$\Delta_m G \equiv G - \sum_{i=0}^{r} n_i \mu_i^o = \sum_{i=0}^{r} n_i \Delta_m \mu_i \qquad (1.4.4)$$

where

$$\Delta_m V_i \equiv V_i - V_i^o \tag{1.4.5}$$

$$\Delta_m H_i \equiv H_i - H_i^o \tag{1.4.6}$$

$$\Delta_m S_i \equiv S_i - S_i^o \tag{1.4.7}$$

$$\Delta_m \mu_i \equiv \mu_i - \mu_i^o \tag{1.4.8}$$

The superscript o refers to a pure liquid, and the subscript m stands for mixing.

If $\Delta_m \mu_i$ is expressed by

$$\Delta_m \mu_i(T, p, x) = RT \ln x_i \tag{1.4.9}$$

over the entire range of composition for all i, the solution is called **ideal**. In Eq. (1.4.9), R is the gas constant [see Eq. (1.1.46)]. The ideal solution is very important, because it provides a reference for studying thermodynamic properties of real solutions. Its properties are easily derived. Substitution of Eq. (1.4.9) into Eqs. (1.2.43) through (1.2.45) gives

$$\Delta_m V_i = (\partial \Delta_m \mu_i / \partial p)_{T,x} = 0 \tag{1.4.10}$$

$$\Delta_m H_i = -T^2 [\partial(\Delta_m \mu_i / T)/\partial T]_{p,x} = 0 \tag{1.4.11}$$

$$\Delta_m S_i = -(\partial \Delta_m \mu_i / \partial T)_{p,x} = -R \ln x_i \tag{1.4.12}$$

It follows that

$$\Delta_m V = 0, \quad \Delta_m H = 0 \tag{1.4.13}$$

$$\Delta_m S = -R \sum_{i=0}^{r} n_i \ln x_i \tag{1.4.14}$$

$$\Delta_m G = RT \sum_{i=0}^{r} n_i \ln x_i \tag{1.4.15}$$

The quantity $\Delta_m S$ given by Eq. (1.4.14) is called the **ideal entropy of mixing.** Since $x_i < 1$, it follows that $\Delta_m S > 0$ and $\Delta_m G < 0$.

a. Raoult's Law. We denote by p_i the partial pressure of component i in the equilibrium vapor phase over an ideal solution, and by p_i^o the vapor pressure of pure liquid component i. If the vapor phase is assumed to behave as an ideal gas mixture, the chemical potential μ_i^g of component i in the vapor phase is given by

$$\mu_i^g = \mu_i^\theta(T) + RT \ln(p_i/p^\theta) \tag{1.4.16}$$

where $\mu_i^\theta(T)$ is the chemical potential of pure gas i at a standard pressure p^θ and temperature T. If this μ_i^g is equated to μ_i for the ideal solution and then μ_i^θ is eliminated between the resulting equation and its particular form for $x_i = 1$, then

$$p_i = p_i^o x_i \exp\left\{[\mu_i^o(T,p) - \mu_i^o(T,p_i^o)]/(RT)\right\} \tag{1.4.17}$$

It can be shown that at ordinary pressures

$$[\mu_i^o(T,p) - \mu_i^o(T,p_i^o)]/(RT) \ll 1 \tag{1.4.18}$$

so that

$$p_i = p_i^o x_i \tag{1.4.19}$$

This is **Raoult's law.**

Equation (1.4.19) for $i = 0$ gives

$$(p_0^o - p_0)/p_0^o = 1 - x_0 = \sum_{i=1}^{r} x_i \equiv x_s \tag{1.4.20}$$

Therefore, the relative vapor pressure lowering of the solvent (component 0) in an ideal solution is equal to the total mole fraction x_s of solute.

b. van't Hoff's Law. Suppose that an $(r + 1)$-component solution is separated from the solvent (component 0) by a membrane permeable only to component 0. The condition for osmotic equilibrium between the solution and solvent phases is

$$\mu_0^o(T,p') = \mu_0(T,p,x), \quad [x \equiv (x_1, x_2, \ldots, x_r)] \tag{1.4.21}$$

The osmotic pressure π is defined as the difference between solution pressure p and solvent pressure p':

$$\pi = p - p' \tag{1.4.22}$$

In the customary osmotic pressure measurement, solvent pressure p' is held fixed and the excess pressure on the solution side at equilibrium is determined as a function of solute concentration. In this case, $\mu_0^o(T, p')$ is kept constant and so is the chemical potential $\mu_0(T, p, x)$ of the solvent component in the solution phase as x is varied. For this reason we shall call π measured at fixed solvent pressure p' the osmotic pressure at constant μ_0. In the other type of measurement, solution pressure p is held fixed and the decrease in solvent pressure needed to attain equilibrium is determined as a function of solute concentration. The osmotic pressure measured in this case is referred to as one at constant p.

We integrate Eq. (1.2.44) with respect to pressure from p' to p and expand the integrand $V_0(T, p, x)$ about a solvent pressure p' held fixed to obtain

$$\mu_0(T,p,x) = \mu_0(T,p',x) + \pi V_0 - \pi^2 V_0 \kappa_0/2 + \ldots \tag{1.4.23}$$

where κ_0 is the isothermal compressibility of the solvent component. Since the compressibility of a liquid is very small, the third and higher terms in the above expansion can be neglected.† For an ideal solution, V_0 is equal to $V_0{}^o$ and $\mu_0(T, p', x)$ is given by Eq. (1.4.9) with p replaced by p', so that Eq. (1.4.23) becomes

$$\mu_0(T, p, x) = \mu_0{}^o(T, p') + RT\,\ell n\, x_0 + \pi V_0{}^o \tag{1.4.24}$$

if the compressibility effect is neglected. Substituting this into Eq. (1.4.21), we obtain for the osmotic pressure at constant μ_0

$$\pi = -(RT/V_0{}^o)\,\ell n\, x_0 = -(RT/V_0{}^o)\,\ell n\,(1 - x_s) \tag{1.4.25}$$

Next we expand $\mu_0{}^o(T, p')$ about a solution pressure p held fixed to obtain

$$\mu_0{}^o(T, p') = \mu_0{}^o(T, p) - \pi V_0{}^o \tag{1.4.26}$$

where the compressibility effect has been neglected. If this is introduced into Eq. (1.4.21), together with Eq. (1.4.9) for $\mu_0(T, p, x)$, we obtain for π the same expression as Eq. (1.4.25). Thus we find that when the solution is ideal, the osmotic pressure at constant μ_0 and the one at constant p have exactly the same dependence on solute concentration.

For dilute solutions in which $x_s \ll 1$, Eq. (1.4.25) reduces to

$$\pi = (RT/V_0{}^o)\, x_s \tag{1.4.27}$$

This limiting form for π is called **van't Hoff's law.**

If x_i is converted to C_i by Eq. (1.1.43), Eq. (1.4.27) is transformed to

$$\pi/RT = \sum_{i=1}^{r} C_i/M_i = C_s/M_n \tag{1.4.28}$$

where

$$C_s \equiv \sum_{i=1}^{r} C_i \tag{1.4.29a}$$

and

$$M_n \equiv \sum_{i=1}^{r} n_i M_i / \sum_{i=1}^{r} n_i = C_s[\Sigma_i(C_i/M_i)]^{-1} = (\Sigma_i \xi_i/M_i)^{-1} \tag{1.4.29b}$$

with

$$\xi_i \equiv C_i/C_s \tag{1.4.29c}$$

†The compressibility of water is about 5×10^{-10} Pa^{-1} (the pascal Pa is the SI unit for pressure; 1 atm = 101.325 kPa = 1.01325×10^5 N m^{-2}). Hence even if π = 200 kPa (about 2 atm) the third term in Eq. (1.4.23) is a factor of 10^{-5} smaller than the second term πV_0.

The quantity M_n defined here represents the molar mass averaged over all solutes in the solution. The molecular weight corresponding to it is usually called the **number-average molecular weight of the solute**; see § 2.2.1.

1.4.2. Activity in Real Solutions

Equation (1.4.15) does not apply to real solutions. Hence we must write in general

$$\Delta_m G = RT \sum_{i=0}^{r} n_i \ln x_i + G^E (T, p, n_0, n_1, \ldots, n_r) \qquad (1.4.30)$$

where G^E is called the excess Gibbs free energy. It follows by using Eqs. (1.2.20), (1.2.21), (1.2.26), and (1.2.38) that

$$\Delta_m S = -(\partial \Delta_m G/\partial T)_{p,n} = -R \sum_{i=0}^{r} n_i \ln x_i - (\partial G^E/\partial T)_{p,n} \qquad (1.4.31)$$

$$\Delta_m V = (\partial \Delta_m G/\partial p)_{T,n} = (\partial G^E/\partial p)_{T,n} \qquad (1.4.32)$$

$$\Delta_m H = -T^2 [\partial(\Delta_m G/T)/\partial T]_{p,n} = -T^2 [\partial(G^E/T)/\partial T]_{p,n} \qquad (1.4.33)$$

$$\Delta_m \mu_i = (\partial \Delta_m G/\partial n_i)_{T,p,n_j} = RT \ln x_i + (\partial G^E/\partial n_i)_{T,p,n_j} \qquad (1.4.34)$$

All the thermodynamic properties of a real solution can be derived from knowledge of its G^E as a function of T, p, and $n_0, n_1, \ldots, n_r$.

Instead of G^E we more often use for the description of real solutions either the **activity coefficient** f_i of component i, defined by

$$\ln f_i \equiv (1/RT)\,(\partial G^E/\partial n_i)_{T,p,n_j} \qquad (1.4.35)$$

or the **activity** a_i (more correctly the **relative activity**):

$$a_i \equiv f_i x_i \qquad (1.4.36)$$

Substitution of these definitions into Eq. (1.4.34) gives

$$\Delta_m \mu_i = RT \ln x_i + RT \ln f_i = RT \ln a_i \qquad (1.4.37a)$$

or

$$\mu_i = \mu_i^o + RT \ln a_i \qquad (1.4.37b)$$

Since $\mu_i \to \mu_i^o$ as $x_i \to 1$, it follows that

$$\lim_{x_i \to 1} a_i = 1, \qquad \lim_{x_i \to 1} f_i = 1 \qquad (1.4.38)$$

Furthermore, referring to Eq. (1.2.39), we get

$$G^E = RT \sum_{i=0}^{r} n_i \ln f_i \tag{1.4.39}$$

Thus $G^E(T, p, n)$ is known if the $f_i(T, p, x)$ $(i = 0, 1, \ldots, r)$ are given, either experimentally or theoretically.

Equation (1.4.37a) has the same form as Eq. (1.4.9) except that x_i is replaced by a_i. Using the procedures described in the previous section, we can derive the following results from Eq. (1.4.37a).

The partial pressure p_i of component i in the equilibrium vapor over a real solution is given by

$$p_i = p_i{}^o a_i(T, p, x) \tag{1.4.40}$$

The osmotic pressure π^* of a real solution at constant μ_0 (i.e., at constant solvent pressure p') is represented by

$$\pi^* = -(RT/V_0) \ln a_0 (T, \mu_0, x) \tag{1.4.41a}$$

and the corresponding osmotic pressure π at constant solution pressure p by

$$\pi = -(RT/V_0{}^o) \ln a_0 (T, p, x) \tag{1.4.41b}$$

The set (T, μ_0, x), rather than (T, p', x), has been taken as the state variables for a_0 in Eq. (1.4.41a). This is permissible, because, by virtue of Eq. (1.4.21), the chemical potential μ_0 of the solvent component in the solution phase at osmotic equilibrium is determined when T and p' are given. It is preferable for thermodynamic analysis that, as in the above equations, the activity is expressed as a function of state properties of the system, not those of its surroundings.

Substitution of Eq. (1.4.37b) into the Gibbs-Duhem relation, Eq. (1.2.42), gives

$$\sum_{i=0}^{r} x_i (\partial \ln a_i / \partial x_j)_{T, p, x_k} = 0, \quad (j = 1, 2, \ldots, r) \tag{1.4.42}$$

For a binary solution this equation reduces to

$$(1 - x_1)(\partial \ln a_0 / \partial x_1)_{T,p} + x_1 (\partial \ln a_1 / \partial x_1)_{T,p} = 0 \tag{1.4.43}$$

These equations impose a restriction on the activities of the components in a solution.

If we measure the solvent vapor pressure p_0 or the osmotic pressure (π^* or π) for a given solution, a_0 can be evaluated from Eq. (1.4.40) or (1.4.41a,b). For binary solutions we can evaluate a_1 from $a_0(x_1)$ by applying Eqs. (1.4.43) and (1.4.38). Multicomponent solutions in which the solutes are mixed in fixed proportions, as is the case with solutions of a polymer heterogeneous in molecular weight, are called **quasibinary** (see § 2.2.3). The thermodynamic properties of such a solution are completely determined if p_0 or π (or π^*) is known as a function of the total solute concentration x_s. It is important to note that the guiding principles for an experimental study of the thermodynamics of solutions can be stated clearly in terms of activity.

1.4.3. Activity Coefficients in Dilute Solution

The thermodynamics of real solutions can be developed further by introducing **Henry's empirical law**, which states that the partial pressure p_i of component i in a vapor phase is proportional to the solubility x_i of the same component in a liquid phase in equilibrium with the vapor:

$$p_i = k_i x_i \tag{1.4.44}$$

The Henry law constant k_i depends only on T and p. Equation (1.4.44) may be considered the first term in the general expansion of p_i in powers of x_i. Hence we might expect that it should be valid in general for $x_i \ll 1$.

Let us carry out a procedure similar to the derivation of Raoult's law from Eq. (1.4.9), this time starting with Eq. (1.4.44). If this equation is inserted into Eq. (1.4.16), we obtain

$$\mu_i^g = \mu_i^\theta (T) + RT \ln (k_i x_i / p^\theta), \quad (i = 1, 2, \ldots, r) \tag{1.4.45a}$$

Because $\mu_i = \mu_i^g$ at equilibrium, this equation gives

$$\mu_i = \mu_i^\infty (T, p) + RT \ln x_i, \quad (i = 1, 2, \ldots, r) \tag{1.4.45b}$$

where

$$\mu_i^\infty (T, p) \equiv \mu_i^\theta (T) + RT \ln (k_i / p^\theta) \tag{1.4.45c}$$

Equation (1.4.45b) is substituted into the Gibbs-Duhem relation, Eq. (1.2.42), to give

$$(\partial \mu_0 / \partial x_j)_{T,p,x_k} = -RT/x_0 = -RT/(1 - \sum_{i=1}^{r} x_i) \tag{1.4.46}$$

If this is integrated with respect to x_j and the limiting condition that $\mu_0 \to \mu_0{}^o$ as $x_0 \to 1$ is invoked, we obtain

$$\mu_0 = \mu_0{}^o(T, p) + RT \ln x_0 \tag{1.4.47}$$

This result indicates that in dilute solutions for which Henry's law holds the chemical potential of the solvent (component 0) is that in an ideal solution. Accordingly, the solvent vapor pressure obeys Raoult's law and the osmotic pressure obeys van't Hoff's law.

As the solution becomes more concentrated, Henry's law breaks down. For such solutions we may put

$$\mu_0 = \mu_0{}^o + RT\ln a_0 = \mu_0{}^o + RT\ln (f_0 x_0) \qquad (1.4.48a)$$

$$\mu_i = \mu_i^\infty + RT\ln a_i = \mu_i^\infty + RT\ln (f_i x_i), \quad (i = 1, 2, \ldots, r) \qquad (1.4.48b)$$

Here, since we have chosen infinite dilution as the standard state for the solutes, we have the limiting conditions

$$\lim_{x_0 \to 1} a_i = 1, \qquad \lim_{x_0 \to 1} f_i = 1, \quad (i = 0, 1, \ldots, r) \qquad (1.4.49)$$

It should be noted that the a_i and f_i $(i = 1, 2, \ldots, r)$ defined by Eq. (1.4.48b) are not the same as the quantities a_i and f_i in the preceding section, the latter being defined with the pure liquid as the standard state of each component. However, we shall not distinguish these activities and activity coefficients by different symbols.

As mentioned in the preceding section, the activity is a concept that is closely associated with experimental work. It is, therefore, expedient to define activity coefficients in different ways to correspond to composition variables that are used in practice.

a. Activity Coefficient γ_i (Molality Scale). At infinite dilution we have the relation $x_i = M_0 m_i$ [Eq. (1.1.43)]. This is inserted into Eq. (1.4.45b) to give

$$\mu_i = \mu_{im}^\theta + RT\ln (m_i/m^\theta), \quad (i = 1, 2, \ldots, r) \qquad (1.4.50a)$$

where

$$\mu_{im}^\theta \equiv \mu_i^\infty + RT\ln (M_0 m^\theta) \qquad (1.4.50b)$$

and m^θ denotes a standard molality, conventionally chosen to be 1 mol kg^{-1}. To express μ_i correctly at finite concentration we introduce a factor γ_i into the argument of the logarithm in Eq. (1.4.50a). Thus

$$\mu_i = \mu_{im}^\theta + RT\ln (\gamma_i m_i/m^\theta), \quad (i = 1, 2, \ldots, r) \qquad (1.4.51)$$

The dimensionless quantity γ_i is called the activity coefficient of component i on the molality scale. The quantity μ_{im}^θ, which may be termed the standard chemical potential of component i on the molality scale, has the following physical significance: for a hypothetical solution in which the chemical potential of every component obeys Eq. (1.4.50a), μ_{im}^θ represents the chemical potential of component i when all components other than component 0 are at the standard concentration m^θ. Eliminating μ_i from Eqs. (1.4.48b) and (1.4.51) by means of Eq. (1.4.50b) and using Eq. (1.1.35) for the relation between x_i and m_i, we obtain

$$\gamma_i = f_i \Big/ \left(1 + M_0 \sum_{j=1}^{r} m_j\right), \quad (i = 1, 2, \ldots, r) \tag{1.4.52}$$

The value of γ_i is not affected by the choice of m^θ. Hence, in place of 1 mol kg^{-1}, we may choose m^θ to be m_0 as defined by Eq. (1.1.34). Then the relative molality m_i/m^θ becomes equal to n_i/n_0, the amount of component i relative to that of component 0, which is often called the **mole ratio**.

b. Activity Coefficient y_i (Volume Molality and Mass Concentration Scale). The relation $x_i = V_0{}^o c_i$, which holds at infinite dilution, is introduced into Eq. (1.4.45b) to obtain

$$\mu_i = \mu_{ic}^\theta + RT\ln\,(c_i/c^\theta), \quad (i = 1, 2, \ldots, r) \tag{1.4.53a}$$

where

$$\mu_{ic}^\theta \equiv \mu_i^\infty + RT\ln\,(V_0{}^o c^\theta) \tag{1.4.53b}$$

The symbol c^θ denotes a standard value of volume molality. The activity coefficient y_i of component i on the volume molality scale is then introduced by

$$\mu_i = \mu_{ic}^\theta + RT\ln\,(y_i c_i/c^\theta), \quad (i = 1, 2, \ldots, r) \tag{1.4.54}$$

Like γ_i, the coefficient y_i is a dimensionless quantity. The physical meaning of the standard chemical potential μ_{ic}^θ of component i on the volume molality scale may be found by considering a hypothetical solution governed by Eq. (1.4.53a). If Eq. (1.4.54) is compared with Eq. (1.4.48b) and then Eq. (1.1.36) is introduced, we obtain the following interrelations among the activity coefficients y_i, f_i, and γ_i:

$$y_i = (V_m/V_0{}^o) f_i = f_i \Big/ \left(V_0{}^o \sum_{i=0}^{r} c_i\right) = (v_M/v_0{}^o)\,\gamma_i \tag{1.4.55}$$

The general relation $C_i = M_i c_i$ allows Eq. (1.4.54) to be expressed as

$$\mu_i = \mu_{iC}^\theta + RT\ln\,(y_i C_i/C^\theta), \quad (i = 1, 2, \ldots, r) \tag{1.4.56a}$$

where

$$\mu_{iC}^\theta \equiv \mu_i^\infty + RT\ln\,(V_0{}^o C^\theta/M_i) \tag{1.4.56b}$$

and C^θ is a standard value of mass concentration. From Eq. (1.4.56a) we see that y_i can also be taken as the activity coefficient consistent with mass concentrations. The value of y_i is not affected by the choice of c^θ and C^θ. Changing these standard concentrations simply alters the values of the standard chemical potentials μ_{ic}^θ and μ_{iC}^θ.

Finally, we note that the activity coefficients γ_i and y_i have the limiting properties

$$\lim_{m_i \to 0} \gamma_i = \lim_{c_i \to 0} y_i = \lim_{C_i \to 0} y_i = 1, \quad (i = 1, 2, \ldots, r) \tag{1.4.57}$$

1.4.4. The Virial Expansion of Osmotic Pressure

For nonelectrolyte solutions all the activity coefficients defined above can be expanded in integral powers of composition variables, and this is also the case with the osmotic pressure.

First, we consider the osmotic pressure measurement at constant temperature and solution pressure p. In this case, since T and p are to be chosen as independent variables, it is convenient to use molalities m_i as the composition variables. We expand $\ell n\, \gamma(T, p, m)$ at fixed T and p in powers m_i as

$$(\ell n\, \gamma_i)_{T,p} = \sum_{j=1}^{r} \beta_{ij} m_j + \tfrac{1}{2} \sum_{j=1}^{r} \sum_{k=1}^{r} \beta_{ijk} m_j m_k + \ldots \tag{1.4.58}$$

where

$$\beta_{ij} \equiv (\partial \ell n\, \gamma_i / \partial m_j)^o_{T,p,m_k} = \beta_{ji} \tag{1.4.59a}$$

$$\beta_{ijk} \equiv (\partial^2 \ell n\, \gamma_i / \partial m_j\, \partial m_k)^o_{T,p,m_\ell} = \beta_{jki} = \beta_{kij} = \ldots \tag{1.4.59b}$$

and the superscript o indicates infinite dilution. The exchangeability of subscripts shown in Eqs. (1.4.59a) and (1.4.59b) follows from Eq. (1.3.42), i.e., $\mu_{ij} = \mu_{ji}$. Inserting Eq. (1.4.58) into Eq. (1.4.51) and differentiating with respect to m_j, we obtain

$$\mu_{ij} = (\partial \mu_i / \partial m_j)_{T,p,m_k} = RT(\delta_{ij} m_i^{-1} + \beta_{ij} + \sum_{k=1}^{r} \beta_{ijk} m_k + \ldots) \tag{1.4.60}$$

where δ_{ij} is the Kronecker delta, i.e.,

$$\delta_{ii} = 1, \quad \delta_{ij} = 0 \quad (i \neq j) \tag{1.4.61}$$

Introducing Eq. (1.4.60) in the Gibbs-Duhem relation, Eq. (1.1.58), to obtain $(\partial \mu_0 / \partial m_j)_{T,p,m_k}$ and combining the result with the partial derivative of Eq. (1.4.41b) with respect to m_j, we get

$$\begin{aligned} V_0{}^o(\partial \pi / \partial m_j)_{T,p,m_k} &= -(\partial \mu_0 / \partial m_j)_{T,p,m_k} = M_0 \sum_{i=1}^{r} m_i \mu_{ij} \\ &= M_0 RT \sum_{i=1}^{r} \left(\delta_{ij} + \beta_{ij} m_i + \sum_{k=1}^{r} \beta_{ijk} m_i m_k + \ldots\right) \end{aligned} \tag{1.4.62}$$

With the condition that $\pi \to 0$ as all m_i values approach zero, integration gives

$$\pi v_0{}^o/RT = \sum_{i=1}^{r} m_i + \tfrac{1}{2}\sum_{i=1}^{r}\sum_{j=1}^{r} \beta_{ij} m_i m_j + \tfrac{1}{3}\sum_{i=1}^{r}\sum_{j=1}^{r}\sum_{k=1}^{r} \beta_{ijk} m_i m_j m_k + \ldots \tag{1.4.63}$$

Comparison with Eq. (1.4.58) shows that the osmotic pressure at constant solution pressure p can be expanded in powers of m_i with the same expansion coefficient as in the m_i-expansion of $(\ell n\, \gamma_i)_{T,p}$ for solute components.

As has been noted above, osmotic pressure measurements are usually made holding fixed the pressure p' of the solvent phase as solute concentration is varied. In this case, the chemical potential μ_0 of the solvent component in the solution phase is held fixed, and the activities of the solute components are treated as functions of T, μ_0, and composition. If we choose mass concentrations C_i as the composition variables and expand $\ell n\, y_i(T, \mu_0, C)$ (y_i being the activity coefficient appropriate for C_i) in powers of C_i, we obtain

$$(\ell n\, y_i)_{T,\mu_0} = M_i\left(\sum_{j=1}^{r} B_{ij} C_j + \tfrac{1}{2}\sum_{j=1}^{r}\sum_{k=1}^{r} B_{ijk} C_j C_k + \ldots\right) \tag{1.4.64}$$

Here M_i has been factored out to make the expansion coefficients B_{ij}, B_{ijk}, etc., invariant with exchange of subscripts [see Eq. (1.4.83)]. Substitution of Eq. (1.4.64) into Eq. (1.4.56a), followed by differentiation with respect to C_j, yields

$$M_i^{-1}(\partial\mu_i/\partial C_j)_{T,\mu_0,C_k} = RT\left[\delta_{ij}(M_i C_i)^{-1} + B_{ij} + \sum_{k=1}^{r} B_{ijk} C_k + \ldots\right] \tag{1.4.65}$$

At constant T and μ_0, Eq. (1.2.41) gives

$$V(dp)_{T,\mu_0} = \sum_{i=1}^{r} n_i (d\mu_i)_{T,\mu_0} \tag{1.4.66}$$

Since solvent pressure p' is fixed in the case under consideration, it follows from Eq. (1.4.22) that $d\pi^*$ equals dp. Here π^* denotes the osmotic pressure at constant μ_0. Thus we obtain from Eq. (1.4.66)

$$(\partial\pi^*/\partial C_j)_{T,\mu_0,C_k} = \sum_{i=1}^{r} (C_i/M_i)(\partial\mu_i/\partial C_j)_{T,\mu_0,C_k} \tag{1.4.67}$$

Substituting Eq. (1.4.65) and integrating with respect to C_j, we get

$$\pi^*/RT = \sum_{i=1}^{r} C_i/M_i + \tfrac{1}{2}\sum_{i=1}^{r}\sum_{j=1}^{r} B_{ij} C_i C_j + \tfrac{1}{3}\sum_{i=1}^{r}\sum_{j=1}^{r}\sum_{k=1}^{r} B_{ijk} C_i C_j C_k + \ldots \tag{1.4.68}$$

where the integration constant has been determined from the physical condition that π^* vanishes at infinite dilution.

When, as in solutions of a polymer which is polydisperse in molecular weight, the solute components are mixed in fixed proportions, Eq. (1.4.68) can be expressed as

$$\pi^*/RT = C_s/M_n + A_2^{OS} C_s^2 + A_3^{OS} C_s^3 + \cdots \tag{1.4.69}$$

where

$$A_2^{OS} = \frac{1}{2} \sum_{i=1}^{r} \sum_{j=1}^{r} \xi_i \xi_j B_{ij} \tag{1.4.70}$$

$$A_3^{OS} = \frac{1}{3} \sum_{i=1}^{r} \sum_{j=1}^{r} \sum_{k=1}^{r} \xi_i \xi_j \xi_k B_{ijk} \tag{1.4.71}$$

Here C_s is the mass concentration of the entire solute, i.e., $C_s = \Sigma_i C_i$, M_n is its number-average molar mass, and ξ_i is the weight fraction of component i in the solute [see Eq. (1.4.29c)]. The C_s-expansion of π^* represented by Eq. (1.4.69) is called the **virial expansion** for osmotic pressure, and A_2^{OS} and A_3^{OS} are termed the **osmotic second and third virial coefficients**, respectively. The coefficient A_2^{OS} is an average of the binary interaction coefficients B_{ij} (which are functions of T and μ_0) for all solute components. It depends on the molecular weight distribution in the solute as well as the values of the B_{ij}.

As can be understood from the argument given above, the situation is clear-cut if we distinguish the osmotic pressure measurements at fixed p and at fixed μ_0 (or p') and use m_i for the former and C_i for the latter as the composition variables. However, in actuality, we often use activity coefficients determined from osmotic measurements at constant μ_0 to describe thermodynamic data obtained at constant solution pressure. In such cases, caution is needed.

The Gibbs free energy is the characteristic function for systems under constant temperature and pressure. It is, therefore, natural to expand $\ln y_i$ for fixed T and p in powers of C_i. For this reason Eq. (1.4.64) must be regarded as a rather special expansion, because it refers to fixed T and μ_0. We can write the C_i-expansion of $\ln y_i$ at T and p as

$$(\ln y_i)_{T,p} = M_i \left(\sum_{j=1}^{r} \mathscr{B}_{ij} C_j + \frac{1}{2} \sum_{j=1}^{r} \sum_{k=1}^{r} \mathscr{B}_{ijk} C_j C_k + \ldots \right) \tag{1.4.72}$$

Before proceeding further, it is worth remarking that the appropriate composition variables for systems at constant T and p are not mass concentrations but molalities. The coefficients $\mathscr{B}_{ij}$, $\mathscr{B}_{ijk}$, etc., are not the same as the B coefficients defined by Eq. (1.4.64). In fact, since

$$M_i^{-1} (\partial \mu_i / \partial C_j)_{T,p,C_k} \neq M_j^{-1} (\partial \mu_j / \partial C_i)_{T,p,C_k} \tag{1.4.73}$$

the $\mathscr{B}$ coefficients are not symmetric with respect to exchange of the first two subscripts.

We can relate the $\mathscr{B}$ and β coefficients by using the conversion relations between C_i and m_i and between y_i and γ_i, i.e., Eqs. (1.1.40) and (1.4.55). Since these expressions contain v_M, we first must obtain the C-expansion of v_M. By use of Eq. (1.1.71) it can be shown that

$$v_M = v_M{}^o + \sum_{j=1}^{r} (\partial v_M/\partial m_j)^o_{T,p,m_k} m_j$$

$$+ \tfrac{1}{2} \sum_{j=1}^{r} \sum_{k=1}^{r} (\partial^2 v_M/\partial m_j \partial m_k)^o_{T,p,m_\ell} m_j m_k + \cdots \tag{1.4.74}$$

$$= v_0{}^o \left\{1 + \Sigma_j v_j{}^o C_j + \Sigma_j \Sigma_k [v_j^o v_k{}^o + \tfrac{1}{2} (\partial v_j/\partial C_k)^o] C_j C_k + \ldots\right\}$$

We note that $(\partial v_j/\partial C_k)^o = (\partial v_k/\partial C_j)^o$ and that the superscript zero here indicates infinite dilution rather than pure state. We introduce Eqs. (1.4.58) and (1.4.72) into Eq. (1.4.55) and, with help of Eqs. (1.1.40) and (1.4.74), rearrange the right-hand side of the resulting equation in powers of C_i. Then it follows from comparison of both sides that

$$\mathscr{B}_{ij} = \beta_{ij} v_0{}^o/(M_i M_j) + v_j^o/M_i \tag{1.4.75a}$$

$$\mathscr{B}_{ijk} \simeq \beta_{ijk} (v_0{}^o)^2/(M_i M_j M_k) + \beta_{ij} v_k{}^o v_0{}^o/(M_i M_j)$$

$$+ \beta_{ik} v_j^o v_0{}^o/(M_i M_k) + v_j^o v_k{}^o/M_i \tag{1.4.75b}$$

Solving for β_{ij} and β_{ijk} gives

$$\beta_{ij} = (M_i M_j/v_0{}^o) (\mathscr{B}_{ij} - v_j^o/M_i) \tag{1.4.76a}$$

$$\beta_{ijk} \simeq (M_i M_j M_k/(v_0{}^o)^2) (\mathscr{B}_{ijk} - \mathscr{B}_{ij} v_k{}^o - \mathscr{B}_{ik} v_j^o + v_j{}^o v_k{}^o/M_i) \tag{1.4.76b}$$

The approximate equality sign was used in Eqs. (1.4.75b) and (1.4.76b), because terms involving $(\partial v_j/\partial C_k)^o$ were neglected. As can be verified with help of Eqs. (1.1.71) and (1.1.40), this approximation, i.e., $(\partial v_j/\partial C_k)^o = 0$, corresponds to neglecting the second term in Eq. (1.1.74), i.e., assuming that the dependence of v_0 on C_i starts with a term of third order in C_i. It can be shown that, in the same approximation, the C_i-expansions of π and π^* agree through the third order in C_i.

We now introduce Eqs. (1.4.76a) and (1.4.76b) into Eq. (1.4.63), convert m_i to C_i, and equate the resulting expression for π to Eq. (1.4.68) to obtain the following relations between the $\mathscr{B}$ and B coefficients

$$B_{ij} = \mathscr{B}_{ij} + v_i{}^o/M_j \tag{1.4.77a}$$

$$B_{ijk} \simeq \mathscr{B}_{ijk} + \mathscr{B}_{jk} v_i^o + v_i^o v_j^o / M_k \tag{1.4.77b}$$

Finally, we write relations between the B and β coefficients:

$$M_i M_j B_{ij} = \beta_{ij} v_0{}^o + V_i^o + V_j^o \tag{1.4.78a}$$

$$\begin{aligned} M_i M_j M_k B_{ijk} &\simeq \beta_{ijk} (v_0{}^o)^2 + v_0{}^o (\beta_{ij} V_k{}^o + \beta_{jk} V_i^o + \beta_{ki} V_j^o) \\ &\quad + (V_i^o V_j^o + V_j^o V_k{}^o + V_k{}^o V_i^o) \end{aligned} \tag{1.4.78b}$$

As would be expected, these two relations are invariant with exchange of subscripts.

Thermodynamic treatments of systems of constant μ_0 are given a detailed account in § 2.3.4. Here we note that the characteristic function of such systems is given by

$$\mathscr{A}(T, V, \mu_0, n_1, \ldots, n_r) \equiv A(T, V, n_0, \ldots, n_r) - \mu_0 n_0 \tag{1.4.79}$$

and has the following properties

$$d\mathscr{A} = -SdT - pdV - n_0 d\mu_0 + \sum_{i=1}^{r} \mu_i dn_i \tag{1.4.80}$$

$$\mu_i = (\partial \mathscr{A}/\partial n_i)_{T, V, \mu_0, n_k}, \quad (i = 1, 2, \ldots, r) \tag{1.4.81}$$

From the latter relations we obtain for the derivatives of μ_i with respect to c_i and C_j

$$\begin{aligned} V^{-1} (\partial \mu_i/\partial c_j)_{\mu_0, c_k} &= (\partial \mu_i/\partial n_j)_{V, \mu_0, n_k} \\ &= \partial^2 \mathscr{A}/\partial n_j \, \partial n_i = V^{-1} (\partial \mu_j/\partial c_i)_{\mu_0, c_k} \end{aligned} \tag{1.4.82}$$

$$\begin{aligned} M_i^{-1} (\partial \mu_i/\partial C_j)_{\mu_0, C_k} &= (M_i M_j)^{-1} (\partial \mu_i/\partial c_j)_{\mu_0, c_k} \\ = (M_i M_j)^{-1} (\partial \mu_j/\partial c_i)_{\mu_0, c_k} &= M_j^{-1} (\partial \mu_j/\partial C_i)_{\mu_0, C_k} \end{aligned} \tag{1.4.83}$$

These relations guarantee exchangeability of the subscripts in the expansion coefficients in the C-expansion of $\ln y_i$ at constant μ_0. This indicates that both c_i and C_i are appropriate composition variables for systems at constant μ_0, just as m_i is so for systems at constant pressure.

1.4.5. Remarks on the Choice of Composition Variables

Now that we have come to the end of our introduction to solution thermodynamics it may be useful to offer a few comments on the features of the composition variables so far introduced. We would hope thus to dispel any confusion arising from the appearance of so many kinds of composition variables.

The conventionally used composition variables can be classified into three groups. The first group includes the mole fraction x_i and weight fraction w_i, which are defined as the amount and mass of each component relative to the amount and mass of the solution, respectively. These quantities are independent of temperature T and pressure p and hence are the appropriate composition variables when T and p are chosen as the other independent variables to describe the state of a system. For a closed system they do not change whatever processes may take place within it. However, in open systems, there is a change in the total amount n of the system when a component, say, component i, is added to or removed from the system. Thus when x_i changes, the mole fractions of all other components simultaneously vary. Therefore, the derivative operator $(\partial/\partial n_i)_{T,p,n_j}$ cannot simply be equated to $(1/n)\,(\partial/\partial x_i)_{T,p,x_j}$, and we need an expression like Eq. (1.3.19) to relate these two operators. Consequently, we have $(\partial Y_i/\partial x_j)_{T,p,x_k} \neq (\partial Y_j/\partial x_i)_{T,p,x_k}$, where Y_i is the partial molar Y of component i.

The molality m_i and mole ratio n_i/n_0 (see § 1.4.3), which are, respectively, the amounts of component i relative to the mass or to the amount of a particular component 0 chosen as the principal solvent, belong to the second group of composition variables. Like x_i and w_i, both m_i and n_i/n_0 are independent of T and p. However, unlike the mole fractions and weight fractions, only m_i and n_i/n_0 change when component i is added or removed from the system. Thus we have the relation $(\partial/\partial n_i)_{T,p,n_j} = (1/n_0\,M_0)\,(\partial/\partial m_i)_{T,p,n_0,m_j}$, where M_0 is the molar mass of the principal solvent. The subscript n_0 can be omitted when the quantity to be differentiated is intensive. Thus there exists the useful relation $(\partial Y_i/\partial m_j)_{T,p,m_k} = (\partial Y_j/\partial m_i)_{T,p,m_k}$ which greatly simplifies thermodynamic calculations. It is also important to appreciate that $(\partial/\partial m_i)_{T,p,n_0,m_j}$ corresponds to experimental procedures, i.e., the derivatives that are directly measurable. This is not the case with $(\partial/\partial x_i)_{T,p,x_j}$, because it is impossible to keep all the mole fractions other than x_i constant when the amount of component i is changed. For these reasons the molality is the most appropriate choice for thermodynamic analysis of systems with fixed n_0 at specified T and p.

The third group of composition variables include the volume molality (or molar concentration) c_i and the mass concentration C_i. Only when the volume V of the system is fixed, can c_i and C_i be changed, with the volume molalities or mass concentrations of all other components held constant. Thus, the relations $(\partial/\partial n_i)_{T,V,n_j} = (1/V)\,(\partial/\partial c_i)_{T,V,c_j} = (M_i/V)\,(\partial/\partial C_i)_{T,V,C_j}$ hold, and we have $(\partial\mu_i/\partial c_j)_{T,V,c_k} = (\partial\mu_j/\partial c_i)_{T,V,c_k}$, where μ_i is the chemical potential of component i. These relations are useful in thermodynamic analysis of isochoric systems. For the formulation of systems at constant μ_0, the choice of V as a state variable becomes almost imperative. Hence c_i and C_i are also appropriate composition variables for such systems.

As far as intensive properties are concerned, we may choose c_i or C_i as a state variable together with T and p. But, in this case, $(\partial/\partial n_i)_{T,p,n_j}$ is no longer equal to $(1/V)\,(\partial/\partial c_i)_{T,p,c_j}$, because a change in n_i, at fixed T, p, and n_j $(j \neq i)$, is accompanied by a change in V and hence by variations of the c_j of all other components. Thus we have $(\partial\mu_i/\partial c_j)_{T,p,c_k} \neq (\partial\mu_j/\partial c_i)_{T,p,c_k}$, which is inconvenient for thermodynamic calculations. It must also be noted that, like $(\partial/\partial x_i)_{T,p,x_j}$, $(\partial/\partial c_i)_{T,p,c_j}$ can be realized only in a thought experiment. The same is true of $(\partial/\partial T)_{p,c}$ and $(\partial/\partial p)_{T,c}$, because V changes with temperature and pressure. Here c is a shorthand notation for the set $(c_0, c_1, \ldots, c_r)$. In spite of these disadvantages, the set of variables T, p, c has been widely used, along with the set of variables T, p, m, for the description of thermodynamic data on solutions. This is probably due to the fact that most experiments on liquid solutions are done at specified T and p. Furthermore, solutions with desired concentrations on the volume basis are easily prepared and volume concentrations can be determined simply, for example, by an appropriate optical device.

1.5. Regular Solutions

1.5.1. Critical Point of Regular Solutions

The concept of a regular solution was first proposed by Hildebrand [6]. The ideal solution is characterized by the three conditions in Eqs. (1.4.13) and (1.4.14). According to Hildebrand, a solution is called **regular** when it obeys two of these conditions except $\Delta_m H = 0$. Thus the regular solution is defined by

$$\Delta_m V = 0 \tag{1.5.1a}$$

$$\Delta_m S = -R \sum_{i=0}^{r} n_i \ln x_i \tag{1.5.1b}$$

Any molecule in the pure state of a substance is surrounded by molecules of the same kind. When it goes into a solution, it comes into contact with molecules of other kinds. It seems reasonable to assume that there arises an energy of mixing $\Delta_m U$ which is proportional to the frequency of pairwise contacts of molecules of different species. The relative volume occupied by a component in a solution is represented by the volume fraction ϕ_i. If it is assumed that the probability of contact between a molecule of component i and one of component j is proportional to $\phi_i\,\phi_j$, then

$$\Delta_m U = V \sum_{i=0}^{r-1} \sum_{j=i+1}^{r} b_{ij}\, \phi_i\, \phi_j \tag{1.5.2}$$

where V is the volume of the solution, and b_{ij} is the proportionality factor depending on the (i, j) pair. When the condition $\Delta_m V = 0$, one of the conditions for a regular solution, holds, then $\Delta_m U = \Delta_m H$, so that Eq. (1.5.2) can be regarded as an expression for $\Delta_m H$. This type of expression for the enthalpy of mixing is usually called the **van Laar equation** [7]. It has been demonstrated that, except for the systems which involve hydrogen bonding or unusual molecular interactions, many nonelectrolyte solutions of small molecules obey approximately the basic conditions of the regular solutions, i.e., Eqs. (1.5.1a) and (1.5.1b), and the van Laar equation for $\Delta_m H$.

For binary regular solutions we have

$$\Delta_m G = RT\,(n_0\, \ell n\, x_0 + n_1\, \ell n\, x_1) + V b_{01}\, \phi_0\, \phi_1 \tag{1.5.3}$$

by introducing Eq. (1.5.1b) and the van Laar equation into the relation $\Delta_m G = \Delta_m H - T\Delta_m S$. It follows from Eq. (1.5.3) that

$$\mu_0 = \mu_0{}^o + RT\, \ell n\, x_0 + V_0{}^o\, b_{01}\, \phi_1{}^2 \tag{1.5.4a}$$

$$\mu_1 = \mu_1{}^o + RT\, \ell n\, x_1 + V_1{}^o\, b_{01}\, \phi_0{}^2 \tag{1.5.4b}$$

The critical point (T_c, x_{1c}) of this binary solution is determined by

$$\frac{1}{x_{1c}} - 2\left(\frac{b_{01}}{RT_c}\right) \frac{(1 - x_{1c})\,(V_0{}^o\, V_1{}^o)^2}{[V_0{}^o + x_{1c}\,(V_1{}^o - V_0{}^o)]^3} = 0 \tag{1.5.5}$$

and

$$-\frac{1}{x_{1c}{}^2} + 2\left(\frac{b_{01}}{RT_c}\right) \frac{[V_1{}^o + 2\,(1 - x_{1c})\,(V_1{}^o - V_0{}^o)]\,(V_0{}^o\, V_1{}^o)^2}{[V_0{}^o + x_{1c}\,(V_1{}^o - V_0{}^o)]^4} = 0 \tag{1.5.6}$$

These equations may be solved for x_{1c} and T_c to give

$$x_{1c} = \frac{V_1{}^o - Q}{V_1{}^o - V_0{}^o} \tag{1.5.7}$$

$$\frac{b_{01}}{RT_c} = \frac{(V_0{}^o + V_1{}^o - Q)^3\,(V_0{}^o + Q)\,(V_1{}^o + Q)}{2\,(V_0{}^o\, V_1{}^o)^3} \tag{1.5.8}$$

where Q is defined by

$$Q \equiv [(V_0{}^o)^2 - V_0{}^o\, V_1{}^o + (V_1{}^o)^2]^{1/2} \tag{1.5.9}$$

When the two components have equal molar volumes, i.e., $V_0^o = V_1^o$, these expressions yield

$$x_{1c} = 1/2, \quad V_0^o b_{01}/RT_c = 2 \tag{1.5.10}$$

Table 1.1 gives measured values of T_c and x_{1c} for binary mixtures of $(C_3H_7COOCH_2)_4C$, a substance of relatively large molar volume, with liquids having much smaller molar volumes. It is seen that Eq. (1.5.7) predicts the experimental values of x_{1c} quite closely even for binary solutions in which V_1^o differs markedly from V_0^o.

As will be shown in the next chapter, solutions of flexible linear polymers have very large excess entropies of mixing. According to the Flory-Huggins theory for such solutions, the chemical potential of the solvent component is given by

$$\mu_0 = \mu_0^o + RT\left\{\ln \phi_0 + [1 - (V_0^o/V_1^o)]\,\phi_1\right\} + V_0^o b_{01}\,\phi_1^2 \tag{1.5.11}$$

and the critical composition x_{1c} is expressed by

$$x_{1c} = (V_0^o)^{3/2}/[(V_0^o)^{3/2} + (V_1^o)^{3/2}] \tag{1.5.12}$$

The values of x_{1c} calculated by this equation are also shown in Table 1.1. Equation (1.5.7) is obviously better than Eq. (1.5.12) for these systems [8].

Since, as is illustrated here, the regular-solution theory applies well to systems composed of low-molecular-weight substances differing in molecular size, it is possible to develop theories applicable to a wide class of solutions by incorporating refinements and corrections [9].

Table 1.1. Critical points for binary solutions of components differing in molar volume* : Solute (component 1) $(C_3H_7COOCH_2)_4C$

Solvent (component 0)	T_c / °C	V_0^o / cm^3	V_1^o / cm^3	x_{1c} (obs)	x_{1c} (calc) Eq. (1.5.7)	x_{1c} (calc) Eq. (1.5.12)
CH_2Cl_2	38.5	65.7	549.5	0.057	0.061	0.040
$CHCl_3$	43.5	82.6	553	0.073	0.077	0.055
n-C_5H_{12}	62.1	123.8	564	0.13	0.115	0.093
CCl_4	72.1	103	571	0.09	0.094	0.079
cyclo-C_5H_{10}	80.8	101	575	0.10	0.092	0.069
i-C_8H_{18}	78.8	178	573	0.17	0.164	0.147
n-C_8H_{18}	119.4	185	603	0.16	0.162	0.145

*Indicated values of V_0^o and V_1^o refer to $T = T_c$.

1.5.2. The Solubility Parameter

An important quantity associated with the regular-solution theory is the solubility parameter introduced by Hildebrand, which has to do with an approximate estimation of the b_{ij} constants involved in the van Laar equation (1.5.2). Once the magnitude of b_{ij} is known, we can predict solubility characteristics of the system at any given temperature. Hence, any method which allows b_{ij} to be estimated has great practical value.

We denote by E_i^o the molar energy of vaporization of component i at a given temperature. The molar energy U_i^o of that component of the pure liquid relative to the gas is written

$$U_i^o = -E_i^o \tag{1.5.13}$$

Hence the total energy U^o of a system of $r + 1$ isolated liquid components is expressed by

$$U^o = \sum_{i=0}^{r} n_i U_i^o = -\sum_{i=0}^{r} n_i E_i^o \tag{1.5.14}$$

The quantity U_i^o/V_i^o is the energy per unit volume of pure liquid i and may be regarded as a measure of the interaction energy for contacts of like molecules. We introduce $U_{ij}/(V_i^o V_j^o)^{1/2}$, an energy per unit volume, as a measure of the energy of interaction between unlike molecules i and j. It is assumed that component i in the solution occupies a volume $n_i V_i^o$ and that the probability of an i, j molecular contact is $\phi_i \phi_j$. Then the total energy U of the solution is represented by

$$U = -V \sum_{i=0}^{r} \left[\phi_i^2 \frac{E_i^o}{V_i^o} - \phi_i \sum_{j \neq i} \phi_j \frac{U_{ij}}{(V_i^o V_j^o)^{1/2}} \right] \tag{1.5.15}$$

The energy of mixing $\Delta_m U$ is, therefore, given by

$$\Delta_m U = V \sum_{i=0}^{r-1} \sum_{j=i+1}^{r} \left[\frac{E_i^o}{V_i^o} - 2 \frac{E_{ij}}{(V_i^o V_j^o)^{1/2}} + \frac{E_j^o}{V_j^o} \right] \phi_i \phi_j \tag{1.5.16}$$

where U_{ij} has been written $-E_{ij}$.

For solutions in which London dispersion forces dominate molecular interactions, the interaction energy between a pair of unlike molecules is approximately equal to the geometric mean of the interaction energies between pairs of like molecules. Thus we assume for such a solution

$$E_{ij} = (E_i^o E_j^o)^{1/2} \tag{1.5.17}$$

Introducing this into Eq. (1.5.16), we obtain the Scatchard-Hildebrand equation [10]

$$\Delta_m U = V \sum_{i=0}^{r-1} \sum_{j=i+1}^{r} (\delta_i - \delta_j)^2 \phi_i \phi_j \qquad (1.5.18)$$

with

$$\delta_i \equiv (E_i{}^o / V_i{}^o)^{1/2} \qquad (1.5.19)$$

The quantity δ_i is called the **solubility parameter** of component i [11, 12]. The energy of vaporization $E_i{}^o$ needed to compute δ_i may be obtained from

$$E_i{}^o = \Delta_v H_i - RT \qquad (1.5.20)$$

where $\Delta_v H_i$ is the molar heat of vaporization of component i.

From comparison of Eqs. (1.5.18) and (1.5.2) we find that the coefficient b_{ij} in the van Laar equation for the enthalpy of mixing is estimated from the solubility parameters of components i and j by

$$b_{ij} = (\delta_i - \delta_j)^2 \qquad (1.5.21)$$

This expression indicates that the value of b_{ij} is either positive or zero. Therefore, mixing of liquids governed by London dispersion forces is either endothermic ($\Delta_m H > 0$) or athermal ($\Delta_m H = 0$) if volume change on mixing is negligible. In general, the greater the difference in δ between two liquids, the larger the $\Delta_m H$ becomes, and thus the two liquids become less miscible. For this reason the best solvent for a substance is the solute itself.

1.5.3. Osmotic Pressure of Regular Solutions

When the chemical potential of the solvent is represented by Eq. (1.5.4a), the second virial coefficient $A_2{}^{OS}$ for a binary solution is given by

$$A_2{}^{OS} = (V_1{}^o/M_1{}^2)\,[1 - (1/2)(V_0{}^o/V_1{}^o) - (V_1{}^o\, b_{01}/RT)] \qquad (1.5.22)$$

In many cases, b_{01} is positive, so that the upper limit of $A_2{}^{OS}$ may be taken as $V_1{}^o/M_1{}^2$. In other words, the upper limit of $A_2{}^{OS}M_1$ would be the specific volume of the solute $v_1{}^o = V_1{}^o/M_1$, which is of the order of 1 cm^3 g^{-1} for ordinary low-molecular-weight substances.

Equation (1.4.69) may be written, for binary solutions, as

$$\pi/RT = (C_1/M_1)(1 + A_2{}^{OS}M_1\, C_1 + \ldots) \qquad (1.5.23)$$

Table 1.2. Osmotic pressure of aqueous sucrose solutions

a. Concentration dependence (15°C)			b. Temperature dependence (C = 0.01 g cm^{-3})		
C	π	$(\pi/C) \times 10^{-2}$	T	π	π/T
g cm^{-3}	mm Hg	g^{-1} cm^3 mm Hg	K	mm Hg	mm Hg K^{-1}
0.01	535	535	280.0	505	1.80
0.02	1016	503	295.2	548	1.85
0.04	2082	521	305.2	544	1.79
0.06	3075	513	309.2	567	1.83

If the upper limit of $A_2{}^{OS}M_1$ is of the order of 1 cm^3 g^{-1}, the second term in the parentheses in this expression does not exceed a few percent of the first term when C_1 is less than 0.02 to 0.03 g cm^{-3}. In such cases, van't Hoff's law will be obeyed within the accuracy of experimental error. Table 1.2 shows part of the historic experimental data of Pfeffer [13] on aqueous solutions of sucrose. No deviation from van't Hoff's law can be seen in these ranges of concentration and temperature.

References

1. "Manual of Symbols and Terminology for Physicochemical Quantities and Units," Pure Appl. Chem. **21**, 1 (1970); see also M. L. McGlashan, "Physicochemical Quantities and Units (The Grammar and Spelling of Physical Chemistry)," 2nd Ed., Royal Institute of Chemistry, London, 1971.
2. J. R. Rowlinson, "Liquids and Liquid Mixtures," Butterworths, London, 1959.
3. E. F. Casassa and H. Eisenberg, Adv. Protein Chem. **19**, 287 (1964).
4. R. Kikuchi and D. de Fontaine, Scr. Metall. **10**, 995 (1976).
5. J. W. Gibbs, "Collected Works," Vol. I, p. 132, Yale Univ. Press, New Haven, CT, 1948.
6. J. H. Hildebrand, J. Am. Chem. Soc. **51**, 66 (1929); J. H. Hildebrand and R. L. Scott, "Regular Solutions," Prentice-Hall, Englewood Cliffs, NJ, 1962.
7. J. J. van Laar, Z. Phys. Chem. **72**, 723 (1910).
8. K. Shinoda and J. H. Hildebrand, J. Phys. Chem. **62**, 481 (1958); **65**, 1885 (1961).
9. J. H. Hildebrand, J. M. Prausnitz, and R. L. Scott, "Regular and Related Solutions," Van Nostrand-Reinhold, New York, 1970.
10. G. Scatchard, Chem. Rev. **8**, 321 (1931); J. H. Hildebrand and S. E. Wood, J. Chem. Phys. **1**, 817 (1933).
11. J. H. Hildebrand and R. L. Scott, "The Solubility of Nonelectrolytes," Appendix I, Reinhold, New York, 1950.

12. H. Burrel, Solubility Parameter Values, in "Polymer Handbook," 2nd Ed., J. Brandrup and E. H. Immergut, Eds., p. IV-337, Wiley, New York, 1975.
13. W. Pfeffer, "Osmotische Untersuchungen," W. Engelmann, 1877. The data cited in Table 1.2 are taken from S. Glasstone, "Elements of Physical Chemistry," 2nd Ed., pp. 245, 246, Van Nostrand, New York, 1960.

Thermodynamics of Polymer Solutions

CHAPTER 2

Thermodynamics of Polymer Solutions

2.1. The Flory-Huggins Theory

2.1.1. Anomalous Osmotic Pressure

In the 1930's, Staudinger and his collaborators found that polymer solutes can be classified into two groups, "normal" and "anomalous," with respect to their osmotic pressure behavior. Glycogen is an example of a normal polymer. Table 2.1 shows osmotic pressure data on glycogen in water and in formamide [1]. In either solvent, van't Hoff's law holds well over the range of concentrations studied (0.005 to 0.03 g cm^{-3}), i.e., π/C is essentially constant and independent of solvent. For this reason, glycogen is considered normal. The data shown in Table 2.1 give about 280,000 for the number-average molecular weight of this substance. It is reported that glycogen is a highly branched polymer and, as a whole, assumes a compact spherical shape. Many globular proteins also exhibit normal osmotic pressure behavior [2].

On the contrary, unbranched chain polymers in solution generally show marked deviations from van't Hoff's law. This is illustrated in Table 2.2, which presents the classic data obtained by Caspari [3] for solutions of natural rubber in benzene. The concentrations C studied are comparable to those in Tables 1.2 and 2.1, but π/C varies appreciably with C. Staudinger and Fisher's [4] data for gutta percha in toluene and in carbon tetrachloride are plotted in Fig. 2.1. We see that π/C depends on both C and the solvent even at these low polymer concentrations, which, in fact, are lower by an order of magnitude than those used in the experiments cited above. However, extrapolation of the plotted points to $C = 0$ yields π/C which does not depend on the solvent.

By the late 1930's, the accumulation of such experimental data clearly indicated that the osmotic pressure of a linear polymer in solution can be expressed by a virial expansion, Eq. (1.4.69). Hence, the "anomaly" in osmotic pressure behavior is simply ascribed to the magnitude of the second virial coefficient A_2^{OS}, not to an anomaly in the state of dispersion of polymer molecules,

Table 2.1. Osmotic pressures of glycogen solutions (20°C)

solvent: water			solvent: formamide		
$100\,C$	$1000\,\pi$	$10\,(\pi/C)$	$100\,C$	$1000\,\pi$	$10\,(\pi/C)$
g cm^{-3}	atm	g^{-1} cm^3 atm	g cm^{-3}	atm	g^{-1} cm^3 atm
0.5	0.43	0.86	0.5	0.44	0.88
1.0	0.88	0.88	1.0	0.86	0.86
2.0	1.74	0.87	2.0	1.74	0.87
3.0	2.58	0.86	3.0	2.58	0.86

Table 2.2. Osmotic pressures of rubber in benzene (25°C)

$100\,C$	$1000\,\pi$	$10\,(\pi/C)$
g cm^{-3}	atm	g^{-1} cm^3 atm
1.01	3.6	3.58
1.43	6.6	4.46
2.10	11.7	5.58
2.92	21.4	7.34
3.59	30.7	8.55
5.26	59.2	11.25

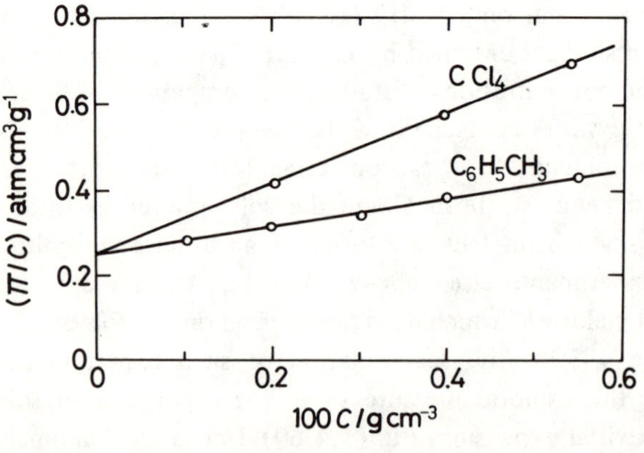

Fig. 2.1. Reduced osmotic pressure as a function of polymer mass concentration for gutta percha in toluene and in carbon tetrachloride.

such as would be caused by association. Furthermore, it was shown that A_2^{OS} for a homologous series of polymer samples is essentially independent of molecular weight.

In conjunction with the characteristic difference in osmotic behavior between globular and linear polymers, it is of interest to consider the effect of molecular shape on the pressure of a nonideal gas. Supposing that a gas obeys the van der Waals equation of state

$$[p + (n^2 a/V^2)]\,(V - nb) = nRT \tag{2.1.1}$$

and expanding p/RT in powers of the density $\rho = nM/V$, we obtain

$$p/RT = \rho/M + A_2{}^g \rho^2 + \cdots \tag{2.1.2}$$

where

$$A_2{}^g = (b/M^2)\,[1 - (T_B/T)] \tag{2.1.3a}$$

with

$$T_B \equiv a/(Rb) \tag{2.1.3b}$$

In these equations, M is the molar mass of the gas, and a and b are parameters characterizing intermolecular attraction and repulsion, respectively. The temperature T_B is called the Boyle temperature, in reference to the fact that at $T = T_B$, $A_2{}^g$ becomes zero and Boyle's law holds up to relatively high density. If the gas molecule is spherical, the parameter b is given by

$$b = 4N_A\omega = (N_A/2)\,(4\pi d^3/3) \tag{2.1.4}$$

where ω is the molecular volume, d is the molecular diameter, and N_A is the Avogadro constant. Thus $4\pi d^3/3$ is the mutually excluded volume between two gas molecules. From Eq. (2.1.4) we see that the factor b/M^2 in Eq. (2.1.3a) for a series of spherical molecules of different diameters but with the same density, $M/(N_A\,\omega)$, is inversely proportional to M.

Next, consider two rodlike molecules of diameter d and length ℓ with fixed orientation in space. The excluded volume is represented by the parallelepiped in Fig. 2.2. This volume depends on the angle θ between the two rods in contact and equals $2d\ell^2 \sin\theta$ for $\ell \gg d$. If this is averaged over all orientations of one rod relative to the other, we obtain for the parameter b

$$b = \frac{N_A \int_0^{\pi}\int_0^{2\pi} (2d\ell^2 \sin\theta) \sin\theta \, d\theta \, d\varphi}{2\int_0^{\pi}\int_0^{2\pi} \sin\theta \, d\theta \, d\varphi} = \left(\frac{\ell}{d}\right) N_A\,\omega \tag{2.1.5}$$

where ω is the molecular volume $\pi d^2 \ell/4$. The b so obtained for the rod is much larger than the b for a sphere of the same volume. The same is true of

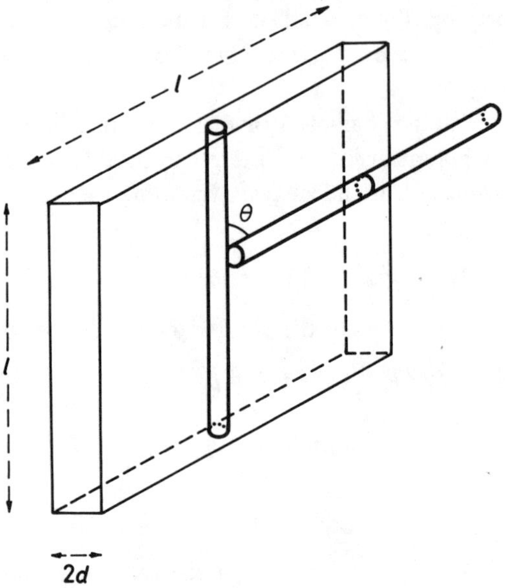

Fig. 2.2. Excluded volume of a rodlike molecule.

the second virial coefficient $A_2{}^g$. For a series of rods of different ℓ with constant d, b is proportional to M^2, so that the factor b/M^2 in Eq. (2.1.3a) is independent of M.

Pursuing the analogy between Boyle's law for gas pressure and van't Hoff's law for osmotic pressure, we may infer that the second virial coefficients for these two pressures can be interpreted in terms of essentially similar concepts. If so, the normal osmotic behavior characteristic of glycogen and globular proteins may be ascribed to the spherical or nearly spherical shapes of the molecules of these substances, while the anomalous osmotic behavior of natural rubber and other linear polymers may be considered due to the thin rodlike shapes of the molecules.

2.1.2. The Flory-Huggins Theory – 1. Osmotic Pressure

In the preceding section, we have pointed out the possibility of correlating the "anomalous" osmotic pressure behavior of chain polymers with their "slenderness." However, ordinary polymer chains, like those of natural rubber, are not only very slender but are also flexible. The Flory-Huggins theory described below is an attempt to explain characteristic solution behavior of linear chain polymers by a molecular model which includes both slenderness and flexibility [5,6].

As illustrated in Fig. 2.3, it is assumed that molecules in a solution are arranged regularly on an array of cells in space, each solvent molecule occupying one cell and each polymer chain occupying a series of P_1 successive cells. For convenience, the term segment is used for a portion of the polymer chain which occupies one cell. If it is assumed that no cell of a lattice of L cells is vacant, we have the relation

$$L = N_0 + P_1 N_1 \tag{2.1.6}$$

where N_0 is the number of solvent molecules, and N_1 is the number of polymer chains.

To compute the number of ways Ω in which these solvent and polymer molecules are arranged on the lattice, we number the solvent molecules $1, 2, \ldots, N_0$, and the polymer chains $1, 2, \ldots, N_1$. Further, we assume that the ends of each polymer chain are distinguishable and count the constituent segments as 1, 2, $\ldots, P_1$ from one end (head) to the other end (tail).

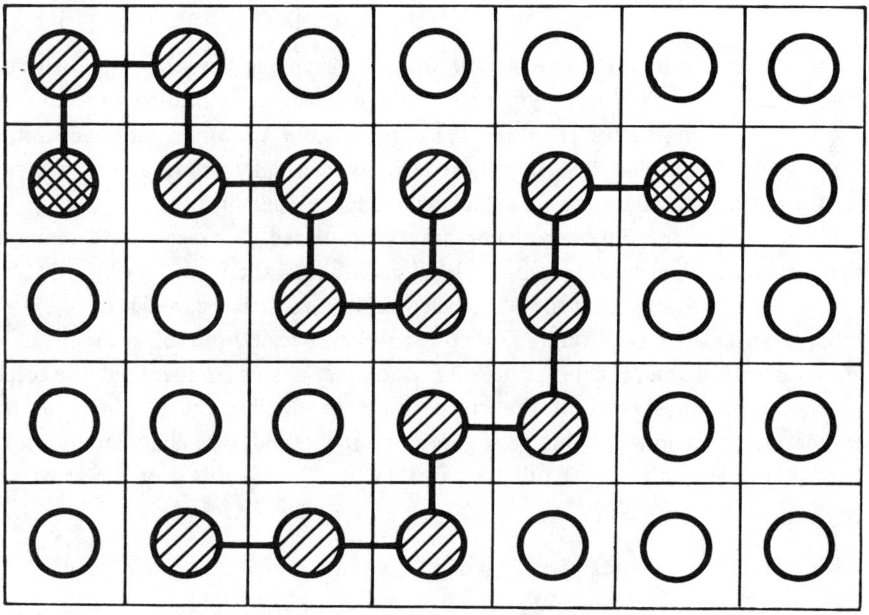

Fig. 2.3. Lattice model of a polymer solution. Shaded circles: polymer segments. Unfilled circles: solvent molecules.

We denote by ν_1 the number of ways of placing a polymer chain on the empty lattice. The first segment may be placed in any one of the L cells. The second segment must be put in one of the z nearest-neighbor cells around the first segment. The number z is the lattice coordination number. The third segment can be placed in any one of the $z - 1$ available cells around the second segment. Continuing this argument up to the P_1-th segment, we have

$$\nu_1 = Lz\,(z-1)^{P_1-2} \tag{2.1.7}$$

Here we have ignored the possibility that there will be fewer than $z - 1$ choices for the placement of a given segment if any of the prospective cells are occupied by previously placed segments.

After we have arranged $(i-1)$ polymer chains on the lattice, there remain $L - (i-1)P_1$ unoccupied lattice cells. The number of ways, ν_i, of placing the i-th polymer chain is given by

$$\begin{aligned}\nu_i &= [L-(i-1)P_1]\left\{z\left[\frac{L-(i-1)P_1}{L}\right]\right\}\left\{(z-1)\left[\frac{L-(i-1)P_1}{L}\right]\right\}^{P_1-2} \\ &= [L-(i-1)P_1]\left[\frac{L-(i-1)P_1}{L}\right]^{P_1-1} z\,(z-1)^{P_1-2}\end{aligned} \tag{2.1.8}$$

if, according to Flory [5], we assume that, on arranging the second and subsequent segments of the i-th polymer chain, the probability of finding a vacant cell can be approximated by $[L - (i-1)P_1]/L$. There are other, more accurate, choices for this probability, but most lead to essentially the same expressions for the entropy of mixing as are calculated on Flory's assumption.

When all the N_1 polymer chains have been placed, N_0 vacant cells remain; these must be filled with solvent molecules. Although there are $N_0\,(N_0 - 1)\ldots 1 = N_0!$ ways of arranging the N_0 labeled molecules, it is meaningless to regard these arrangements as different from one another, because the solvent molecules are indistinguishable. Further, since all N_1 polymer chains are identical, the total number of arrangements of the chains must be divided by $N_1!$ in order to enumerate the distinguishable configurations. If the ends of a chain are indistinguishable, this result must be divided further by 2^{N_1}. In this way, we arrive at

$$\Omega = (N_1!\,\sigma^{N_1})^{-1}\prod_{i=1}^{N_1}\nu_i \tag{2.1.9}$$

where ν_i is given by Eq. (2.1.8) and σ is a symmetry number (unity for a chain with distinguishable head and tail, two otherwise). Substituting Eq. (2.1.8) into Eq. (2.1.9) and applying Stirling's formula

$$x! \simeq (x/e)^x \quad (x \gg 1) \tag{2.1.10}$$

we obtain

$$\Omega = \frac{L!}{N_0!\,(P_1N_1)!}\left(\frac{P_1N_1}{L}\right)^{(P_1-1)N_1}(\omega_R)^{N_1} \tag{2.1.11}$$

where

$$\omega_R = P_1\, z\,(z-1)^{P_1-2}/(\sigma e^{P_1-1}) \tag{2.1.12}$$

The configurational entropy $S_{\text{conf}}(N_0, N_1)$ of the solution is thus given by

$$S_{\text{conf}}(N_0,N_1) = k\,\ell n\,\Omega = -k(N_0\,\ell n\,\phi_0 + N_1\,\ell n\,\phi_1) + kN_1\,\ell n\,\omega_R \tag{2.1.13}$$

where k is the Boltzmann constant, and ϕ_0 and ϕ_1 are defined by

$$\phi_0 = N_0/(N_0 + P_1N_1), \quad \phi_1 = P_1N_1/(N_0+P_1N_1) \tag{2.1.14}$$

i.e., ϕ_0 is the volume fraction of the solvent and ϕ_1 is that of the polymer solute. The configurational entropy S^o_{conf} of unmixed components is given by

$$S^o_{\text{conf}} = S_{\text{conf}}(N_0, 0) + S_{\text{conf}}(0, N_1) = kN_1\,\ell n\,\omega_R \tag{2.1.15}$$

Hence the entropy of mixing, $\Delta_m S$, is

$$\begin{aligned}\Delta_m S &= S_{\text{conf}}(N_0,N_1) - S^o_{\text{conf}} \\ &= -k(N_0\,\ell n\,\phi_0 + N_1\,\ell n\,\phi_1) \\ &= -R(n_0\,\ell n\,\phi_0 + n_1\,\ell n\,\phi_1)\end{aligned} \tag{2.1.16}$$

where n_0 and n_1 are the amounts of solvent and polymer solute, respectively. The quantity $k\,\ell n\,\omega_R$, which does not appear in the expression for $\Delta_m S$, is called the **disorientation entropy** per polymer chain, i.e., the entropy of a randomly kinked, disoriented polymer molecule relative to a perfect crystal of completely stretched-out polymer molecules.

Next, we denote by u_{oo}, u_{ss}, and u_{os} the interaction energies which arise when a pair of nearest cells are occupied by two solvent molecules, by two polymer segments, and by one solvent molecule and one polymer segment, respectively, and define Δu by

$$\Delta u \equiv u_{os} - (1/2)(u_{oo} + u_{ss}) \tag{2.1.17}$$

The quantity Δu is one-half the energy increase accompanying the creation of two nearest-neighbor solvent (O) and segment (S) pairs from a nearest-neighbor O-O pair and a nearest-neighbor S-S pair by an exchange of O and S. The exchange process may be written schematically:

$$(O - O) + (S - S) \rightarrow (O - S) + (S - O) \tag{2.1.18}$$

Hence, if there are zX O-S pairs in the solution, the energy of mixing $\Delta_m U$ is represented by

$$\Delta_m U = zX \, \Delta u \tag{2.1.19}$$

There are altogether $zL/2$ pairs of nearest-neighbor lattice cells. The probability that any cell is occupied by a solvent molecule is ϕ_0, and the corresponding probability for a polymer segment is ϕ_1. Therefore, if the mixing of segments and solvent molecules is completely random, the total number of O-S pairs should be given by

$$zX = (zL/2)2 \, \phi_0 \, \phi_1 = zL \, \phi_0 \, \phi_1 \tag{2.1.20}$$

Furthermore, if we can neglect the volume change on mixing $\Delta_m V$, the energy of mixing $\Delta_m U$ can be equated to the enthalpy of mixing $\Delta_m H$. Thus we obtain for $\Delta_m H$ an equation of the van Laar type [see Eq. (1.5.2)] :

$$\Delta_m H = L \, (z \Delta u) \, \phi_0 \, \phi_1 \tag{2.1.21}$$

With Eqs. (2.1.16) and (2.1.21) substituted into the general relation $\Delta_m G = \Delta_m H - T \Delta_m S$, we obtain

$$\Delta_m G = RT \, [n_0 \, \ell n \, \phi_0 + n_1 \, \ell n \, \phi_1 + (n_0 + P_1 n_1) \, \chi \, \phi_0 \, \phi_1] \tag{2.1.22}$$

which gives

$$\mu_0 = \mu_0{}^o + RT \, [\ell n \, (1 - \phi_1) + (1 - P_1{}^{-1}) \, \phi_1 + \chi \, \phi_1{}^2] \tag{2.1.23}$$

$$\mu_1 = \mu_1{}^o + RT \, [\ell n \, \phi_1 - (P_1 - 1)(1 - \phi_1) + P_1 \, \chi \, (1 - \phi_1)^2] \tag{2.1.24}$$

where

$$\chi = \chi_H \equiv z \, \Delta u / kT \tag{2.1.25}$$

The osmotic pressure π is obtained by deriving the activity a_0 from Eq. (2.1.23) and introducing it into Eq. (1.4.41b):

$$\begin{aligned} \pi &= - (RT/V_0{}^o) \, [\ell n \, (1 - \phi_1) + (1 - P_1{}^{-1}) \, \phi_1 + \chi \, \phi_1{}^2] \\ &= (RT/V_0{}^o) \, [P_1{}^{-1} \, \phi_1 + (\tfrac{1}{2} - \chi) \, \phi_1{}^2 + (\tfrac{1}{3}) \, \phi_1{}^3 + \cdots] \end{aligned} \tag{2.1.26}$$

Converting ϕ_1 to C_1 by Eq. (1.1.43), we obtain for the second and third virial coefficients in the C_1-expansion of π

$$A_2^{OS} = [(v_1^o)^2/V_0^o]\,(\tfrac{1}{2} - \chi_H) \tag{2.1.27}$$

$$A_3^{OS} = (v_1^o)^3/3V_0^o \tag{2.1.28}$$

where v_1^o is the specific volume of the polymer solute. From the definition of the segment we may equate the molar volume V_1^o of the polymer solute to $V_0^o P_1$. Hence, the term $(v_1^o)^2/V_0^o$ can be written $V_1^o P_1/M_1^2$, where M_1 is the molar mass of the polymer solute. Comparison with Eq. (1.5.22) shows that the A_2^{OS} given by Eq. (2.1.27) is $P_1/2$ times larger than that given by Eq. (1.5.22) if we can ignore the exchange terms in both expressions. This explains why solutions of flexible polymers have anomalously large A_2^{OS} in comparison with regular solutions. Further, since $V_1^o P_1$ is proportional to P_1^2 and hence to M_1^2, we see that values of A_2^{OS} for a homologous series of such polymers are independent of molecular weight.

2.1.3. The Flory-Huggins Theory – 2. Phase Equilibrium

In the theory described above, the exchange of O and S merely involved an energy increase Δu. In real systems, it will also influence thermal vibrations and rotations of both solvent and polymer segments. Though it is not easy to formulate rigorously, this effect may be treated as an entropy change accompanying the formation of nearest-neighbor O-S pairs. We may assume that the formation of such a molecular pair increases the free energy of the solution by $\Delta u - T\Delta s$. Then, we obtain the same expression as Eqs. (2.1.22) through (2.1.24) for $\Delta_m G$, μ_0, and μ_1, with χ redefined as

$$\chi = \chi_H + \chi_S \tag{2.1.29}$$

where

$$\chi_S \equiv z\Delta s/k \tag{2.1.30}$$

The parameters χ_H and χ_S may be replaced by a new set of parameters ψ and Θ defined by

$$\psi \equiv \tfrac{1}{2} - \chi_S \tag{2.1.31}$$

$$\Theta \equiv \chi_H T/(\tfrac{1}{2} - \chi_S) \tag{2.1.32}$$

Then χ can be expressed as

$$\chi = \tfrac{1}{2} - \psi\left(1 - \frac{\Theta}{T}\right) \tag{2.1.33}$$

Hence, in place of Eq. (2.1.27), we obtain for A_2^{OS}

$$A_2^{OS} = \frac{(v_1^{o})^2}{V_0^{o}} \left(\tfrac{1}{2} - \chi\right) = \frac{(v_1^{o})^2}{V_0^{o}} \psi \left(1 - \frac{\Theta}{T}\right) \tag{2.1.34}$$

At $T = \Theta$, A_2^{OS} vanishes, so that van't Hoff's law holds up to relatively high polymer concentration. This characteristic temperature Θ is called the **theta temperature** or **Flory temperature**. It corresponds to the Boyle temperature T_B of real gases, at which Boyle's law for pressure holds up to relatively high density.

Now, we proceed to the problem of liquid-liquid phase equilibrium. Converting from x_1 to ϕ_1 in Eq. (1.3.25) and substituting Eq. (2.1.24) for μ_1, we obtain for the spinodal

$$\phi_1^2 - \left(1 - \frac{P_1 - 1}{2P_1\chi}\right)\phi_1 + \frac{1}{2P_1\chi} = 0 \tag{2.1.35}$$

This equation gives the dashed lines in Fig. 2.4. For $P_1 = \infty$ the spinodal consists of the vertical line at $\phi_1 = 0$ and the branch $\phi_1 = 1 - (2\chi)^{-1}$.

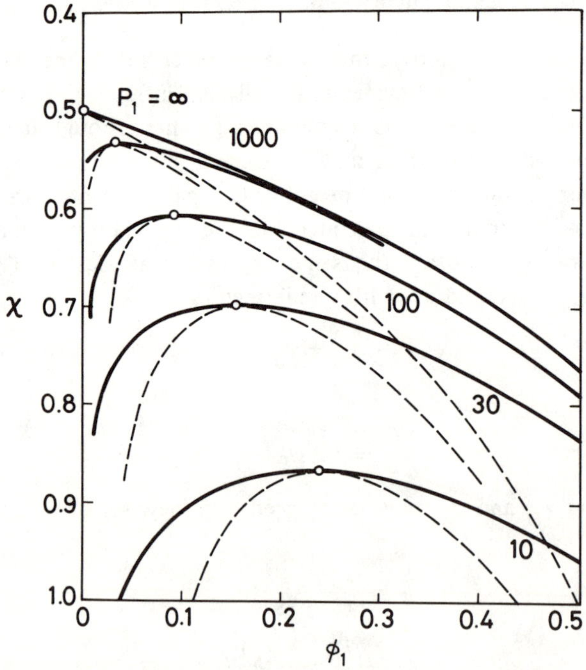

Fig. 2.4. Calculated binodals (solid lines) and spinodals (dashed lines) of binary solutions of a polymer as a function of relative chain length P_1. Unfilled circles: critical points.

The critical point, determined by Eq. (1.3.26), is calculated from the relations

$$(1 - \phi_{1c})^{-1} - (1 - P_1^{-1}) - 2\chi_c \phi_{1c} = 0 \tag{2.1.36}$$

$$(1 - \phi_{1c})^{-2} - 2\chi_c = 0 \tag{2.1.37}$$

where χ_c denotes the value of χ at the critical point: temperature T_c and composition ϕ_{1c}. Solution of these equations gives

$$\phi_{1c} = (1 + P_1^{1/2})^{-1} \tag{2.1.38}$$

$$\chi_c = \tfrac{1}{2}(1 + P^{-1/2})^2 \tag{2.1.39}$$

The unfilled circles in Fig. 2.4 indicate the critical points calculated from these expressions. With Eq. (2.1.33), Eq. (2.1.39) may be rewritten

$$T_c^{-1} = \Theta^{-1} + (\Theta\psi)^{-1}\,[P_1^{-1/2} + (2P_1)^{-1}] \tag{2.1.40}$$

Phase separation can take place only if $\chi \geqslant \chi_c$. Denoting the compositions of two phases by ϕ_1' and ϕ_1'', we have the condition for phase equilibrium

$$\mu_i(\phi_1') = \mu_i(\phi_1''), \quad (i = 0, 1) \tag{2.1.41}$$

Introduction of Eqs. (2.1.23) and (2.1.24) yields

$$\begin{aligned} &\ln(1 - \phi_1') + (1 - P_1^{-1})\phi_1' + \chi(\phi_1')^2 \\ &= \ln(1 - \phi_1'') + (1 - P_1^{-1})\phi_1'' + \chi(\phi_1'')^2 \end{aligned} \tag{2.1.42}$$

$$\begin{aligned} &\ln \phi_1' - (P_1 - 1)(1 - \phi_1') + P_1\chi(1 - \phi_1')^2 \\ &= \ln \phi_1'' - (P_1 - 1)(1 - \phi_1'') + P_1\chi(1 - \phi_1'')^2 \end{aligned} \tag{2.1.43}$$

The solid lines in Fig. 2.4 show numerical solutions of this set of equations for ϕ_1' and ϕ_1'', i.e., binodals for different P_1 values. Each binodal touches the corresponding spinodal tangentially at the critical point.

Figure 2.5 shows T versus ϕ_1 phase diagrams for cyclohexane solutions of polystyrene [7]. Each curve corresponds to a different molecular weight. Usually, when a homogeneous solution is cooled, the onset of phase separation is marked by the appearance of turbidity at a temperature called the **cloud-point** or **precipitation temperature**. A graph representing the cloud-point temperature as a function of the solution composition, such as any of the solid curves in Fig. 2.5, is called the **cloud-point curve**, and the maximum of this curve is referred to as the **threshold cloud-point**. For a monodisperse polymer sample

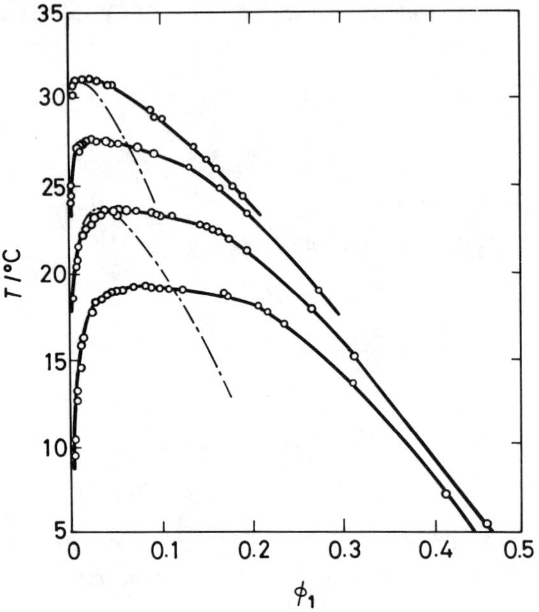

Fig. 2.5. Cloud-point curves of polystyrene fractions of different molecular weights in cyclohexane. Solid lines: experimental [7]. Dot-dash lines: theoretical.

dissolved in a single solvent (hence a binary system), the cloud-point curve coincides with the binodal and the threshold cloud-point is the critical point. It should be noted that these features do not hold for solutions of polymers heterogeneous in molecular weight [8]. Discussion of this point is deferred to a later section; here we attempt an analysis of experimental data, ignoring for the moment the effect of molecular weight polydispersity.

Figure 2.6 shows the molecular weight dependence (more correctly, the P_1 dependence) of T_c, which has been derived from the data of Fig. 2.5 by equating T_t to T_c; here T_t denotes the **threshold temperature**, i.e., the temperature at the threshold cloud-point. As Eq. (2.1.40) predicts, T_c^{-1} varies linearly with $P_1^{-1/2} + (2P_1)^{-1}$, and the intercept and slope of the straight line for polystyrene yield

$$\Theta = 307.2 \text{ K}, \quad \psi = 1.056 \tag{2.1.44}$$

This Θ is close to the value 307.7 K obtained by analyzing temperature dependence of A_2^{OS} in terms of Eq. (2.1.34). The straight line for polyisobutylene in Fig. 2.6 can also be analyzed according to Eq. (2.1.40) to evaluate Θ and ψ.

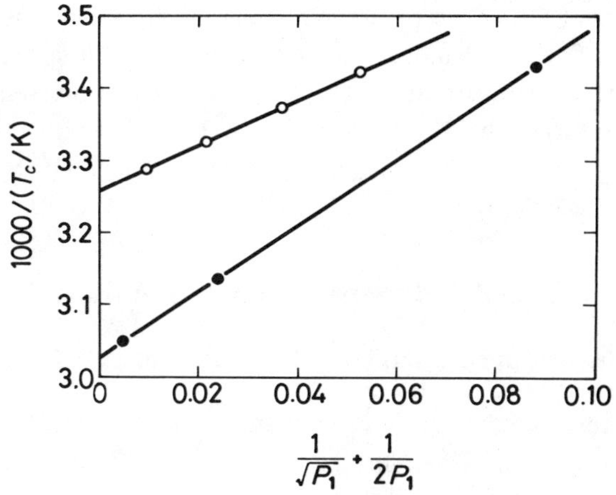

Fig. 2.6. Chain length dependence of critical point T_c (actually, threshold temperature T_t) for polystyrene in cyclohexane (unfilled circles) and for polyisobutylene in diisobutyl ketone (filled circles). Data of Shultz and Flory [7].

If the empirical values of Θ and ψ obtained in this way are inserted into Eq. (2.1.33) and the binodals shown in Fig. 2.4 are converted to T versus ϕ_1 curves by use of the resulting $\chi(T)$, we obtain the dot-dash curves illustrated in Fig. 2.5. These theoretical binodals display the general character of the experimental curves in that the cloud-point curve is markedly skewed toward low concentration; but inspection reveals that the computed critical concentration ϕ_{1c} is lower than the measured threshold concentration. Furthermore, the solubility gap is much narrower than the observed cloud-point curve. These discrepancies are discussed further in § 2.2.6.

As has been noted, the second virial coefficient of a solution of a polymer is usually quite large, and this is also true of the higher virial coefficients. Hence, for example, when P_1 is larger than 100, van't Hoff's law should be obeyed only at concentrations below 10^{-4} g cm^{-3}. However, osmotic pressures at such low concentrations become too small to measure with precision. Therefore, the van't Hoff region may not be accessible by dilution. The only way to obtain the "ideal" condition (usually called the theta state) is to choose a solvent which makes the second virial coefficient zero at a temperature convenient for experimental work. For this reason, the theta state plays a very important role in both experimental and theoretical studies of polymers in solution.

From Eq. (2.1.40) it follows that

$$\lim_{P_1 \to \infty} T_c = \Theta \qquad (2.1.45)$$

which suggests that Θ is close to the experimental threshold temperature for a polymer sample of very high molecular weight and hence the solvent at Θ should be a poor solvent. Determination of Θ by the extrapolation illustrated in Fig. 2.6 is usually referred to as the **Shultz-Flory method**. However, as will be shown in § 2.2.5, caution is needed in using this procedure for polymers with a broad distribution of molecular weight.

2.2. Polydisperse Polymer Solutions

2.2.1. Molecular Weight Distribution and Average Molecular Weight

So far we have considered a binary system consisting of a monodisperse polymer solute and a single solvent. However, since synthetic polymers are always heterogeneous in molecular weight, an understanding of their solutions requires a theoretical treatment of systems having a distribution of molecular weight.

Given a mixture of r polymers which differ only in molecular weight, the total amount n_s is represented by

$$n_s = \sum_{i=1}^{r} n_i \tag{2.2.1}$$

where n_i is the amount of the i-th polymer component with molar mass M_i. We define ξ_i by

$$\xi_i \equiv n_i M_i / \sum_{j=1}^{r} n_j M_j, \quad \left(\sum_{i=1}^{r} \xi_i = 1 \right) \tag{2.2.2}$$

which represents the weight fraction of component i in the polymer sample. Different average molar masses of the sample may be defined, e.g.,

$$M_n \equiv (\Sigma_i n_i M_i)/n_s = 1/\Sigma_i (\xi_i/M_i) \tag{2.2.3}$$

$$M_w \equiv \Sigma_i n_i M_i^2 / \Sigma_i n_i M_i = \Sigma_i \xi_i M_i \tag{2.2.4}$$

$$M_z \equiv \Sigma_i n_i M_i^3 / \Sigma_i n_i M_i^2 = (M_w)^{-1} \Sigma_i \xi_i M_i^2 \tag{2.2.5}$$

These are, in order, the number-average, the weight-average, and the z-average molar masses. Letting M_{ri} be the molecular weight of component i, we also define the **number-average**, **weight-average**, and ***z*-average molecular weights** of the sample by

$$M_{rn} \equiv 1/\Sigma_i (\xi_i/M_{ri}) \tag{2.2.6}$$

$$M_{rw} \equiv \Sigma_i \xi_i M_{ri} \tag{2.2.7}$$

$$M_{rz} \equiv (M_{rw})^{-1} \Sigma_i \xi_i M_{ri}^2 \tag{2.2.8}$$

Besides the molecular weight, the **degree of polymerization** Z_i is often used to express the size of a polymer. In terms of the molar mass M_u of an appropriately chosen repeat unit of the polymer chain, we define Z_i by

$$Z_i \equiv M_i/M_u \tag{2.2.9}$$

Then the number-average, weight-average, and z-average degrees of polymerization are

$$Z_n \equiv M_n/M_u, \quad Z_w \equiv M_w/M_u, \quad Z_z \equiv M_z/M_u \tag{2.2.10}$$

Furthermore, we denote by P_i the molar volume of a polymer relative to a suitable reference volume and call it the **relative chain length**. The quantity P_1 that appeared in the Flory-Huggins theory is an example. In that case, the reference volume was the molar volume of the solvent, so that

$$P_i = V_i^*/V_0^o = v_s^* M_i/(v_0^o M_0) \tag{2.2.11}$$

with M_0, V_0^o, and v_0^o denoting the molar mass, molar volume, and specific volume of the pure solvent, and V_i^* and v_s^* the molar volume of polymer component i and specific volume of the polymer in the amorphous state. These latter values must be extrapolated from appropriate data when crystallization or vitrification prevents attainment of the amorphous state. It is important to specify them when P_i is employed for data analysis. Also note that, as is evident from the definition of P_i, it is meaningless to assign different v_s^* values to terminal and "interior" repeat units of a polymer chain. Finally, we define the number-average, weight-average, and z-average relative chain lengths by

$$P_n \equiv 1/\Sigma_i(\xi_i/P_i), \quad P_w \equiv \Sigma_i \xi_i P_i, \quad P_z \equiv P_w^{-1} \Sigma_i \xi_i P_i^2 \tag{2.2.12}$$

When, as is usually the case, we are concerned with a polymer of high molecular weight it is advantageous to represent the molecular weight distribution by a continuous function of M_r. The weight fraction of polymer molecules having molecular weights between M_r and $M_r + dM_r$ is denoted by $\xi(M_r)dM_r$, and $\xi(M_r)$ is called the weight distribution of molecular weight. From this definition it follows that

$$\int_0^\infty \xi(M_r)dM_r = 1 \tag{2.2.13}$$

We denote the k-th moment of $\xi(M_r)$ by λ_k, i.e.,

$$\lambda_k \equiv \int_0^\infty M_r^k \xi(M_r)dM_r \tag{2.2.14}$$

with λ_0 being unity.

In terms of the λ_k, the average molecular weights defined above can be written

$$M_{rn} = \lambda_0/\lambda_{-1}, \quad M_{rw} = \lambda_1/\lambda_0, \quad M_{rz} = \lambda_2/\lambda_1 \qquad (2.2.15)$$

For the standard deviation S_w of $\xi(M_r)$ we have

$$S_w^{\,2} \equiv \int_0^\infty (M_r - M_{rw})^2\, \xi(M_r)\, dM_r = \lambda_2 - \lambda_1^{\,2} \qquad (2.2.16)$$

Hence

$$S_w/M_{rw} = [(M_{rz}/M_{rw}) - 1]^{1/2} \qquad (2.2.17)$$

Next, we write the mole fraction of polymer molecules with molecular weights between M_r and $M_r + dM_r$ as $f(M_r)\,dM_r$ and refer to $f(M_r)$ as the number distribution of molecular weight. The standard deviation S_n of this distribution satisfies the relation

$$S_n/M_{rn} = [(M_{rw}/M_{rn}) - 1]^{1/2} \qquad (2.2.18)$$

These results indicate that spread of the molecular weight distribution $f(M_r)$ may be estimated by the ratio M_{rw}/M_{rn} and that of $\xi(M_r)$ by the ratio M_{rz}/M_{rw}.

The **exponential distribution** and the **logarithmic-normal distribution** defined below are often adopted as analytic forms for $\xi(M_r)$. The former is expressed by

$$\xi(M_r) = \frac{y^{h+1}}{\Gamma(h+1)} M_r^{\,h} \exp(-yM_r) \qquad (2.2.19)$$

where

$$y = h/M_{rn} = (h+1)/M_{rw} = (h+2)/M_{rz} = \ldots \qquad (2.2.20)$$

and $\Gamma(h+1)$ is the gamma function. When $h = 1$, the ratio M_{rw}/M_{rn} becomes 2. The exponential distribution for this special case is often called the **most probable distribution** (see § 3.3.6).

The logarithmic-normal distribution has the form

$$\xi(M_r) = \frac{1}{\beta(\pi)^{1/2} M_r^{\,o}} \exp\left[-\frac{1}{\beta^2}\left(\ln \frac{M_r}{M_r^{\,o}}\right)^2\right] \qquad (2.2.21)$$

where

$$\beta^2 \equiv 2 \ln (M_{rw}/M_{rn}), \quad M_r^{\,o} \equiv (M_{rw} M_{rn})^{1/2} \qquad (2.2.22)$$

This distribution satisfies the characteristic relations

$$M_{rw}/M_{rn} = M_{rz}/M_{rw} = M_{r,z+1}/M_{rz} = \ldots \qquad (2.2.23)$$

and is often used to approximate a distribution with a long tail at high molecular weight.

2.2.2. Interaction Parameters in Real Polymer Solutions

The Flory-Huggins theory for a binary solution can be extended in straightforward fashion to a general solution consisting of a single solvent and a mixture of homologous polymers different in chain length. We designate the solvent as component 0 and the polymers as components 1 through r. The relative chain length of polymer i is denoted by P_i, its volume fraction in the solution by ϕ_i, and the total volume fraction of the polymer solute by ϕ_s. Then we have

$$\phi_s = \sum_{i=1}^{r} \phi_i = 1 - \phi_0 \tag{2.2.24}$$

We assume that the polymer species are uniform in specific volume. This assumption holds very well for any series of homologous polymers, except for its members of very short chain length. With this assumption, the weight fraction ξ_i of polymer i in the polymer mixture can be represented by ϕ_i/ϕ_s, i.e.,

$$\xi_i = \phi_i/\phi_s, \quad (i = 1, 2, \ldots, r) \tag{2.2.25}$$

The Gibbs free energy of mixing for the solution is given by

$$\Delta_m G = RT\left[n_0 \ln \phi_0 + \sum_{i=1}^{r} n_i \ln \phi_i + \left(n_0 + \sum_{i=1}^{r} P_i n_i\right)\chi \phi_0 \phi_s\right] \tag{2.2.26}$$

which is an extension of Eq. (2.1.22). Here the reference state is the separated pure components. Accordingly, Eq. (2.2.26) contains a contribution from mixing the polymer species to a bulk amorphous phase, and gives

$$\mu_0 = \mu_0{}^o + RT\,[\ln(1 - \phi_s) + (1 - P_n{}^{-1})\,\phi_s + \chi \phi_s{}^2] \tag{2.2.27a}$$

$$\begin{aligned}\mu_i = \mu_i{}^o + RT\,[&\ln \phi_i - (P_i - 1) + P_i (1 - P_n{}^{-1})\,\phi_s \\ &+ P_i \chi (1 - \phi_s)^2], \quad (i = 1, 2, \ldots, r)\end{aligned} \tag{2.2.27b}$$

where P_n is the number-average relative chain length of the polymer mixture:

$$P_n = \phi_s / \sum_{i=1}^{r} (\phi_i/P_i) = \left(\sum_{i=1}^{r} \xi_i/P_i\right)^{-1} \tag{2.2.28}$$

For $\phi_s = 1$, Eq. (2.2.27b) gives

$$\mu_i = \mu_i{}^o + RT\,[\ln \xi_i + 1 - (P_i/P_n)] \tag{2.2.29}$$

the chemical potential of polymer i in the pure polymer mixture. The second term on the right-hand side is the contribution from the entropy of mixing polymer components with different chain lengths.

The Flory-Huggins theory is founded on a number of assumptions. Therefore, quantitative agreement with data on real polymer solutions is not to be expected. The theory might be improved by reducing the number of assumptions and/or by using a more realistic molecular model. However, we shall not pursue this line of study, but rather adopt a phenomenological approach to the thermodynamics of polymer solutions.

As noted in Chapter 1, real solutions can be described thermodynamically in terms of the activity coefficients with the ideal solution or the infinitely dilute solution as the reference state. This formalism, of course, can be applied to polymer solutions. However, since polymer solutions exhibit appreciable deviations from ideal-solution behavior even at very low concentration, these reference states are inconvenient, especially when we are concerned with phenomena, such as phase separation, which occur in relatively concentrated solutions.

Despite its shortcomings, the Flory-Huggins theory does predict the main features of thermodynamic properties of polymer solutions. Therefore it seems attractive to attempt a phenomenological description of such solutions, taking the Flory-Huggins solution as the reference state. In this approach, we simply treat the parameter χ as a phenomenological function of the state variables, not as a constant relating to exchange of a pair of nearest-neighbor solvent and polymer segment. To define this phenomenological χ, which is hereafter called the **mutual interaction parameter**, it is appropriate to resort to Eq. (2.2.27a), because $\Delta_m \mu_0$ is directly related to osmotic pressure and solvent vapor pressure which are the most often measured quantities in thermodynamic studies of polymer solutions. Thus we define χ by

$$\Delta_m \mu_0 \equiv RT\,[\ell n\,(1-\phi_s) + (1-P_n^{-1})\,\phi_s + \chi\phi_s^2] \tag{2.2.30}$$

In this approach, it is also appropriate to replace the χ in Eq. (2.2.27b) by a species-dependent mutual interaction parameter χ_{Pi} according to

$$\Delta_m \mu_i \equiv RT\,[\ell n\,\phi_i - (P_i - 1) + P_i(1-P_n^{-1})\,\phi_s + P_i\chi_{Pi}(1-\phi_s)^2] \tag{2.2.31}$$

The parameters χ and χ_{Pi} are, in general, functions of the state variables $T, p, \phi_1, \phi_2, \ldots, \phi_r$, but thermodynamics affords no information about these functions. The conventional assumption so far adopted in the phenomenological approach to polymer solutions is that both χ and χ_{Pi} do not depend on the volume fractions of individual polymer components but on the total volume fraction ϕ_s. In the paucity of a theoretical treatment based on another choice of the concentration dependence of these parameters, we proceed with the subsequent analysis, adopting the conventional assumption mentioned above.

We resolve χ to enthalpic component χ_H and entropic component χ_S, which are defined respectively by

$$\chi_H = -T(\partial\chi/\partial T)_{p,\phi_s}, \quad \chi_S = \chi + T(\partial\chi/\partial T)_{p,\phi_s} \tag{2.2.32}$$

Then it follows from Eq. (2.2.30) that

$$\Delta_m H_0 = -T^2 \, [\partial(\Delta_m \mu_0/T)/\partial T]_{p,\phi_s} = RT\,\chi_H\,\phi_s^2 \tag{2.2.33}$$

and

$$\Delta_m S_0 = -(\partial\Delta_m \mu_0/\partial T)_{p,\phi_s} = -R\,[\ln(1-\phi_s) + (1-P_n^{-1})\,\phi_s + \chi_S\,\phi_s^2] \tag{2.2.34}$$

which can be regarded as defining χ_H and χ_S. We define χ_P by

$$\chi_P \equiv \sum_{i=1}^{r} \xi_i\,\chi_{Pi} \tag{2.2.35}$$

which is an average of χ_{Pi} over the polymer solute. Then, by use of Eq. (1.1.57) and the condition $V_i^o = P_i V_0^o$, we find that χ_P is related to χ by

$$\chi_P\,(1-\phi_s)^2 = -\chi\,\phi_s\,(1-\phi_s) + \int_{\phi_s}^{1} \chi\,d\phi_s \tag{2.2.36}$$

This relation allows χ_P to be computed if χ is determined as a function of ϕ_s, and also allows $\Delta_m G$ to be expressed in the form

$$\begin{aligned}\Delta_m G &= n_0\,\Delta_m \mu_0 + \sum_{i=1}^{r} n_i\,\Delta_m \mu_i \\ &= RT\left(n_0 + \sum_{i=1}^{r} n_i P_i\right)[(1-\phi_s)\ln(1-\phi_s) \\ &\quad + \sum_{i=1}^{r} (\phi_i/P_i)\ln\phi_i + g(1-\phi_s)\,\phi_s]\end{aligned} \tag{2.2.37}$$

Here g is a new mutual interaction parameter defined by

$$g \equiv \frac{1}{1-\phi_s}\int_{\phi_s}^{1} \chi\,d\phi_s \tag{2.2.38}$$

This relation is converted to express χ in terms of g, yielding

$$\chi = g - (1-\phi_s)\,(\partial g/\partial\phi_s)_{T,p} \tag{2.2.39}$$

The phenomenological formalism described above was initiated by Huggins [9] and Tompa [10] and elaborated by Koningsveld [11]. We remark that it is founded on the important assumption that both χ and χ_{Pi} and hence g depend on ϕ_s only.

Like the logarithm of an activity coefficient, χ and g can be expanded in integral powers of ϕ_s:

$$\chi = \chi_0 + \chi_1 \phi_s + \chi_2 \phi_s^2 + \dots \tag{2.2.40}$$

$$g = g_0 + g_1 \phi_s + g_2 \phi_s^2 + \dots \tag{2.2.41}$$

where the coefficients χ_i and g_i ($i = 0, 1, \dots$) are functions of T and p. Substitution of these series into Eqs. (2.2.38) and (2.2.39) yields

$$g_i = \sum_{k=i}^{\infty} \chi_k/(k+1) \tag{2.2.42}$$

$$\chi_i = (i+1)(g_i - g_{i+1}) \tag{2.2.43}$$

With Eq. (2.2.30) for $\Delta_m \mu_0$ and with Eq. (2.2.40) for χ, we obtain for the second and third virial coefficients

$$A_2^{OS} = [(v_s^*)^2/V_0^o](\tfrac{1}{2} - \chi_0) \tag{2.2.44}$$

$$A_3^{OS} = [(v_s^*)^3/V_0^o](\tfrac{1}{3} - \chi_1) \tag{2.2.45}$$

where v_s^* is the specific volume of the amorphous polymer. Denoting the enthalpic and entropic components of χ_0 by χ_{0H} and χ_{0S} and redefining the parameters ψ and Θ by

$$\psi \equiv \tfrac{1}{2} - \chi_{0S} \tag{2.2.46}$$

$$\Theta \equiv \chi_{0H}\, T/(\tfrac{1}{2} - \chi_{0S}) \tag{2.2.47}$$

we can express χ_0 as

$$\chi_0 = \tfrac{1}{2} - \psi\left(1 - \frac{\Theta}{T}\right) \tag{2.2.48}$$

which allows Eq. (2.2.44) to be put in the same form as Eq. (2.1.34).

We have so far developed the theory using ϕ_i as the composition variables. But, as has been noted, ϕ_i cannot escape from the arbitrariness associated with the choice of V_i^o. Therefore, for an unequivocal description of observations it is desirable to use weight fractions w_i as the composition variables, because they are free from arbitrariness. Following Scholte [12], we summarize below the basic phenomenological equations to be used when the composition is expressed in terms of w_i.

The mutual interaction parameters χ_w and χ_{Mi}, which correspond, respectively to χ and χ_{Pi}, are defined by

$$\Delta_m \mu_0 \equiv RT\left[\ln(1 - w_s) + \left(1 - \frac{M_0}{M_n}\right) w_s + \chi_w w_s^2\right] \tag{2.2.49}$$

$$\Delta_m \mu_i \equiv RT\left[\ln w_i - \left(\frac{M_i}{M_0} - 1\right) + \frac{M_i}{M_0}\left(1 - \frac{M_0}{M_n}\right) w_s + \left(\frac{M_i}{M_0}\right) \chi_{Mi} (1 - w_s)^2\right], \quad (i = 1, 2, \ldots, r) \tag{2.2.50}$$

These parameters are assumed to depend on T, p, and w_s, where w_s is the total weight fraction of the polymer solute. The relationships corresponding to Eqs. (2.2.35), (2.2.36), and (2.2.38) are

$$\chi_M \equiv \sum_{i=1}^{r} \xi_i \chi_{Mi} = -\frac{w_s}{1 - w_s} \chi_w + \frac{1}{(1 - w_s)^2} \int_{w_s}^{1} \chi_w \, dw_s \tag{2.2.51}$$

$$g_w \equiv \frac{1}{1 - w_s} \int_{w_s}^{1} \chi_w \, dw_s = w_s \chi_w + (1 - w_s) \chi_M \tag{2.2.52}$$

The Gibbs free energy of mixing per unit mass of solution, denoted here by $\Delta_m G_w$, is given by

$$\begin{aligned} \Delta_m G_w &= \Delta_m G/\rho V = [(1 - w_s)/M_0]\, \Delta_m \mu_0 + \sum_{i=1}^{r} (w_i/M_i)\, \Delta_m \mu_i \\ &= (RT/M_0)\, [(1 - w_s) \ln (1 - w_s) \\ &\quad + \sum_{i=1}^{r} (M_0/M_i)\, w_i \ln w_i + g_w w_s (1 - w_s)] \end{aligned} \tag{2.2.53}$$

This expression can be used to define g_w. Once g_w is determined, χ_w and χ_M can be evaluated from

$$\chi_w = g_w - (1 - w_s) (\partial g_w / \partial w_s)_{T,p} \tag{2.2.54a}$$

$$\chi_M = g_w + w_s (\partial g_w / \partial w_s)_{T,p} \tag{2.2.54b}$$

We expand χ_w in powers of w_s as

$$\chi_w = \chi_{0w} + \chi_{1w} w_s + \chi_{2w} w_s^2 + \ldots \tag{2.2.55}$$

It can be shown that if M_n of the polymer mixture is sufficiently large, the second virial coefficient is represented by

$$A_2{}^{OS} = (v_0{}^o/M_0) (\tfrac{1}{2} - \chi_{0w}) \tag{2.2.56}$$

Equating this to Eq. (2.2.44) shows that χ_0 and χ_{0w} are related by

$$\chi_0 = \tfrac{1}{2} - \left(\frac{v_0{}^o}{v_s{}^*}\right)^2 (\tfrac{1}{2} - \chi_{0w}) \tag{2.2.57}$$

A contribution from the concentration dependence of solution density ρ appears in the relation between χ_{1w} and the third virial coefficient.

Another composition variable called the **base mole fraction** φ_i is sometimes used. It is defined by

$$\varphi_i \equiv x_i Z_i / \sum_{j=0}^{r} x_j Z_j = \frac{w_i (M_0/M_u)}{1 - w_s + w_s (M_0/M_u)} \tag{2.2.58}$$

where Z_i is the degree of polymerization of polymer i defined by Eq. (2.2.9), with $Z_0 = 1$. The thermodynamic relations involving φ_i as the composition variable can be derived by the substitutions $\phi_i \to \varphi_i$ and $P_i \to Z_i$ from those involving ϕ_i as the composition variable. It is advantageous to use the base mole fraction when we wish to compare thermodynamic data on different polymers with the same backbone structure, since this composition variable does not depend on side-chain substitutions.

2.2.3. Spinodal and Critical Point

The spinodal is determined by Eq. (1.3.68), and the critical point by Eqs. (1.3.68) and (1.3.69). In application of these equations, G_m must be replaced by G_ϕ, the Gibbs free energy per unit volume of solution, if the composition is represented by volume fractions ϕ_i. Substitution for $\Delta_m G$ from Eq. (2.2.37) gives

$$G_\phi = \frac{n_0 \mu_0{}^o + \Sigma_i n_i \mu_i{}^o + \Delta_m G}{(n_0 + \Sigma_i n_i P_i) V_0{}^o}$$

$$= \frac{RT}{V_0{}^o} \left[(1 - \phi_s) \frac{\mu_0{}^o}{RT} + \sum_{i=1}^{r} \frac{\phi_i}{P_i} \left(\frac{\mu_i{}^o}{RT} \right) + (1 - \phi_s) \ln (1 - \phi_s) + \sum_{i=1}^{r} \frac{\phi_i}{P_i} \ln \phi_i + g (1 - \phi_s) \phi_s \right] \tag{2.2.59}$$

Unless $\Delta_m V$ is zero, the denominator $(n_0 + \Sigma_i P_i n_i) V_0{}^o$ is not exactly equal to the solution volume V, so that G_ϕ does not precisely represent G/V.

With Eq. (2.2.59) we obtain at constant T and p

$$\frac{V_0{}^o}{RT} \left(\frac{\partial G_\phi}{\partial \phi_i} \right) = \frac{1}{RT} \left(- \mu_0{}^o + \frac{\mu_i{}^o}{P_i} \right) - \ln (1 - \phi_s) + \frac{1}{P_i} \ln \phi_i - \left(1 - \frac{1}{P_i} \right) + g (1 - 2\phi_s) + \left(\frac{\partial g}{\partial \phi_s} \right) (1 - \phi_s) \phi_s \tag{2.2.60}$$

$$\frac{V_0{}^o}{RT}\left(\frac{\partial^2 G_\phi}{\partial\phi_i\,\partial\phi_j}\right) = \frac{1}{1-\phi_s} - 2\chi - \left(\frac{\partial\chi}{\partial\phi_s}\right)\phi_s \equiv J, \quad (i \neq j) \tag{2.2.61a}$$

$$\frac{V_0{}^o}{RT}\left(\frac{\partial^2 G_\phi}{\partial\phi_i{}^2}\right) = J + \frac{1}{P_i\phi_i} \equiv J + J_i \tag{2.2.61b}$$

where Eq. (2.2.39) has been used. Again, we note that χ and g are here treated as functions of ϕ_s, the total volume fraction of the polymer solute. Introducing Eqs. (2.2.61a) and (2.2.61b) into Eq. (1.3.68) for the spinodal, with G_{ij} replaced by $G_{ij}{}^{(\phi)}$, we obtain

$$D \equiv \begin{vmatrix} J+J_1 & J & \cdots & J \\ J & J+J_2 & \cdots & J \\ \cdots & \cdots & \cdots & \cdots \\ J & J & \cdots & J+J_r \end{vmatrix} = 0 \tag{2.2.62}$$

By use of the formula

$$\begin{vmatrix} A+a_1 & B+b_1 & \cdots \\ A+a_2 & B+b_2 & \cdots \\ \cdots & \cdots & \cdots \\ A+a_r & B+b_r & \cdots \end{vmatrix} = \begin{vmatrix} a_1 & b_1 & \cdots \\ a_2 & b_2 & \cdots \\ \cdots & \cdots & \cdots \\ a_r & b_r & \cdots \end{vmatrix} + A\begin{vmatrix} 1 & b_1 & \cdots \\ 1 & b_2 & \cdots \\ \cdot & \cdots & \cdots \\ 1 & b_r & \cdots \end{vmatrix} + B\begin{vmatrix} a_1 & 1 & \cdots \\ a_2 & 1 & \cdots \\ \cdot & \cdots & \cdots \\ a_r & 1 & \cdots \end{vmatrix} + \cdots \tag{2.2.63}$$

the determinant D can be expanded to give

$$D = J\left(\prod_{i=1}^{r} J_i\right)\left(\sum_{i=1}^{r} J_i{}^{-1}\right) + \prod_{i=1}^{r} J_i \tag{2.2.64}$$

Hence Eq. (2.2.62) gives

$$J + \left(\sum_{i=1}^{r} J_i{}^{-1}\right)^{-1} = 0 \tag{2.2.65}$$

Thus the spinodal is described by

$$(1-\phi_s)^{-1} - 2\chi - (\partial\chi/\partial\phi_s)_{T,p}\,\phi_s + (P_w\,\phi_s)^{-1} = 0 \tag{2.2.66}$$

where P_w is the weight-average relative chain length of the polymer mixture:

$$P_w = \sum_{i=1}^{r} P_i \phi_i / \phi_s = \sum_{i=1}^{r} P_i \xi_i \tag{2.2.67}$$

When a solution is prepared by dissolving a polydisperse polymer sample in a solvent, it may be considered as a **quasibinary** system. The weight fractions ξ_i are characteristic parameters describing the molecular weight distribution of the polymer sample and hence do not vary with ϕ_s. Thus, for a quasibinary solution, the P_w [and the P_n defined by Eq. (2.2.28)] is independent of ϕ_s, and the spinodal represented as a curve on the χ versus ϕ_s diagram depends only on the weight-average relative chain length of the polymer solute. It is important to note that, though the proof is omitted here, this conclusion on the spinodal of a quasibinary solution does not hold when χ depends separately on the volume fractions of individual polymer components, rather than their sum ϕ_s only.

Differentiating Eq. (2.2.64) with respect to ϕ_i and using Eq. (2.2.65), we obtain

$$\begin{aligned} D_i &\equiv (\partial D / \partial \phi_i)_{T, p, \phi_j} \\ &= \left(\prod_{j=1}^{r} J_j \right) \left[(\partial J / \partial \phi_s) \left(\sum_{j=1}^{r} J_j^{-1} \right) - P_i \left(\sum_{j=1}^{r} J_j^{-1} \right)^{-1} \right] \end{aligned} \tag{2.2.68}$$

Equation (1.3.69) for the critical point can be written in terms of J, J_i, and D_i as

$$Q \equiv \begin{vmatrix} D_1 & D_2 & \dots & D_r \\ J & J + J_2 & \dots & J \\ \dots & \dots & \dots & \dots \\ J & J & \dots & J + J_r \end{vmatrix} = 0 \tag{2.2.69}$$

This determinant can be expanded to give

$$\begin{aligned} Q = {} & (D_1/J_1) \left(\prod_{i=1}^{r} J_i \right) \left(1 + J \sum_{i=1}^{r} J_i^{-1} \right) \\ & - (J/J_1) \left(\prod_{i=1}^{r} J_i \right) \left(\sum_{i=1}^{r} D_i J_i^{-1} \right) \end{aligned} \tag{2.2.70}$$

The first term on the right-hand side vanishes by virtue of Eq. (2.2.65). Hence Eq. (2.2.69) gives

$$\sum_{i=1}^{r} D_i J_i^{-1} = 0 \tag{2.2.71}$$

Substituting for J_i and D_i from Eqs. (2.2.61b) and (2.2.68), we obtain [11]

$$(1-\phi_s)^{-2} - 3(\partial\chi/\partial\phi_s)_{T,p} - (\partial^2\chi/\partial\phi_s^2)_{T,p}\,\phi_s - (P_z/P_w^2\,\phi_s^2) = 0 \quad (2.2.72)$$

where P_z is defined by

$$P_z \equiv \sum_{i=1}^{r} P_i^2\,\phi_i / \sum_{i=1}^{r} P_i\phi_i = P_w^{-1} \sum_{i=1}^{r} P_i^2\,\xi_i \quad (2.2.73)$$

A set of Eqs. (2.2.66) and (2.2.72) determines the critical point. If χ is independent of ϕ_s, this point is given by

$$\phi_{sc} = 1/(1+P_w P_z^{-1/2}) \quad (2.2.74a)$$

$$\chi_c = \tfrac{1}{2}(1+P_z^{1/2}P_w^{-1})(1+P_z^{-1/2}) \quad (2.2.74b)$$

where the subscript c indicates the critical point. These expressions for ϕ_{sc} and χ_c were first derived in a different way by Stockmayer [13]. In contrast to the spinodal, the critical point for a quasibinary solution depends on the z-average relative chain length P_z as well as on the weight-average one P_w. If the polymer solute is monodisperse in molecular weight, Eqs. (2.2.74a) and (2.2.74b) reduce to Eqs. (2.1.38) and (2.1.39), respectively.

From what we have shown above, we see that quasibinary solutions of polydisperse polymers with the same P_w have an identical spinodal on the χ versus ϕ_s diagram. Their critical points are all located on this quasibinary spinodal. In particular, the critical point for the true binary solution appears at the maximum of this curve. Equation (2.2.74a) indicates that the critical point moves along the spinodal toward higher concentration as the polymer becomes more polydisperse, i.e., as P_z departs more from P_w. Thus, on the χ versus ϕ_s diagram, the critical point of a polydisperse polymer sample in a single solvent appears to the right of that for a monodisperse one with the same P_w. This conclusion holds also when χ depends on the total concentration of the polymer, as was shown by Gordon et al. [14].

2.2.4. Cloud-Point Curves – 1. Ternary System

Phase separation in a binary solution composed of a single solvent and a monodisperse polymer has been discussed in § 2.1.3. In extending the discussion of this phenomenon to a quasibinary solution, it is relevant to start with the simplest of such solutions, i.e., a ternary system consisting of a solvent and two monodisperse polymers (components 1 and 2) with different relative chain lengths P_1 and P_2. We have shown in Fig. 1.10 the composition dependence of the Gibbs free energy for a ternary solution which exhibits separation into two phases. From that diagram we expect that if P_2 is larger than P_1, a phase diagram like that shown in Fig. 2.7 obtains on the composition triangle for the ternary system. Since, in general, the solubility of a polymer decreases with increasing chain length, there will be situations in which a binary solution composed of polymer 2 and solvent 0 undergoes phase separation at a temperature T_1, but a binary solution of polymer 1 and solvent 0 remains homogeneous at the same temperature. The thick solid curve P_1' Q_1' Q_1'' P_1'' in Fig. 2.7 illustrates the binodal at T_1. The open circle on the binodal and the thick dashed curve touching it denote the critical point and the spinodal. For any composition inside the solid curve at $T = T_1$ the solution separates into two phases. The coexisting phases are marked by pairs of points connected by thin solid or dashed lines, e.g., P_1' and P_1'', R_1' and R_1'', Q_1' and Q_1'', etc. Here the polymer-poor (dilute) phase is indicated by a single prime and the polymer-rich (concentrated) phase by a double prime. As the temperature is raised from T_1 to T_2, T_3, etc., the region enclosed by the binodal shrinks and eventually converges to a point which corresponds to the critical point of a binary system composed of polymer 2 and solvent 0. Above the temperature T_{2c}, at which this convergence occurs, the ternary system remains a single phase, no matter what the composition is.

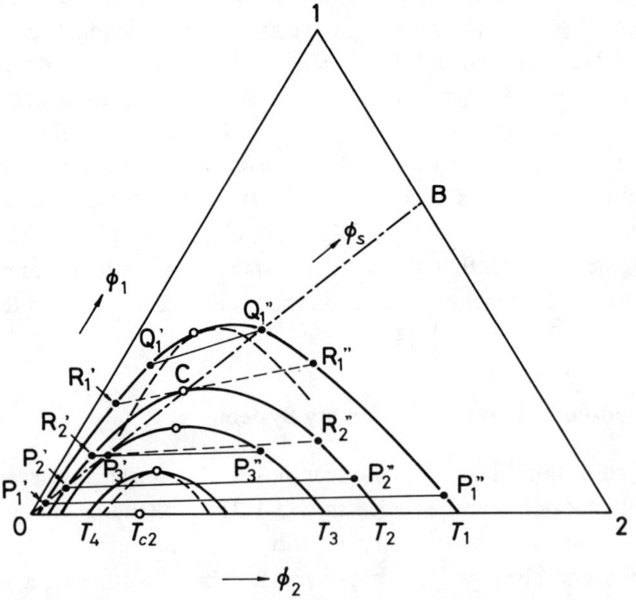

Fig. 2.7. Phase diagram of a ternary system (solvent 0 + polymer 1 + polymer 2). Thick solid lines: binodals. Thick broken lines: spinodals. Thin solid and broken lines: tie lines. Dot-dash line: composition line.

Now, suppose that we make up a binary mixture of homologous polymers 1 and 2 at the composition represented by point B in Fig. 2.7 and dilute it successively by addition of solvent 0. The process follows the straight line connecting B to the apex 0. For convenience we call 0B the composition line. If the total volume fractions of polymer, ϕ_s, at the points of intersection of the composition line 0B with the binodals are plotted against T for the respective binodals, we obtain a relation indicated by the thick solid line $P_1{}'$ $P_2{}'$ $P_3{}'$ $CQ_1{}''$ in Fig. 2.8. This line is a quasibinary cross section of the binodal surface and is called the (quasibinary) cloud-point curve for the ternary solution. For example, the solution represented by point $P_1{}'$ in Fig. 2.8 does not undergo phase separation at T_3 or T_2, but becomes turbid when it is cooled to T_1. However, the concentrated phase separating out at this cloud point is not the solution represented by point $Q_1{}''$ in Fig. 2.7, but the one related to point $P_1{}'$ by the tie line (the thin solid line in Fig. 2.7), i.e., point $P_1{}''$. Therefore, the coexisting concentrated phase does not fall on the cloud-point curve. In this respect, the cloud-point curve is basically different from that for a binary solution. We

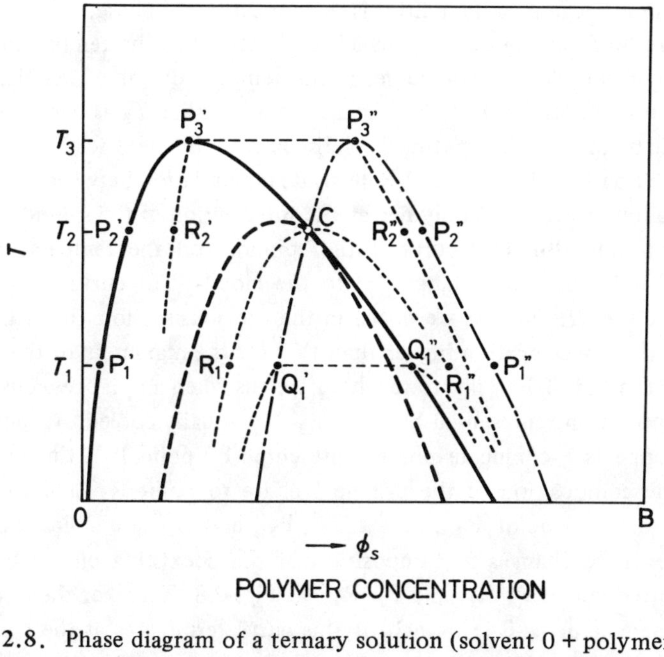

Fig. 2.8. Phase diagram of a ternary solution (solvent 0 + polymer 1 + polymer 2). Solid line: cloud-point curve. Dot-dash line: shadow curve. Thick broken line: spinodal. Thin broken line: coexistence curve. C: critical point.

denote the volumes of concentrated and dilute phases by V'' and V', respectively, and define the **phase-volume ratio** R by

$$R \equiv V''/V' \tag{2.2.75}$$

Thus, R is zero when solution P_1' begins to separate into two phases at $T = T_1$. The phase having the larger volume at the inception of phase separation is called the **principal phase**, and the corresponding separated phase the **conjugate phase**. Thus, when turbidity appears at point P_1', the principal phase is given by point P_1', and the conjugate phase by point P_1''. On the other hand, when turbidity appears at point Q_1'', R is infinite; point Q_1'' represents the principal phase and point Q_1' the conjugate phase. In short, the cloud-point curve represents the relation between temperature and composition of the principal phase. The dot-dash line Q_1' CP_3'' P_2'' P_1'' in Fig. 2.8 shows the relation between temperature and composition of the conjugate phase. This curve is called the **shadow curve**. In actuality, it is very difficult to measure shadow curves directly, because the volume of the conjugate phase is theoretically zero. As is shown in Fig. 2.8, the cloud-point curve and the corresponding shadow curve intersect at the critical point C. In the region to the left of point C the principal phase is the dilute phase and in the region to the right it is the concentrated phase.

The threshold temperature T_t, which was defined as the temperature at the maximum of the cloud-point curve, is the temperature at which the binodal touches the composition line. In Fig. 2.7, we see that T_3 is just equal to T_t. An analytic method for computing T_t is discussed in the next section.

The thick dashed line in Fig. 2.8 illustrates the relation between temperature and the volume fraction of polymer at the intersection of the spinodal with the composition line 0B. This curve is the spinodal on the temperature versus composition diagram. It is tangential to the cloud-point curve at the critical point; see Eq. (2.2.130). As we noted in the previous section, the critical point is located at a concentration higher than that for the maximum of the spinodal.

We can use Fig. 2.7 to illustrate what happens when a homogeneous solution of the composition represented by point P_3' is gradually cooled. At temperature T_3 there appears a conjugate phase represented by point P_3''. Further cooling changes the composition of the principal phase to point R_2' and that of the conjugate phase to point R_2''. Points R_2', P_3', and R_2'' are colinear and form a tie line. These changes in composition of the coexisting phases follow the pair of dotted curves $P_3'R_2'$ and $P_3''R_2''$ in Fig. 2.8. Together they constitute the **coexistence curve**. One branch of this curve terminates at the cloud-point curve, and the other terminates at the shadow curve. This property can be utilized to locate the shadow curve, which is not amenable to direct observation [15]. A series of solutions of different ϕ_s is prepared and the coexistence curve for each of them is determined. The end of each principal-phase branch can be easily located, because it is a cloud point. This information allows us to infer the corresponding terminus of the conjugate-phase branch. Connecting the terminal points so obtained at different ϕ_s gives the desired shadow curve.

If a given solution happens to have the critical composition ϕ_{sc}, the two branches of the coexistence curve merge, as is illustrated in Fig. 2.8; and the volume ratio R defined by Eq. (2.2.75) may remain of the order of unity even as the temperature T approaches T_c. This behavior is contrasted to the fact that for solutions below ϕ_{sc}, R tends to zero as T approaches the cloud-point temperature, while the solutions above ϕ_{sc}, R diverges to infinity in the same process. These features of R can be used to locate the critical point [15].

Returning to Fig. 2.7, we consider what happens when the temperature is lowered below T_1. The region enclosed by the binodal is progressively enlarged, and ultimately the critical point reaches the side $\overline{01}$ of the composition triangle at the critical temperature T_{c1} of the binary system composed of polymer 1 and solvent 0. Figure 2.9(a) shows the binodal at $T = T_{c1}$. For $T < T_{c1}$, the binodal is more expanded, as shown in Fig. 2.9(b), and has no critical point.

When P_1 and P_2 differ greatly, separation into three phases can take place. Figures 2.9(c) and 2.9(d) show binodals involving a three-phase separation. A solution of any composition inside the shaded triangular region splits into three phases indicated by triangular marks, called here **three-phase points**. We note that the binodals in these diagrams consist of several branches intersecting at three-phase points.

According to Tompa [16] and Šolc [17], when separation into three phases occurs, the phase diagram of a ternary solution changes in order (a) through (d) of Fig. 2.10 as T is lowered. In these diagrams, the regions where significant changes in phase relationships occur are extremely magnified to facilitate understanding. Thus, the dot-dashed line representing a composition line is drawn parallel to the side $\overline{01}$. Critical points and three-phase points are indicated by unfilled circles and triangles, respectively, and binodals and spinodals by solid and dashed lines, respectively. The thick solid line defines the boundary between homogeneous and inhomogeneous phases. The fact that all these lines and points are located very close to the side $\overline{01}$ implies that separation into three phases occurs at very low concentrations of polymer 2.

Until T reaches T_d, the phase diagram consists of a binodal containing one critical point C_1. At $T = T_d$, a new critical point, often called the double critical point, evolves on the spinodal associated with the C_1-binodal, and as T is lowered, this point splits into discrete critical points C_2 and C_3 and two new binodals grow on both sides of C_2 and C_3, as shown in (a) of Fig. 2.10. Point C_3 eventually moves away from the region enclosed by the C_1-binodal, and

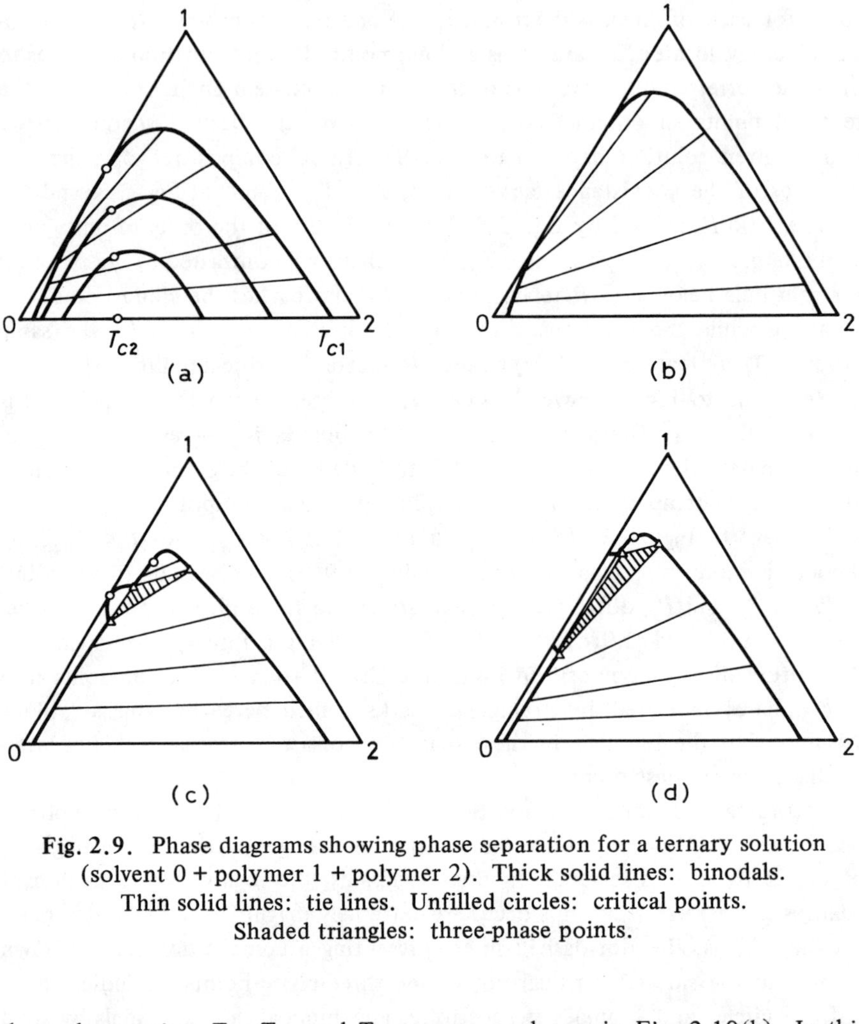

Fig. 2.9. Phase diagrams showing phase separation for a ternary solution (solvent 0 + polymer 1 + polymer 2). Thick solid lines: binodals. Thin solid lines: tie lines. Unfilled circles: critical points. Shaded triangles: three-phase points.

three-phase points T_1, T_2, and T_3 appear, as shown in Fig. 2.10(b). In this situation, a system with any composition inside the triangle T_1 T_2 T_3 separates into three phases represented by T_1, T_2, and T_3. With a further decrease in T, the C_3-binodal expands more, point C_3 moves out of the composition triangle, and points T_1 and T_2 approach, as shown in (c) and (d) of Fig. 2.10. Eventually, points T_1 and T_2 merge with point C_1, and the C_1-binodal retires into the metastable region enclosed by the C_3-binodal. At the same time, point T_3 disappears. This is the final stage for the occurrence of a three-phase equilibrium. After this, point C_1 leaves the C_3-binodal and moves toward point C_2 until both merge. The double critical point then disappears from the phase diagram.

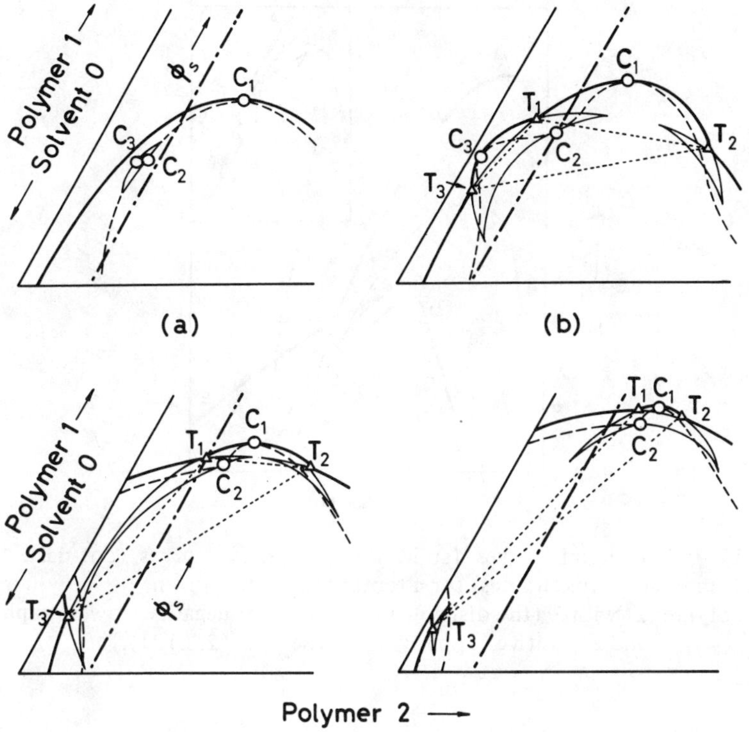

Fig. 2.10. Change with temperature, in the order (a), (b), (c), (d), in the phase diagram for a ternary solution (solvent 0 + polymer 1 + polymer 2) in which separation into three phases occurs. Solid lines: binodals. Broken lines: spinodals. Dot-dash line: composition line. C_1, C_2, C_3 : critical points. T_1, T_2, T_3 : three-phase points.

If we trace the above evolution of the phase relationship along the composition line indicated by the dot-dashed line and project the result on the T versus ϕ_s plane, we obtain a quasibinary cloud-point curve as shown in Fig. 2.11. This curve consists of three branches AT_1B, BC_2D, and DT_1E. These correspond to the binodals associated with the critical points C_1, C_2, and C_3 in Fig. 2.10. Either the branch AT_1 or T_1E of the cloud-point curve has no critical point, and point T_1 is a three-phase point. The critical point C_2 in Fig. 2.10 appears on branch BD of the cloud-point curve in Fig. 2.11. This branch extends from point C_2 into the unstable region bounded by the dashed line representing the spinodal. The dot-dash lines $\bar{A}T_2\bar{B}$, $\bar{B}C_2\bar{D}$, and $\bar{D}T_3\bar{E}$ are the branches of the

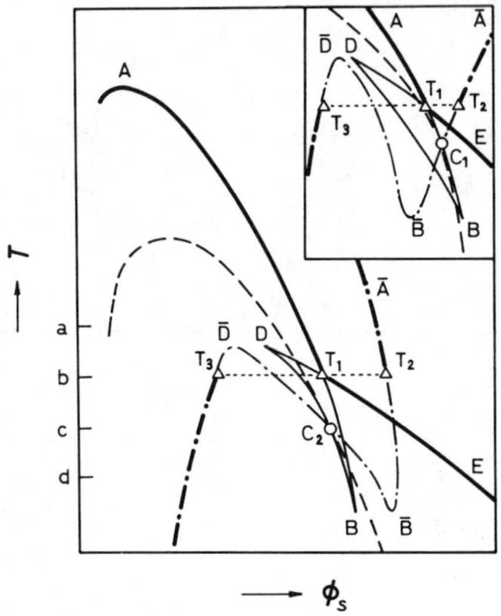

Fig. 2.11. Cloud-point curves (solid lines), shadow curves (dot-dash lines), and spinodals (broken lines) for a ternary solution (solvent 0 + polymer 1 + polymer 2) with Z (the distribution parameter) negative (lower graph) and Z positive (upper graph). [See Eq. (2.2.131).]

shadow curve corresponding, respectively. to the branches AT_1B, BC_2D, and DT_1E of the cloud-point curve. The three-phase points T_2 and T_3 on the shadow curve lie on the horizontal line passing through the three-phase point T_1.

Šolc [17] showed for the system with $P_1 = 1$ and $P_2 = 25$ that a phase diagram of the type described above is obtained when ξ_2 is in the range $0.00042 < \xi_2 < 0.0094$. Here ξ_2 is the volume fraction of polymer 2 in the polymer mixture. In this system, if ξ_2 is slightly larger than 0.0094, the phase diagram is of the type illustrated in the insert of Fig. 2.11. The critical point C_1 appears on the AB branch of the cloud-point curve, and moves upward along this branch with increasing ξ_2. On the other hand, if ξ_2 is smaller than 0.00042, the cloud-point curve has the critical point C_3 on the DE branch. Thus the quasibinary cross section of the phase diagram for a ternary system consisting of a solvent and two homologous polymers can display various kinds of critical points, depending on the mixing ratio of the polymer components.

To discuss the distribution of critical points on the phase diagram of such a system we follow Tompa [16]. In his theory, χ is assumed to be independent of ϕ_s. Then, Eqs. (2.2.66) and (2.2.72) for the critical point reduce to

$$(1 - \phi_s)^{-1} - 2\chi + (P_w \phi_s)^{-1} = 0 \tag{2.2.76}$$

$$(1 - \phi_s)^{-2} - (P_w \phi_s)^{-2} P_z = 0 \tag{2.2.77}$$

where the subscript c to indicate the critical condition has been omitted for simplicity. For the ternary system under consideration, Eqs. (2.2.67) and (2.2.73) are written

$$P_w = P_1 + (P_2 - P_1)\xi_2 \tag{2.2.78}$$

and

$$P_z = P_w^{-1} [P_1^2 + (P_2^2 - P_1^2)\xi_2] \tag{2.2.79}$$

where ξ_2 is equal to ϕ_2/ϕ_s, with $\phi_s = \phi_1 + \phi_2$. Eliminating ξ_2 from these two equations, we obtain

$$P_z = P_1 + P_2 - (P_1 P_2/P_w) \tag{2.2.80}$$

With this equation, we can evaluate ϕ_s and P_w for a given value of $\chi(T)$ from Eqs. (2.2.76) and (2.2.77). The result can be used to determine the position (ϕ_1, ϕ_2) of a critical point inside the composition triangle. It is necessary for the occurrence of separation into three phases that the set of Eqs. (2.2.76) and (2.2.77) has at least two sets of positive solutions.

We introduce a new parameter u defined by

$$u \equiv 2\chi(1 - \phi_s) - 1 \tag{2.2.81}$$

This is combined with Eq. (2.2.76) to give

$$u = (1 - \phi_s)/(P_w \phi_s) \tag{2.2.82}$$

which, with Eq. (2.2.80), transforms Eq. (2.2.77) to

$$u = [P_1 + P_2 - (P_1 P_2/P_w)]^{-1/2} \tag{2.2.83}$$

Since P_w must satisfy the physical requirement $P_1 < P_w < P_2$, the physically allowable range of u is found from Eq. (2.2.83) to be

$$P_2^{-1/2} \leqslant u \leqslant P_1^{-1/2} \tag{2.2.84}$$

Eliminating ϕ_s and P_w from Eqs. (2.2.82) and (2.2.83), with the help of Eq. (2.2.81), we obtain an algebraic equation for u:

$$f(u) \equiv P_1 P_2 u^4 - \lambda u^3 + (P_1 + P_2)u^2 - u - 1 = 0 \tag{2.2.85}$$

where

$$\lambda = P_1 P_2 (2\chi - 1) - P_1 - P_2 \tag{2.2.86}$$

For given P_1 and P_2 we can solve Eq. (2.2.85) for u as a function of χ. The positive root falling in the range specified by Eq. (2.2.84) corresponds to a critical point inside the composition triangle. We denote it by u_c. It follows from Eqs. (2.2.84) and (2.2.85) that if the values of χ corresponding to the upper and lower bounds of u_c are denoted by χ_{c1} and χ_{c2}, they are given by

$$\chi_{c1} = (1 + P_1^{-1/2})^2/2, \quad \chi_{c2} = (1 + P_2^{-1/2})^2/2 \tag{2.2.87}$$

Thus, for χ in the range $\chi_{c2} < \chi < \chi_{c1}$, all the positive roots of Eq. (2.2.85) can be u_c.

First, we consider the case in which P_1 and P_2 are not much different. In this case, as χ is increased, the function $f(u)$ changes in order (a), (b), (c) of Fig. 2.12. The thick segments marked on the abscissa axes indicate the range of u specified by Eq. (2.2.84). Thus the root indicated by an open circle in (b) is the only u_c for the case under consideration, giving a phase diagram like (a) of Fig. 2.9.

Next, if P_2 is much larger than P_1, $f(u)$ changes in order (a), (d), (e), (f), (c) of Fig. 2.12 as χ is increased. For χ in the range $\chi_{c2} < \chi < \chi_d < \chi_{c1}$, $f(u)$ has only one u_c, as in (d); this u_c is designated $u_c(1)$. At $\chi = \chi_d$, a double root u_d appears in the range specified by the thick segment, and as χ exceeds χ_d, it splits into two distinct u_c, denoted $u_c(2)$ and $u_c(3)$, as in (e). For χ larger than χ_{c1}, $u_c(3)$ moves out of the upper edge of the thick segment, as in (f), and when χ reaches $\chi_{d'}$, $u_c(1)$ and $u_c(2)$ merge to a single value $u_{d'}$. With further increase in χ, this double root disappears, and $f(u)$ assumes a shape like that shown in (c). In correspondence with this change in $f(u)$, the phase diagram in the composition triangle varies in order (c), (d), and (b) of Fig. 2.9 or in order (a) through (d) of Fig. 2.10. Roots $u_c(1)$, $u_c(2)$, and $u_c(3)$ correspond to critical points C_1, C_2, and C_3 of Fig. 2.10, respectively.

Double roots u_d and $u_{d'}$ must satisfy Eq. (2.2.85) and the condition

$$df/du = 4P_1P_2u^3 - 3\lambda u^2 + 2(P_1 + P_2)u - 1 = 0 \tag{2.2.88}$$

For example, when $P_1 = 1$ and $P_2 = 25$, the roots and associated λ are as follows: $u_d = 0.8946$, $\lambda_d = 48.78$; $u_{d'} = 0.4225$, $\lambda_{d'} = 53.24$. The corresponding values of χ, ϕ_1, and ϕ_2 are [17]

$$\chi_d = 1.9956, \quad \phi_{1d} = 0.5251, \quad \phi_{2d} = 0.0002$$

and

$$\chi_{d'} = 2.0848, \quad \phi_{1d'} = 0.6526, \quad \phi_{2d'} = 0.0062$$

These values indicate that separation into three phases occurs in a region very close to the side $\overline{01}$ of the composition triangle, as illustrated in Figs. 2.9 and 2.10.

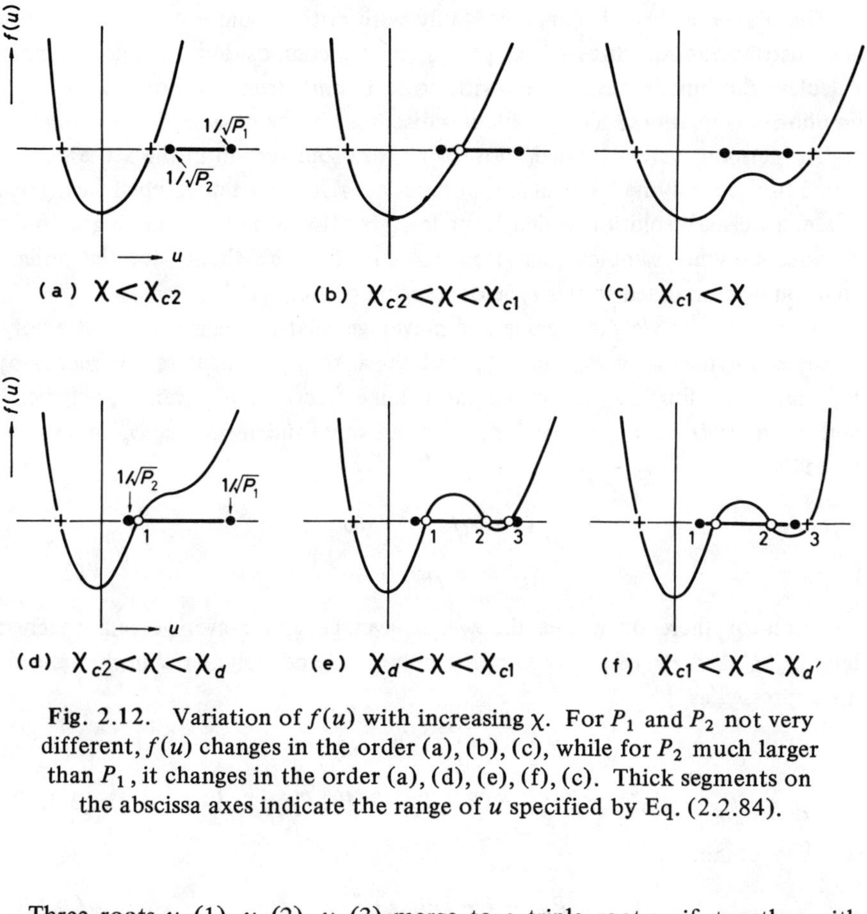

Fig. 2.12. Variation of $f(u)$ with increasing χ. For P_1 and P_2 not very different, $f(u)$ changes in the order (a), (b), (c), while for P_2 much larger than P_1, it changes in the order (a), (d), (e), (f), (c). Thick segments on the abscissa axes indicate the range of u specified by Eq. (2.2.84).

Three roots $u_c(1)$, $u_c(2)$, $u_c(3)$ merge to a triple root u_t if, together with Eqs. (2.2.85) and (2.2.88), the condition

$$d^2f/du^2 = 12P_1P_2u^2 - 6\lambda u + 2(P_1 + P_2) = 0 \tag{2.2.89}$$

is satisfied. Elimination of u and λ from these three equations yields

$$3(P_1{}^2 - 10P_1P_2 + P_2{}^2)^2 + (P_1 + P_2)(P_1{}^2 - 106P_1P_2 + P_2{}^2) - 27P_1P_2 = 0 \tag{2.2.90a}$$

If both P_1 and P_2 are sufficiently large, the first term dominates the equation and yields $P_2/P_1 = 9.9$ as the relevant root. This result implies that if $P_2 < 10P_1$, $f(u)$ can have only one positive root, so that separation into three phases does not occur. For $P_1 = 1$, Eq. (2.2.90a) reduces to

$$(P_2 - 2)^2(3P_2{}^2 - 47P_2 + 1) = 0 \tag{2.2.90b}$$

which gives $P_2 = 15.645$ if $P_2 \neq 2$.

The above analysis is concerned only with critical points. In order to know the distribution of three-phase points in the composition triangle we must calculate the binodals associated with critical points from solutions to the phase equilibrium equations. This problem is discussed in the next section.

The ternary system treated above is an idealized system in the sense that no monodisperse polymer is available in practice. What can actually be investigated is a **quasiternary** solution which is made up by dissolving two homologous polydisperse polymer samples in a single solvent. Tompa's theory for the ternary solution was extended to this type of solution by Šolc [17].

We denote the weight-average and z-average relative chain lengths of a polydisperse polymer I by P_{Iw} and P_{Iz} and those of a polydisperse polymer II by P_{IIw} and P_{IIz}. Further, we denote the volume fractions of solvent 0, polymer I, and polymer II by ϕ_0, ϕ_I, and ϕ_{II}, respectively, and introduce ϕ_s and ξ_{II} defined by

$$\phi_s \equiv \phi_I + \phi_{II} = 1 - \phi_0 \qquad (2.2.91)$$

$$\xi_{II} \equiv \phi_{II}/\phi_s \qquad (2.2.92)$$

In terms of these quantities the weight-average and z-average relative chain lengths, P_w and P_z, of the mixture of polymer I and polymer II in the solution are expressed as

$$P_w = P_{Iw} + (P_{IIw} - P_{Iw})\,\xi_{II} \qquad (2.2.93a)$$

$$P_z = P_w^{-1}\,[P_{Iz}P_{Iw} + (P_{IIz}P_{IIw} - P_{Iz}P_{Iw})\,\xi_{II}] \qquad (2.2.93b)$$

Then we obtain

$$P_z = S - (P/P_w) \qquad (2.2.94)$$

where

$$S \equiv (P_{IIz}P_{IIw} - P_{Iz}P_{Iw})/(P_{IIw} - P_{Iw}) \qquad (2.2.95a)$$

$$P \equiv P_{Iw}P_{IIw}\,(P_{IIz} - P_{Iz})/(P_{IIw} - P_{Iw}) \qquad (2.2.95b)$$

Equation (2.2.94) corresponds to Eq. (2.2.80) for the ternary solution. In place of Eq. (2.2.85) we get

$$f(u) \equiv Pu^4 - \lambda u^3 + Su^2 - u - 1 = 0 \qquad (2.2.96)$$

where

$$\lambda \equiv P(2\chi - 1) - S \qquad (2.2.97)$$

Without loss of generality, we may assume P_{IIw} to be larger than P_{Iw}. Further, if the molecular weight distributions of polymer I and polymer II are not very different, we may assume that P_{IIz} is larger than P_{Iz}. Then it follows from Eqs. (2.2.95a) and (2.2.95b) that both P and S are positive. In this case, we have the relation

$$P_{IIz}^{-1/2} \leqslant u \leqslant P_{Iz}^{-1/2} \tag{2.2.98}$$

which corresponds to Eq. (2.2.84) for the ternary solution. Therefore, Eq. (2.2.96) predicts a distribution of critical points which is essentially similar to that predicted by Eq. (2.2.85). However, when the molecular weight distribution of polymer I is much broader than that of polymer II, there is a possibility that the condition $P_{IIw} > P_{Iw}$ is compatible with the conditions $P_{IIz} < P_{Iz}$. In such a case, Eq. (2.2.95b) gives $P < 0$, and Eq. (2.2.96) can have two or four positive roots.

2.2.5. Cloud-Point Curves – 2. Multicomponent System

The phase separation behavior of a solution consisting of an arbitrary number of homologous polymers in a single solvent is similar in many respects to that of the ternary solution discussed in the preceding section. The solution usually separates into two phases when it is cooled, but separation into three (or even four) phases can occur if the molecular weight distribution of the polymer mixture is extremely broad and asymmetric. Thus we may observe for quasi-binary solutions of polymers a variety of cloud-point curves including curves as illustrated in Figs. 2.8 and 2.11. Here we describe a typical procedure for calculating cloud-point and shadow curves from the conditions for phase equilibrium.

For simplicity we discuss only the case in which the solution separates into two phases, one dilute in polymer and the other concentrated in polymer, respectively. We denote the volume fraction of component i ($i = 0, 1, \ldots, r$) in the former by ϕ_i' and in the latter by ϕ_i''. The conditions that the two phases are at thermodynamic equilibrium are given by

$$\Delta_m\mu_i\,(T, p, \phi_1', \phi_2', \ldots, \phi_r') = \Delta_m\mu_i\,(T, p, \phi_1'', \phi_2'', \ldots, \phi_r''), \quad (i = 0, 1, \ldots, r) \tag{2.2.99}$$

If Eqs. (2.2.30) and (2.2.31) are substituted and the chain length dependence of the χ_{Pi} is ignored, we obtain

$$\begin{aligned} &\ell n\,(1 - \phi_s') + \sum_{i=1}^{r} (1 - P_i^{-1})\,\phi_i' + \chi\,(\phi_s')\,\phi_s'^2 \\ &= \ell n\,(1 - \phi_s'') + \sum_{i=1}^{r} (1 - P_i^{-1})\,\phi_i'' + \chi\,(\phi_s'')\,\phi_s''^2 \end{aligned} \tag{2.2.100}$$

$$P_i^{-1} \ln \phi_i' + \sum_{i=1}^{r} (1 - P_i^{-1}) \phi_i' - \chi(\phi_s') \phi_s' (1 - \phi_s') + \int_{\phi_s'}^{1} \chi(\phi_s) \, d\phi_s$$
$$= P_i^{-1} \ln \phi_i'' + \sum_{i=1}^{r} (1 - P_i^{-1}) \phi_i'' - \chi(\phi_s'') \phi_s'' (1 - \phi_s'') + \int_{\phi_s''}^{1} \chi(\phi_s) \, d\phi_s, \quad (i = 1, 2, \ldots, r) \tag{2.2.101}$$

Here P_n for each phase has been expressed in terms of the P_i and ϕ_i, using Eq. (2.2.28), and χ_P has been transformed to χ by Eq. (2.2.36). Subtracting Eq. (2.2.101) from Eq. (2.2.100), we get

$$\ln (1 - \phi_s') - P_i^{-1} \ln \phi_i' + \chi(\phi_s') \phi_s' - \int_{\phi_s'}^{1} \chi d\phi_s$$
$$= \ln (1 - \phi_s'') - P_i^{-1} \ln \phi_i'' + \chi(\phi_s'') \phi_s'' - \int_{\phi_s''}^{1} \chi d\phi_s, \tag{2.2.102}$$

$$(i = 1, 2, \ldots, r)$$

We may use these equations, in place of Eqs. (2.2.101), to calculate the equilibrium distributions of the polymer components in the two phases. It is convenient to define an important quantity called the **separation factor** σ by

$$\phi_i''/\phi_i' \equiv \exp(P_i \sigma), \quad (i = 1, 2, \ldots, r) \tag{2.2.103}$$

Equations (2.2.102) can then be put in the form:

$$\sigma = \ln (1 - \phi_s'') - \ln (1 - \phi_s') + [\chi(\phi_s'') \phi_s'' - \chi(\phi_s') \phi_s'] + \int_{\phi_s'}^{\phi_s''} \chi d\phi_s \tag{2.2.104}$$

If, as we assumed above, the chain length dependence of the χ_{Pi} can be ignored, both χ_P and χ are independent of the chain length distribution. Then, as can be seen from Eq. (2.2.104), the same is true of the separation factor. Further, if σ is independent of chain length, σ must be positive, because otherwise we have $\phi_i'' < \phi_i'$ for all i other than zero, which contradicts the assignment of the double prime to the phase concentrated in polymer. On introduction of Eq. (2.2.40), Eq. (2.2.104) becomes

$$\sigma = \ln (1 - \phi_s'') - \ln (1 - \phi_s') + 2\chi_0 (\phi_s'' - \phi_s') + \tfrac{3}{2} \chi_1 (\phi_s''^2 - \phi_s'^2) + \tfrac{4}{3} \chi_2 (\phi_s''^3 - \phi_s'^3) + \ldots \tag{2.2.105}$$

We let the volume of the original solution be denoted by V and the volume fraction of polymer component i in it by ϕ_i. Conservation of mass requires

$$V\phi_i = V'\phi_i' + V''\phi_i'', \quad (i = 1, 2, \ldots, r) \tag{2.2.106}$$

where V' and V'' are the volumes of the phases dilute and concentrated in polymer, respectively. Using Eq. (2.2.103) and the phase-volume ratio $R = V''/V'$, we can transform Eq. (2.2.106) to

$$\phi_i = \xi_i \phi_s = \frac{1 + R \exp(P_i \sigma)}{1 + R} \phi_i', \quad (i = 1, 2, \ldots, r) \tag{2.2.107a}$$

or

$$\phi_i = \frac{R + \exp(-P_i \sigma)}{1 + R} \phi_i'', \quad (i = 1, 2, \ldots, r) \tag{2.2.107b}$$

where ξ_i is the weight fraction of polymer component i in the polymer mixture before phase separation. We note that since we have assumed throughout the derivations that the polymer species are uniform in specific volume, the volume fraction ϕ_i/ϕ_s can be equated to ξ_i.

With T, p, and ϕ_i $(i = 1, 2, \ldots, r)$ or T, p, ϕ_s, and ξ_i $(i = 2, 3, \ldots, r)$ specified, there are $2r + 1$ unknown quantities to be determined for the description of a two-phase equilibrium. They are the compositions ϕ_i' and ϕ_i'' $(i = 1, 2, \ldots, r)$ of the coexisting phases and the phase volume ratio R. We have $2r + 1$ conditions for these variables, i.e., Eqs. (2.2.100), (2.2.107a), and (2.2.107b). The computed total volume fractions of polymer, ϕ_s' and ϕ_s'', in the dilute and concentrated phases as functions of T give coexistence curves as we have explained in Fig. 2.8. In the theory of fractionation described in § 2.2.7, the compositions in the coexisting phases are computed, taking the phase-volume ratio, rather than T, as an experimental variable.

We now proceed to the calculation of cloud-point and shadow curves. As already shown, for a solution of a polymer with a dispersion of molecular weight, usually the phase dilute in polymer becomes the principal phase when ϕ_s is lower than the critical composition ϕ_{sc}, and, in this case, the phase-volume ratio R vanishes on the cloud-point curve. Therefore, if $\phi_s < \phi_{sc}$, it follows from Eqs. (2.2.107a) and (2.2.107b) that the cloud-point curve is described by

$$\phi_i' = \xi_i \phi_s, \quad \phi_s' = \phi_s, \quad P_n' = P_n = \left(\sum_{i=1}^{r} \xi_i / P_i \right)^{-1} \tag{2.2.108a}$$

and the corresponding shadow curve by

$$\phi_i'' = \xi_i'' \phi_s, \quad \phi_s'' = \left(\sum_{i=1}^{r} \xi_i'' \right) \phi_s, \quad P_n'' = \sum_{j=1}^{r} \xi_j'' \left(\sum_{i=1}^{r} \xi_i''/P_i \right)^{-1} \tag{2.2.108b}$$

where

$$\xi_i'' \equiv \xi_i \exp(P_i \sigma), \quad (i = 1, 2, \ldots, r) \tag{2.2.108c}$$

Conversely, if $\phi_s > \phi_{sc}$, the polymer-concentrated phase becomes the principal phase and R diverges to infinity on the cloud-point curve. In this case, the cloud-point curve is described by

$$\phi_i' = \xi_i' \phi_s, \quad \phi_s' = (\Sigma_i \xi_i') \phi_s, \quad P_n' = \Sigma_j \xi_j' (\Sigma_i \xi_i'/P_i)^{-1} \tag{2.2.109a}$$

and the corresponding shadow curve by

$$\phi_i'' = \xi_i \phi_s, \quad \phi_s'' = \phi_s, \quad P_n'' = P_n \tag{2.2.109b}$$

where

$$\xi_i' \equiv \xi_i \exp(-P_i \sigma), \quad (i = 1, 2, \ldots, r) \tag{2.2.109c}$$

If we define

$$K \equiv \begin{cases} 1 & (\phi_s < \phi_{sc}) \\ -1 & (\phi_s > \phi_{sc}) \end{cases} \tag{2.2.110}$$

and designate the total volume fraction of polymer, the number-average relative chain length, and the weight fraction of polymer component i in the conjugate phase by $\bar{\phi}_s$, $\bar{P}_n$, and $\bar{\xi}_i$, respectively, the expressions for the shadow curve in Eqs. (2.2.108) and (2.2.109) may be written

$$\bar{\phi}_i = \bar{\xi}_i \phi_s, \quad \bar{\phi}_s = (\Sigma_i \bar{\xi}_i) \phi_s, \quad \bar{P}_n = \Sigma_j \bar{\xi}_j (\Sigma_i \bar{\xi}_i/P_i)^{-1} \tag{2.2.111}$$

where

$$\bar{\xi}_i \equiv \xi_i \exp(KP_i \sigma), \quad (i = 1, 2, \ldots, r) \tag{2.2.112}$$

and ϕ_i, ϕ_s, and ξ_i specify the principal phase, i.e., the original solution.

For a polymer sample with a continuous distribution of relative chain length P, Eq. (2.2.112) may be written

$$\bar{\xi}(P) = \xi(P) \exp(KP\sigma) \tag{2.2.113}$$

where $\xi(P)$ is the normalized distribution of P in the principal phase or in the original solution and $\bar{\xi}(P)$ is a relative (i.e., unnormalized) distribution of P in the conjugate phase. We define

$$\lambda_k \equiv \int_0^\infty P^k \xi(P) dP, \quad (\lambda_0 = 1) \tag{2.2.114a}$$

$$\bar{\lambda}_k \equiv \int_0^\infty P^k \bar{\xi}(P) dP \tag{2.2.114b}$$

where k is an integer. In terms of λ_k and $\bar{\lambda}_k$, the number-average, weight-average, and z-average relative chain lengths in the principal and conjugate phases can be expressed as

$$P_n = 1/\lambda_{-1},\ P_w = \lambda_1,\ P_z = \lambda_2/\lambda_1, \ldots \tag{2.2.115a}$$

$$\bar{P}_n = \bar{\lambda}_0/\bar{\lambda}_{-1},\ \bar{P}_w = \bar{\lambda}_1/\bar{\lambda}_0,\ P_z = \bar{\lambda}_2/\bar{\lambda}_1, \ldots \tag{2.2.115b}$$

and, with Eq. (2.2.111), we have

$$\bar{\phi}_s = \bar{\lambda}_0\,\phi_s \tag{2.2.116}$$

If Eq. (2.2.113) is substituted into Eq. (2.2.114b), we get

$$\begin{aligned}\bar{\lambda}_k &= \int_0^\infty \xi(P) \sum_{j=0}^{\infty} [(K\sigma)^j P^{k+j}/j!]\, dP \\ &= \lambda_k + \sum_{j=1}^{\infty} [(K\sigma)^j/j!]\, \lambda_{k+j}\end{aligned} \tag{2.2.117}$$

which gives

$$d^n \bar{\lambda}_k/d\sigma^n = K^n \bar{\lambda}_{k+n} \tag{2.2.118}$$

For simplicity, we assume that χ does not depend on concentration, and apply the cloud-point condition to Eqs. (2.2.104) and (2.2.100). The results, expressed in terms of λ_k and $\bar{\lambda}_k$, are

$$2\chi\phi_s(\bar{\lambda}_0 - 1) = K\sigma + \ln[(1 - \phi_s)/(1 - \bar{\lambda}_0\phi_s)] \tag{2.2.119}$$

and

$$\begin{aligned}F(\sigma, \phi_s) \equiv{}& K\sigma[(1 + \bar{\lambda}_0)/2] + (\bar{\lambda}_0 - 1) - (\bar{\lambda}_{-1} - \lambda_{-1}) \\ &+ [\phi_s^{-1} - (1/2)(1 + \bar{\lambda}_0)] \ln[(1 - \bar{\lambda}_0\phi_s)/(1 - \phi_s)] = 0\end{aligned} \tag{2.2.120}$$

For a given polymer sample the λ_k are known. Hence, P_w and P_z are also known, and the critical composition ϕ_{sc} can be calculated from Eq. (2.2.74a). Thus once the concentration ϕ_s of the original solution (or of the principal phase) is assigned, K is also determined from the definition, Eq. (2.2.110). In these circumstances, the $\bar{\lambda}_k$, and also $F(\sigma, \phi_s)$, can vary only with σ. Unless the distribution function $\xi(P)$ is extremely asymmetric, Eq. (2.2.120) has a single nontrivial root which is always positive and tends to zero as ϕ_s approaches ϕ_{sc}. The χ value obtained by substituting this σ into Eq. (2.2.119) is denoted χ_{cloud}. The values of χ_{cloud} so obtained for a series of chosen ϕ_s values give the desired cloud-point curve, and the χ_{cloud} plotted against $\bar{\phi}_s = \bar{\lambda}_0\phi_s$ yields the corresponding shadow curve. This method for calculating cloud-point and shadow curves was proposed by Šolc [18]. By an ingenious application of it he clarified various interesting features of the cloud-point curve.

The derivative $d\chi_{\text{cloud}}/d\phi_s$, related closely to the slope of a cloud-point curve, can be calculated from

$$\frac{d\chi_{\text{cloud}}}{d\phi_s} = \left(\frac{\partial \chi_{\text{cloud}}}{\partial \phi_s}\right)_\sigma + \left(\frac{\partial \chi_{\text{cloud}}}{\partial \sigma}\right)_{\phi_s} \frac{d\sigma}{d\phi_s} \tag{2.2.121}$$

From Eq. (2.2.119) we can derive

$$\phi_s \left(\frac{\partial \chi_{\text{cloud}}}{\partial \phi_s}\right)_\sigma = \frac{1}{2(1-\phi_s)(1-\bar{\lambda}_0\phi_s)} - \chi_{\text{cloud}} \tag{2.2.122a}$$

and

$$\phi_s(\bar{\lambda}_0 - 1)\left(\frac{\partial \chi_{\text{cloud}}}{\partial \sigma}\right)_{\phi_s} = \frac{K}{2}\left[1 + \frac{\bar{\lambda}_1\phi_s}{1-\bar{\lambda}_0\phi_s} - 2\bar{\lambda}_1\phi_s\chi_{\text{cloud}}\right] \tag{2.2.122b}$$

Also, from Eqs. (2.2.120) and (A.7) we obtain

$$\frac{d\sigma}{d\phi_s} = -\frac{(\partial F/\partial\phi_s)_\sigma}{(\partial F/\partial\sigma)_{\phi_s}}$$

$$= \frac{2\sigma\phi_s^{-1}(\bar{\lambda}_0 - 1)^{-1} + K\left[(1-\bar{\lambda}_0\phi_s)^{-1} + (1-\phi_s)^{-1} - 4\chi_{\text{cloud}}\right]}{2\chi_{\text{cloud}}\bar{\lambda}_1\phi_s^2 - \bar{\lambda}_1\phi_s^2(1-\bar{\lambda}_0\phi_s)^{-1} - \phi_s} \tag{2.2.123}$$

where we have used Eqs. (2.2.118) and (2.2.119). With Eqs. (2.2.122a) through (2.2.123), Eq. (2.2.121) gives

$$\phi_s \frac{d\chi_{\text{cloud}}}{d\phi_s} = \frac{1}{\phi_s(\bar{\lambda}_0 - 1)}\left[\frac{1}{\phi_s(\bar{\lambda}_0 - 1)} \ln\left(\frac{1-\phi_s}{1-\bar{\lambda}_0\phi_s}\right) - \frac{1}{1-\phi_s}\right] - \chi_{\text{cloud}} \tag{2.2.124}$$

At the threshold cloud-point we have $d\chi_{\text{cloud}}/d\phi_s = 0$, so that Eq. (2.2.124) gives

$$\chi_{\text{cloud}}\phi_s(\bar{\lambda}_0 - 1) - \frac{1}{\phi_s(\bar{\lambda}_0 - 1)} \ln\left(\frac{1-\phi_s}{1-\bar{\lambda}_0\phi_s}\right) + \frac{1}{1-\phi_s} = 0 \tag{2.2.125}$$

A set of Eqs. (2.2.119), (2.2.120), and (2.2.125) is equivalent to the threshold conditions derived in a different way by Tompa [8].

For the special case in which χ depends on temperature according to Eq. (2.1.33) and the polymer solute has a molecular weight distribution of the exponential type, Eq. (2.2.19), Shultz [19] showed that the threshold temperature T_t is represented by

$$\frac{1}{T_t} = \frac{1}{\Theta} + \frac{1}{\Theta\psi}\left(\frac{Q}{P_w^{1/2}} + \frac{1}{2P_w}\right) \tag{2.2.126a}$$

where

$$Q \equiv 1 - \frac{1}{4(h+1)^2}\left[1 - \frac{0.184}{(h+1)^{7/6}}\right] \qquad (2.2.126b)$$

The parameter Q is constant for a series of homologous samples which have exponential molecular weight distributions with the same dispersion parameter h. According to Eq. (2.2.126a), a plot of T_t^{-1} against $P_w^{-1/2}$ for such a series of samples becomes linear for large P_w and the Θ can be determined from the intercept for $P_w = \infty$. However, the slope of the linear region does not allow evaluation of the parameter ψ unless it is corrected for Q.

At the critical point the principal and conjugate phases merge into a single phase, giving $\sigma = 0$, $\bar{\lambda}_k = \lambda_k$, and $\bar{\phi}_s = \phi_s$. Thus, at this point, both $F_\phi \equiv (\partial F/\partial \phi_s)_\sigma$ and $F_\sigma \equiv (\partial F/\partial \sigma)_{\phi_s}$ vanish, and $d\sigma/d\phi_s$ in Eq. (2.2.123) becomes indeterminate. By application of the well-known rule in calculus it can be shown that

$$\left(\frac{d\sigma}{d\phi_s}\right)_{\text{crit}} = -4K \frac{P_w P_z^{-1/2}(1 + P_z^{1/2} P_w^{-1})^2}{3P_z + 2P_z^{1/2} - P_{z+1}} \qquad (2.2.127)$$

We also find from Eq. (2.2.122b) that $(\partial \chi_{\text{cloud}}/\partial \sigma)_{\phi_s}$ vanishes at the critical point. Hence we obtain from Eq. (2.2.121), with Eqs. (2.2.122a), (2.2.74a), and (2.2.74b),

$$\left(\frac{d\chi_{\text{cloud}}}{d\phi_s}\right)_{\text{crit}} = \tfrac{1}{2}(1 + P_z^{1/2} P_w^{-1})^2 (1 - P_w P_z^{-1}) \geqslant 0 \qquad (2.2.128)$$

On the other hand, if χ is independent of concentration, Eq. (2.2.66) for the spinodal is

$$(1 - \phi_s)^{-1} - 2\chi_{\text{sp}} + (P_w \phi_s)^{-1} = 0 \qquad (2.2.129)$$

where the subscript sp indicates the spinodal. Eliminating P_w from this equation and its derivative with respect to ϕ_s, we find

$$\left(\frac{d\chi_{\text{sp}}}{d\phi_s}\right)_{\text{crit}} = \left(\frac{d\chi_{\text{cloud}}}{d\phi_s}\right)_{\text{crit}} \qquad (2.2.130)$$

From these results we can draw the following conclusions: (1) A cloud-point curve and its corresponding spinodal meet tangentially at the critical point. (2) From $(d\chi_{\text{cloud}}/d\phi_s)_{\text{crit}} \geqslant 0$, as can be seen from Eq. (2.2.128), it follows that $(dT_{\text{cloud}}/d\phi_s)_{\text{crit}} \leqslant 0$ for systems in which $d\chi/dT < 0$, and $(dT_{\text{cloud}}/d\phi_s)_{\text{crit}} \geqslant 0$ for systems in which $d\chi/dT > 0$. In either case, the critical point appears on the high concentration side of the threshold cloud-point. The cloud-point curve, critical point, and spinodal shown in Fig. 2.8 are consistent with these conclusions.

Equations (2.2.119) and (2.2.120) contain K and σ as a product $K\sigma$. Therefore, it is convenient to treat $K\sigma$, rather than σ, as a variable to be determined from these equations for the calculation of a cloud-point curve. When we move along the cloud-point curve, $K\sigma$ varies continuously and its sign changes at the critical point. We divide the cloud-point curve into subcritical and supercritical branches at the critical point, defining the former as the branch in which $K\sigma > 0$ and the latter as the branch in which $K\sigma < 0$. For the cloud-point curve shown in the main part of Fig. 2.11, the branch from A to C_2 via T_1 and B is subcritical, and that from C_2 to E via D and T_1 is supercritical.

For the approach from the subcritical branch, the derivative $[d(K\sigma)/d\phi_s]_{\text{crit}}$ is positive or negative, depending on whether the cloud-point curve passes through the critical point in the direction of decreasing or increasing ϕ_s. According to Eq. (2.2.127), this derivative is negative if the distribution parameter Z defined by [18]

$$Z \equiv 3P_z + 2P_z^{1/2} - P_{z+1} \tag{2.2.131}$$

is positive, while it is positive if Z is negative. Thus we see that Z gives important information on the behavior of the cloud-point curve in the vicinity of the critical point. The value of Z is always positive for polymers with molecular weight distributions of the exponential type. Hence, the critical point can never be that of type C_2 (see Fig. 2.11), and the cloud-point curve turns out to be smooth as depicted in Fig. 2.8. If the molecular weight distribution is logarithmic-normal, there is a possibility that Z becomes negative so that a cloud-point curve of type C_2 is obtained. Šolc [18] has discussed the important role of the distribution parameter Z in the theoretical analysis of phase equilibria of polydisperse polymer solutions.

2.2.6. Comparison between Theory and Experiment

Figure 2.13 presents the phase diagram of a polystyrene sample in cyclohexane as determined by Rehage et al. [20]. The average molecular weights of the sample are M_{rn} = 210,000, M_{rw} = 346,000, and M_{rz} = 550,000, which give M_{rw}/M_{rn} = 1.65 and M_{rz}/M_{rw} = 1.59. The thick solid line shows the cloud-point curve. The thin solid lines fitting differently marked points are the coexistence curves for different concentrations of the original solution. The dot-dash line is the shadow curve estimated from the coexistence curves. Finally, the open circle at the intersection of the cloud-point curve and the shadow curve

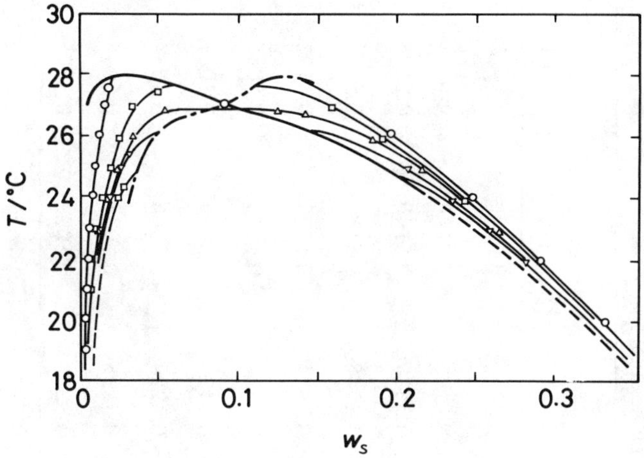

Fig. 2.13. Cloud-point curve (thick solid line), shadow curve (dot-dash line), and coexistence curve (thin solid line) for a polystyrene sample in cyclohexane. Data of Rehage et al. [20].

marks the critical point for this polymer + solvent system (a quasibinary system). On the whole, this phase diagram is similar to the schematic one for a ternary solution as shown in Fig. 2.8.

Table 2.3 summarizes critical point data determined by Koningsveld et al. [21] for a series of polystyrene samples in cyclohexane. Here M_{rn}, M_{rw}, and M_{rz} were evaluated from osmotic pressure, light scattering, and sedimentation equilibrium measurments, respectively. Also ϕ_{sc}, P_w, and P_z were calculated from w_{sc}, M_w, and M_z, with $v_0{}^o = 1.292$ cm^3 g^{-1} and $v_s{}^o = v_s{}^* = 0.9336$ cm^3 g^{-1} [see Eqs. (1.1.30) and (2.2.11)].

First, we compute M_{rz} for the first row sample of Table 2.3 by inserting the measured values of P_w and ϕ_{sc} into Eq. (2.2.74a). The result is 5.5×10^6, which is 10 times larger than the directly measured M_{rz}. If we compute ϕ_{sc} by using the experimental P_w and P_z, we obtain 0.023, which is about one-third the experimental value 0.068. This discrepancy is comparable with the observed differences in ϕ_{sc} between the theoretical binodals and the cloud-point curves for narrow-distribution polystyrene samples in cyclohexane (Fig. 2.5).

Since these comparisons between theory and experiment were made using equations which neglect concentration dependence of χ, we wish to examine what happens when this dependence is taken into account according to Eq. (2.2.40). Introduction of this expansion into Eqs. (2.2.72) and (2.2.66) gives

$$X_c \equiv \frac{1}{3}\left[\frac{1}{(1-\phi_{sc})^2} - \frac{P_z}{P_w{}^2\,\phi_{sc}{}^2}\right] = \chi_{1c} + \frac{8}{3}\chi_{2c}\,\phi_{sc} + \ldots \qquad (2.2.132)$$

Table 2.3. Critical point data for polystyrene in cyclohexane

$\frac{M_{rn}}{10^5}$	$\frac{M_{rw}}{10^5}$	$\frac{M_{rz}}{10^5}$	$\frac{M_{rw}}{M_{rn}}$	$\frac{M_{rz}}{M_{rw}}$	$\frac{T_c}{°C}$	$100\,\phi_{sc}$	X_c	χ_{0c}	
								$\chi_{2c} = 0$	$\chi_{2c} = 0.1554$
2.1	3.46	5.5 ± 0.4	1.6	1.6	27.0	6.8 ± 1	0.345	0.504	–
0.27	0.354	0.455	1.3	1.3	11.45	13.6	0.3703	0.5153	0.5212
0.49	0.51	0.55	1.04	1.08	14.7	11.0	0.3527	0.5140	0.5178
0.55	0.615	0.705	1.12	1.15	17.3	10.7	0.3548	0.5118	0.5155
0.91	0.93	0.96	1.02	1.03	20.5	8.75	0.3441	0.5099	0.5123
1.54	1.66	1.81	1.08	1.09	23.45	7.35	0.3411	0.5068	0.5085
2.00	2.86	4.38	1.4	1.5	25.55	7.0	0.3429	0.5046	0.5061
3.75	3.94	4.23	1.05	1.07	27.55	5.1	0.3294	0.5046	0.5054
4.90	5.27	5.93	1.08	1.13	28.0	4.7	0.3295	0.5038	0.5045
12.5	15.0	17.0	1.20	1.13	30.05	3.05	0.3230	0.5022	0.5025

$$\chi_{0c} = \frac{1}{2}\left(\frac{1}{1-\phi_{sc}} + \frac{1}{P_w\,\phi_{sc}}\right) - \frac{3}{2}X_c\,\phi_{sc} + 2\chi_{2c}\,\phi_{sc}^{\;2} + \ldots \qquad (2.2.133)$$

The quantity X_c can be evaluated from experimental values of M_{rw}, M_{rz}, and ϕ_{sc}. If we neglect the second and higher-order terms in Eq. (2.2.132) and the third and higher-order terms in Eq. (2.2.133), we get $\chi_{0c} = 0.504$ and $\chi_{1c} = 0.345$ from these equations and the data given in the first row of Table 2.3.

The values of X_c listed in Table 2.3 fall in the vicinity of 0.34, but closer inspection reveals a systematic variation with molecular weight of the sample. Following Koningsveld [21], we first ignore the temperature dependence of χ_{1c} and χ_{2c}, and thus regard X_c as a function of ϕ_{sc} only. The filled triangles of Fig. 2.14 indicate X_c plotted against ϕ_{sc}. The straight line fitting them gives $\chi_{1c} = 0.3092$ and $\chi_{2c} = 0.1554$ when analyzed by Eq. (2.2.132). With this χ_{2c} value inserted into Eq. (2.2.133), we get χ_{0c} values in the tenth column of Table 2.3. The filled circles in Fig. 2.14 denoting these values of χ_{0c} plotted against $1/T$ are fitted by a straight line. In this way, we derive [21]

$$\chi = 0.2035 + 90.65\,T^{-1} + 0.3092\,\phi_s + 0.1554\,\phi_s^{\;2} \qquad (2.2.134a)$$

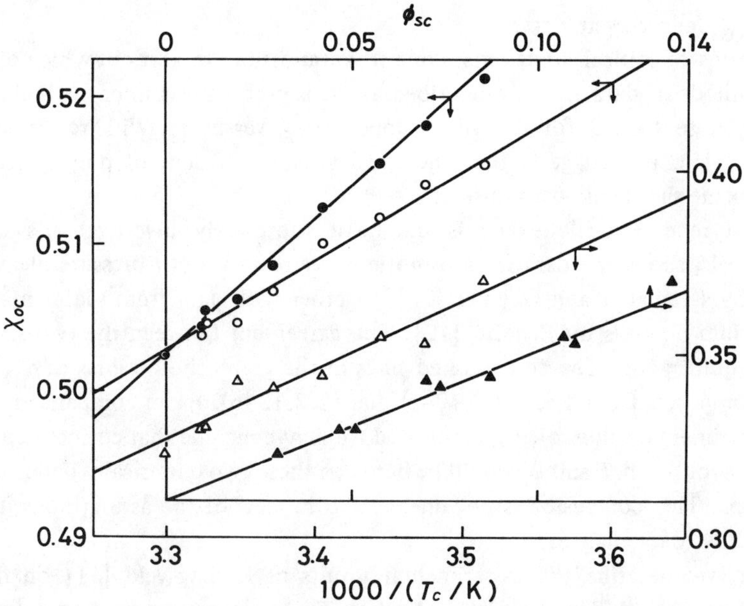

Fig. 2.14. Analysis of experimental data from Table 2.3. Filled triangles: X_c versus χ_{sc}. Filled circles: χ_{0c} versus $T_c^{\;-1}$. Unfilled triangles: X_c versus $T_c^{\;-1}$. Unfilled circles: χ_{0c} (for $\chi_{2c} = 0$) versus $T_c^{\;-1}$.

With the condition that $\chi_0 = 1/2$ at $T = \Theta$, this expression gives $\Theta = 305.7$ K, which is in fair agreement with the value of Θ estimated from the chain length dependence of threshold temperature T_t [see Eq. (2.1.44)].

Next, we attribute the variation of X_c with molecular weight of the polymer to a temperature dependence of χ_{1c}. This means that we neglect the second and higher-order terms in Eq. (2.2.132). The unfilled triangles in Fig. 2.14 show X_c plotted against $1/T_c$. These points are fitted well by a straight line, yielding $\chi_{1c} = -0.300 + 190\ T_c^{-1}$. In agreement with the approximation imposed on Eq. (2.2.132), we ignore the third and higher-order terms in Eq. (2.2.133). The values of χ_{0c} computed on the basis of this approximation are given in the ninth column of Table 2.3, and they are plotted against $1/T_c$. These points are also fitted by a straight line. In this way, we obtain

$$\chi = 0.3015 + 61.0\,T^{-1} + (-0.300 + 190\,T^{-1})\,\phi_s \qquad (2.2.134b)$$

which leads to $\Theta = 307.5$ K. Equations (2.2.134a) and (2.2.134b) give, respectively,

$$\chi_\Theta = 0.5000 + 0.3092\,\phi_s + 0.1554\,\phi_s^2 \qquad (2.2.134c)$$

and

$$\chi_\Theta = 0.5000 + 0.318\,\phi_s \qquad (2.2.134d)$$

where χ_Θ denotes χ at $T = \Theta$.

In terms of critical point data alone it is impossible to sort out which of the two methods of data analysis described above is preferable. Either method gives a value close to 1/3 for the initial slope of χ_Θ versus ϕ_s. With reference to Eq. (2.2.45), this suggests that the third virial coefficient of polystyrene in cyclohexane should almost vanish at $T = \Theta$.

Polystyrene + cyclohexane is the most thoroughly investigated system. Figure 2.15 shows χ_Θ data from osmotic pressure and vapor pressure measurements by Krigbaum and Geymer [22], together with data from sedimentation equilibrium analysis by Scholte [12]. The agreement between the two sets of data is quite good. The four dashed lines in the graph show values of χ_Θ and $\chi_{H\Theta}$ computed from Eqs. (2.2.134a) and (2.2.134b). From comparison with the experimental values of $\chi_{H\Theta}$ indicated we may conclude that correct dependence of χ on both T and ϕ_s would lie between these two empirically determined equations. This conclusion is not unexpected in view of the assumptions made in the above data analysis.

Next, we describe the experimental results of Koningsveld [11] on three samples of polyethylene in diphenyl ether. These are quoted here to illustrate the behavior of polymers which are extremely heterogeneous in molecular weight. Table 2.4 summarizes the critical point data on samples designated A,

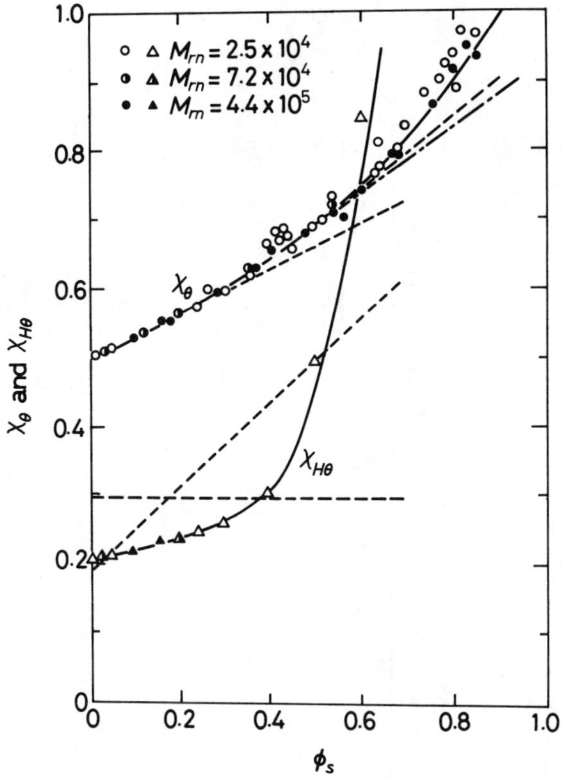

Fig. 2.15. Concentration dependence of χ_Θ and $\chi_{H\Theta}$ for polystyrene in cyclohexane. Points from osmotic pressure and vapor pressure measurements [22]; dot-dash line from sedimentation equilibrium data [12]; dashed lines calculated from Eqs. (2.2.134a) and (2.2.134b).

B, and C. The last column gives the values of M_{rz} computed from Eq. (2.2.74a) with the measured values of M_{rw} and ϕ_{sc}. The agreement with the directly measured values is fairly good, unlike the polystyrene case discussed above. This suggests that the concentration dependence of χ may be ignored for polyethylene in diphenyl ether. Data analysis on the basis of this assumption yields

$$\chi = -0.6068 + 482.2\, T^{-1} \tag{2.2.135}$$

The condition that $\chi_0 = 1/2$ at $T = \Theta$ gives $\Theta = 161.8°C$. The agreement with the temperature, 157 ± 1°C, at which the second virial coefficient from light scattering vanishes is not as good as in polystyrene + cyclohexane. Furthermore, unlike the polystyrene case, the entropic component of χ derived from Eq. (2.2.135) is negative and large.

Table 2.4. Critical point data for polyethylene in diphenyl ether

Sample	$M_{rn}/10^5$	$M_{rw}/10^5$	$M_{rz}/10^5$	M_{rw}/M_{rn}	M_{rz}/M_{rw}	$T_c/°C$	$\phi_{sc}/10^{-2}$	M_{rz} (calc)/10^5
A	0.12	1.53	9.0 ± 2	13	6	143.2 ± 0.2	8.2 ± 0.1	13.0 ± 2
B	0.92	1.40	3.3 ± 0.5	1.5	1.9	147.9 ± 0.1	5.0 ± 0.1	4.0 ± 0.9
C	0.086	0.55	3.0 ± 0.5	6.4	5.5	137.0 ± 0.3	9.7 ± 0.1	2.6 ± 0.3

Figure 2.16 illustrates the cloud-point curves for the three samples. Sample A is an unfractionated polymer with a very broad distribution of molecular weight. Sample B is a fraction with M_{rw} comparable to that of sample A. Sample C is also a fraction, but it has a relatively low molecular weight. Comparison of the data for samples A and B thus may provide information about the effect of molecular weight distribution, and the data for samples B and C may allow estimation of the effect of molecular weight. Koningsveld [11] fitted logarithmic-normal distributions of molecular weight to these samples and, by use of a computer, calculated the cloud-point curves, shadow curves, and spinodals. The computed curves are shown in Fig. 2.17, where A′, B′, and C′ denote the molecular weight distributions fitted to samples A, B, and C, respectively. The theoretical cloud-point curves (thick solid lines) reproduce the positions

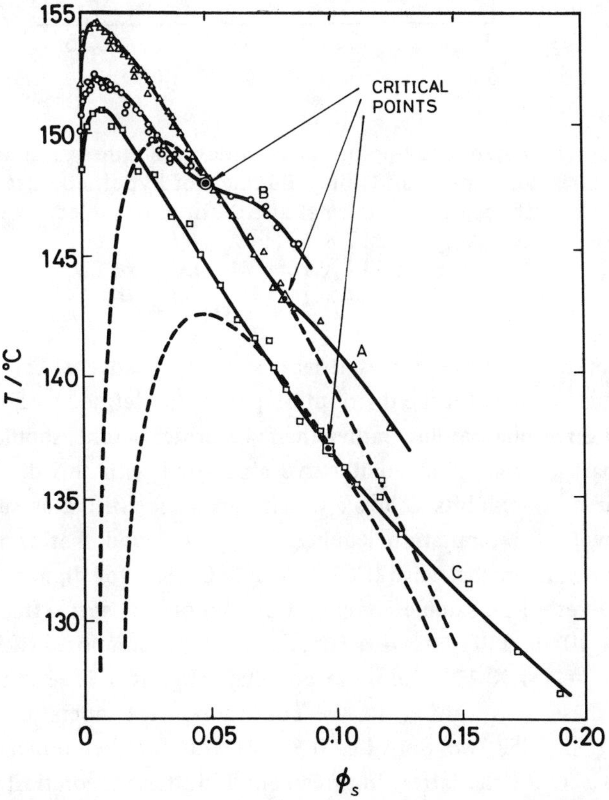

Fig. 2.16. Cloud-point curves (solid lines) and critical points (⊙, ▲, ▣) for samples A, B, C of polyethylene in diphenyl ether. Data of Koningsveld [11].

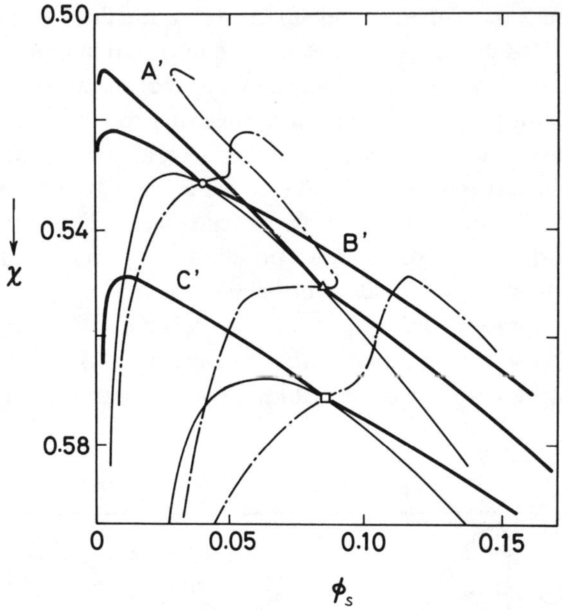

Fig. 2.17. Calculated cloud-point curves (thick solid lines), shadow curves (dot-dash lines), and spinodals (thin solid lines) of hypothetical samples A′, B′, and C′, with logarithmic normal distributions of molecular weight.
A′: $M_{rw} = 1.317 \times 10^5$, $M_{rw}/M_{rn} = 10$
B′: $M_{rw} = 1.317 \times 10^5$, $M_{rw}/M_{rn} = 2$
C′: $M_{rw} = 2.634 \times 10^5$, $M_{rw}/M_{rn} = 2$

and shapes of the observed curves at least semiquantitatively. The distribution A′ gives a negative value for the distribution parameter defined by Eq. (2.2.131). Therefore, a three-phase point, rather than the critical point, should appear on the stable part of the cloud-point curve associated with this distribution. In fact, the curve A′ exhibits a break which can be identified as such a point.

By an involved computation Koningsveld [11] found that separation into three phases occurs in the range $0.6210 > \chi > 0.6124$ for diphenyl ether solutions of a polyethylene sample composed of two monodisperse fractions whose $M_{r1} = 1.2 \times 10^4$ and $M_{r2} = 5.4 \times 10^5$. For a sample composed of $M_{r1} = 2.5 \times 10^4$ and $M_{r2} = 5.4 \times 10^5$ the corresponding range is $0.5798 > \chi > 0.5769$. Converting these ranges of χ to the corresponding temperature bounds by use of Eq. (2.2.135), we find 119.0 – 121.7°C for the former sample and 132.6 – 133.5°C for the latter. In agreement with these theoretical predictions, Koningsveld et al. [15] observed the occurrence of a three-phase equilibrium in the ranges 119.4 – 124°C and 131.1 – 131.6°C for the former and latter

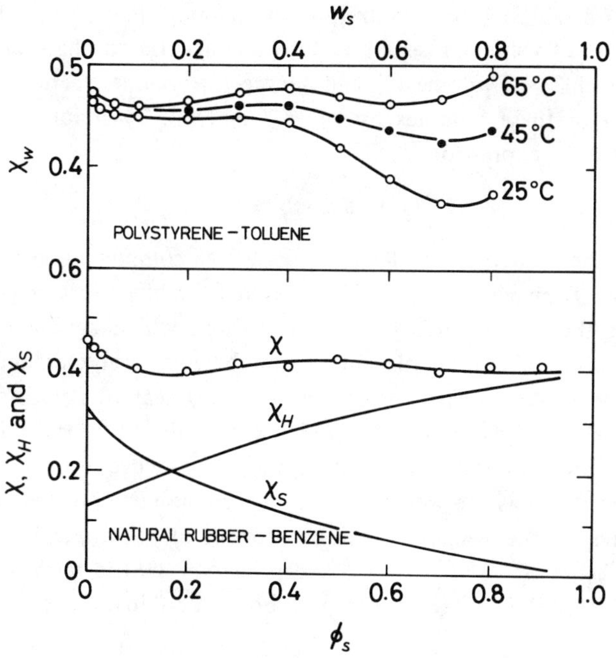

Fig. 2.18. Concentration dependence of χ, χ_H, and χ_S for polystyrene in toluene (upper) [12] and for natural rubber in benzene at 25°C [23].

samples, respectively, though the component samples were narrow fractions rather than strictly monodisperse polymers. This finding confirms the validity of the empirical relation Eq. (2.2.135).

Finally, Fig. 2.18 illustrates the behavior of χ in good-solvent systems, with data of Scholte [12] on polystyrene in toluene and of Gee et al. [23] on natural rubber in benzene. The former were derived from sedimentation equilibrium, and the latter from vapor pressure and osmotic pressure measurements. It can be seen that χ and χ_S in dilute solutions decrease with increasing polymer concentration, while χ_H is a monotonically increasing function of polymer concentration. These trends have been observed in various good-solvent systems and are accepted as fairly general features of such systems.

2.2.7. Fractionation According to Molecular Weight

By taking advantage of liquid-liquid phase separation (usually effected by cooling) we may divide a polymer sample into fractions with narrower distributions of molecular weight. This operation is called **molecular weight fractionation.**

Its basis was established by Schulz [24]. We suppose that an initially homogeneous solution of volume V separates into a concentrated phase of volume V'' and a dilute phase of volume V', and denote the volume fractions of polymer component i in these volumes by ϕ_i, ϕ_i'', and ϕ_i'. By simple considerations Schulz derived the expression

$$\phi_i''/\phi_i' = \exp(P_i \sigma) \tag{2.2.136}$$

where P_i is the relative chain length of polymer component i. This equation has the same form as Eq. (2.2.103) derived from the Flory-Huggins theory. However, in the Schulz theory of fractionation, which antedates the Flory-Huggins theory, only energetic interaction between polymer and solvent is considered to govern phase separation. For this reason, Schulz's separation factor σ consists only of the bracketed terms in Eq. (2.2.104). However, the basic theory can be formulated without explicit knowledge of σ.

We define $K(P_i)$ as the weight of polymer component i in the concentrated phase relative to the weight of the component in the original solution, i.e., $K(P_i) = V''\phi_i''/V\phi_i$. The ratio V''/V can be expressed in terms of the phase-volume ratio R as $R/(1+R)$. Hence, with Eq. (2.2.107b), we get

$$K(P_i) = \frac{1}{1 + R^{-1}\exp(-P_i \sigma)} \tag{2.2.137}$$

The quantity $K(P)$ is called the **precipitation ratio** of the polymer with relative chain length P. Denoting by P^* the value of P at the inflection in K, we can write

$$P^* = -(\ln R)/\sigma \tag{2.2.138}$$

Figure 2.19 shows $K(P)$ as a function of $\ln(P/P^*)$ for different values of R. It is easily shown that

$$(dK/d\ln P)_{P=P^*} = -(1/4)\ln R \tag{2.2.139}$$

When $R \ll 1$, most of a polymer component with $P > P^*$ goes to the concentrated phase, while most of a component with $P < P^*$ remains in the dilute phase, even though every polymer component is concentrated to some degree in the concentrated phase. Such fractionation efficiency may, to a first approximation, be measured by the slope of K versus $\ln P$ at $P = P^*$, i.e., the quantity given by Eq. (2.2.139). Thus the lower the value of R, the higher the fractionation efficiency becomes. The partitioning criterion P^* is controlled by σ, as can be seen from Eq. (2.2.138). Evidently we would need detailed physical information on σ to correlate it with temperature and polymer concentration.

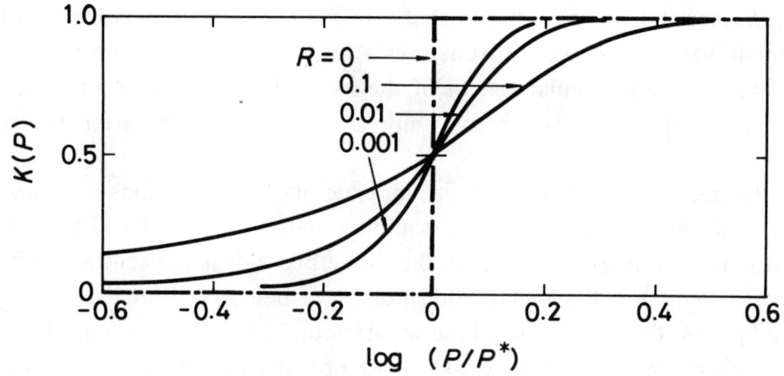

Fig. 2.19. Chain length dependence of precipitation ratio $K(P)$ at different volume ratios R.

There are two principal methods for molecular weight fractionation: **precipitation** and **extraction**. First, considering precipitation, we denote the normalized weight distribution of P in the original polymer sample by $\xi(P)$, and express the precipitation ratio κ of the sample after separation into two phases by

$$\kappa = \int_0^\infty \xi(P) K(P) dP = \int_0^\infty \frac{\xi(P)}{1 + R^{-1} \exp(-P\sigma)} dP \tag{2.2.140}$$

When it is intended to divide the original sample into q **fractions** of approximately equal weights by simple successive precipitations, R and σ must be adjusted by proper choice of initial concentration and temperature so as to make κ for the first phase separation approximately $1/q$. The polymer recovered from the concentrated phase after the first phase separation is called fraction $F1$. The dilute phase is subjected to the second phase separation by readjusting the experimental conditions. The polymer extracted from the resulting concentrated phase is taken as fraction $F2$. By repeating such operations we extract fraction $F(q-1)$ from the concentrated phase in the $(q-1)$-th phase separation and the final fraction Fq from the corresponding dilute phase. The chain length distributions of the fractions separated in this series of operations are represented by

$$\xi_{F1}(P) = \xi(P) K_1(P), \quad \xi_{F2}(P) = \xi(P)[1 - K_1(P)] K_2(P), \ldots$$

$$\xi_{F(q-1)}(P) = \xi(P) \left[\prod_{j=1}^{q-2} (1 - K_j) \right] K_{q-1} \tag{2.2.141a}$$

$$\xi_{Fq}(P) = \xi(P) \prod_{j=1}^{q-1} (1 - K_j)$$

where K_j denotes the precipitation ratio in the j-th phase separation. It should be noted that the distribution functions ξ_{Fj} ($j = 1, 2, \ldots, q$) are not normalized. The average molecular weight of the fraction recovered is less at each step. Figure 2.20 (a) illustrates schematically a successive fractionation process of this type.

In the successive extraction method, experimental conditions are chosen so that κ in the first phase separation is approximately equal to $1 - (1/q)$; now the first fraction $F1$ is extracted from the dilute phase. The concentrated phase is heated to redissolve the precipitated polymer; then it is diluted with solvent, and subjected to the second phase separation. This process is repeated, as is shown schematically in Fig. 2.20 (b), to obtain fraction $F(q - 1)$ from the dilute phase in the $(q - 1)$-th phase separation and the final fraction Fq from the remaining concentrated phase. In this case, the chain length distributions (unnormalized) of the extracted fractions are written

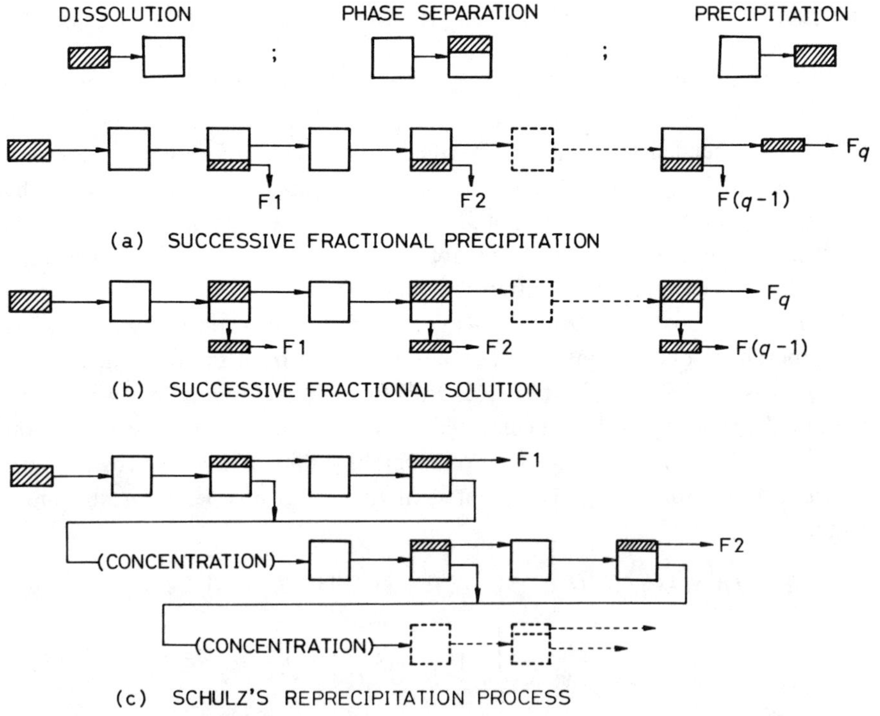

Fig. 2.20. Schematic representation of three typical procedures for fractionation according to molecular weight.

$$\xi'_{F1}(P) = \xi(P)\,[1 - K_1(P)], \quad \xi'_{F2}(P) = \xi(P)\,K_1\,(1 - K_2), \ldots$$

$$\xi'_{F(q-1)} = \xi(P)\left(\prod_{j=1}^{q-2} K_j\right)(1 - K_{q-1}) \tag{2.2.141b}$$

$$\xi'_{Fq} = \xi(P)\prod_{j=1}^{q-1} K_j$$

The average molecular weight of the fractions increases with the fraction number, as would be expected.

Schulz [24] used Eq. (2.2.141a) to calculate ξ_{Fj} (j = 1, 2, . . . , 8) for a sample having an exponential chain length distribution characterized by P_n = 500 and P_w/P_n = 1.5. The calculation was carried out with R fixed at 10^{-3} and σ varied so that P^*, given by Eq. (2.2.138), decreased in order 1200, 1000, 800, 600, 450, 300, 200 as j increased from 1 to 7. The calculated distributions of the eight fractions, together with the distribution of the original sample, are shown in Fig. 2.21. Notwithstanding the choice of a very small R to enhance fractionation efficiency, the distributions of the fractions have tails toward low chain length and there is appreciable overlap. This phenomenon, characteristic of the precipitation method, is referred to as the **tailing effect**. A measure of the effect is the height of the tail of the highest molecular weight fraction relative to that of the original sample as both approach zero at P = 0. For precipitation fractionation this can be expressed by

$$\lim_{P\to 0} \xi_{F1}(P)/\xi(P) = \lim_{P\to 0} K_1(P) = R_1 \tag{2.2.142a}$$

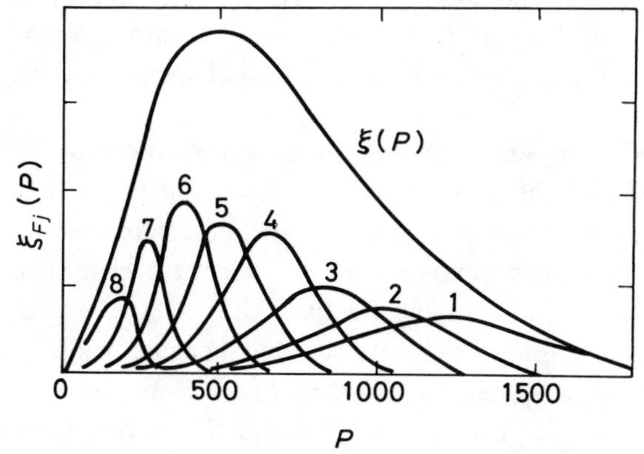

Fig. 2.21. Calculated chain length distributions in fractions separated by successive fractional precipitation from a sample with chain length distribution $\xi(P)$.

For extraction fractionation, since the highest fraction is Fq, the corresponding quantity is

$$\lim_{P \to 0} \xi'_{Fq}(P)/\xi(P) = \lim_{P \to 0} \prod_{j=1}^{q-1} K_j(P) = \prod_{j=1}^{q-1} R_j \tag{2.2.142b}$$

which is much smaller than R_1, because $R_j \ll 1$. Thus fractions obtained by extraction are essentially free of tailing.

To suppress the tailing effect Schulz proposed the **reprecipitation method** [Fig. 2.20 (c)]. The concentrated phase formed in the first phase separation is removed from the system and heated to dissolve the precipitate completely. After the concentration is properly adjusted, the homogeneous solution is cooled and separates into two phases, and the polymer reprecipitated from the concentrated phase is collected. This is taken as fraction $F1$. The dilute phase is combined with the dilute phase left after the first phase separation, adequately concentrated by evaporation, and then subjected to a second phase separation. If R and P^* (hence σ) for the reprecipitation are chosen equal to those for the first phase separation, ξ_{F1} is given by ξK_1^2. Hence the right-hand side of Eq. (2.2.142a) becomes R_1^2, so that tailing is suppressed by a factor R_1 ($\ll 1$). In short, for elimination of this effect it is essential to precipitate the desired fraction after an intermediate step or steps to remove low molecular weight species. Various procedures other than the Schulz method have been proposed to insert such steps [25]. In practice, however, phase separation is more often (and more conveniently) controlled by adjusting the mixing ratio of a solvent and a precipitant rather than changing temperature.

The Schulz theory is based only on Eq. (2.2.136). However, the complete conditions for phase equilibrium must include Eq. (2.2.100); and the solution of the problem of fractionation must also take into account Eq. (2.2.104) describing σ and Eq. (2.2.140) defining κ [26]. Tung's numerical solution [27] is outlined below.

The assigned parameters for numerical computation are the chain length distribution $\xi(P)$ of the given polymer sample, the polymer concentration ϕ_s of the initial single-phase solution, and the parameter κ which controls the size of the fraction to be extracted. Thus, when the chain length distribution is treated as discrete, the volume fractions $\phi_i = \xi_i \phi_s$ ($i = 1, 2, \ldots, r$) of the polymer components in the initial solution become known quantities. The numerical process is started by computing ϕ_i'' from Eq. (2.2.107b) and ϕ_i' from Eq. (2.2.103) for a set of trial R and σ values. Substitution of the computed ϕ_i'' and ϕ_i' into Eq. (2.2.100) allows calculation of χ, which, in turn, allows σ to be computed from Eq. (2.2.104). If the σ value obtained does not agree with the trial value, another value for σ is assumed and the calculation is repeated, keeping R at the initial value. A convergent value of σ from this iteration process, together with the assigned R, gives values of σ and R which are consistent with the given composition of the initial solution. These values of σ and R are then used in Eq. (2.2.140) to check whether the corresponding κ agrees with the κ originally assigned. If it does not, a new value of R is assigned and the whole process is repeated. The final set of σ and R values, which leads to agreement of the assigned and calculated κ, is inserted into Eq. (2.2.137) to compute $K_F(P)$, and then the chain length distribution of the fraction $\xi_F(P)$ is obtained from $\xi_F(P) = \xi(P)K_F(P)$. In order to simulate the entire sequence of fractionations it is necessary to iterate a similar calculation at each stage. Tung used this method, assuming the so-called Tung distribution for $\xi(P)$. The results are summarized in Table 2.5. It is seen that, in agreement with experience, fractionation efficiency can be enhanced by lowering the concentration of the initial solution and doing reprecipitations.

Subsequent to Tung's work, Koningsveld and Staverman [28] and Kamide et al. [29] simulated a number of fractionations and obtained much valuable information. The curves shown in Fig. 2.22 are results computed by Koningsveld and Staverman for a sample with the same very broad logarithmic-normal distribution as that used to calculate the cloud-point curve A′ in Fig. 2.17. The solid lines show the polydispersity index P_w''/P_n'' of the sample in the concentrated phase as a function of the precipitation ratio κ for different initial concentrations ϕ_s, and the dashed lines are the corresponding functions for the polymer in the dilute phase. Polydispersity of the first fraction isolated when a sample is divided into q equal fractions may be estimated from the solid lines at $\kappa = 1/q$ for precipitation and the dashed lines at $\kappa = 1 - q^{-1}$ for extraction. As expected, the solid and dashed lines converge to the polydispersity of the original sample ($P_w/P_n = 10$) at $\kappa = 1$ and $\kappa = 0$, respectively. According to Eq. (2.2.74a), the critical concentration ϕ_{sc} of this system is 0.084. Therefore, we see that all the curves in Fig. 2.22 refer to initial concentrations below ϕ_{sc}. It is apparent from Fig. 2.22 that precipitation fractionation is very inefficient in this case, giving a fraction which is even broader in molecular weight distribution than the original sample, especially if the initial concentration is relatively high and κ is small. On the contrary, extraction fractionation is free from such disadvantages.

In the foregoing we have suggested the use of $(dK/d\,\ln P)_{P=P*}$ as a measure of fractionation efficiency. If this suggestion is always valid, fractionation should become more efficient as R is made smaller, whatever the value of σ. Comparing the numerical data for ϕ_s equal to 0.005 and 0.02 in Table 2.5, we see that R is smaller and the fractionation efficiency is higher at the former than at the latter ϕ_s, which implies that the suggestion is valid. However, it may lead to a wrong

Table 2.5. Simulated fractionations of a sample with chain length distribution $\xi(P) = (1.429 \times 0.003256) \exp(-1.429P^{0.003256})$ and $P_w/P_n = 2.3$

Fraction	successive precipitation										reprecipitation	
	($\phi_s = 0.005$)						($\phi_s = 0.02$)				($\phi_s = 0.01$)	
	σ	R^{-1}	χ	P_w	P_n	P_w/P_n	σ	R^{-1}	P_w	P_w/P_n	P_w	P_w/P_n
1	0.060	792	0.71	111.9	93.0	1.20	0.042	161	101.5	1.55	115.1	1.09
2	0.075	853	0.74	86.1	75.1	1.13	0.052	177	83.8	1.33	86.0	1.06
3	0.092	925	0.77	71.0	63.4	1.12	0.064	189	71.3	1.27	69.4	1.05
4	0.113	1073	0.81	59.9	54.4	1.10	0.080	195	60.5	1.22	57.2	1.05
5	0.140	1069	0.85	50.2	46.1	1.09	0.100	244	51.4	1.18	47.1	1.05
6	0.182	1113	0.91	41.1	38.1	1.08	0.129	256	42.8	1.15	38.3	1.06
7	0.244	1292	1.00	32.7	30.5	1.07	0.177	298	34.6	1.13	30.7	1.07
8	0.369	1342	1.15	24.2	22.6	1.07	0.266	323	26.2	1.11	23.6	1.03
9	0.667	2000	–	16.0	14.9	1.08	0.500	500	17.7	1.11	15.9	1.09
10	–	–	–	7.0	4.6	1.54	–	–	7.7	1.58	7.0	1.55

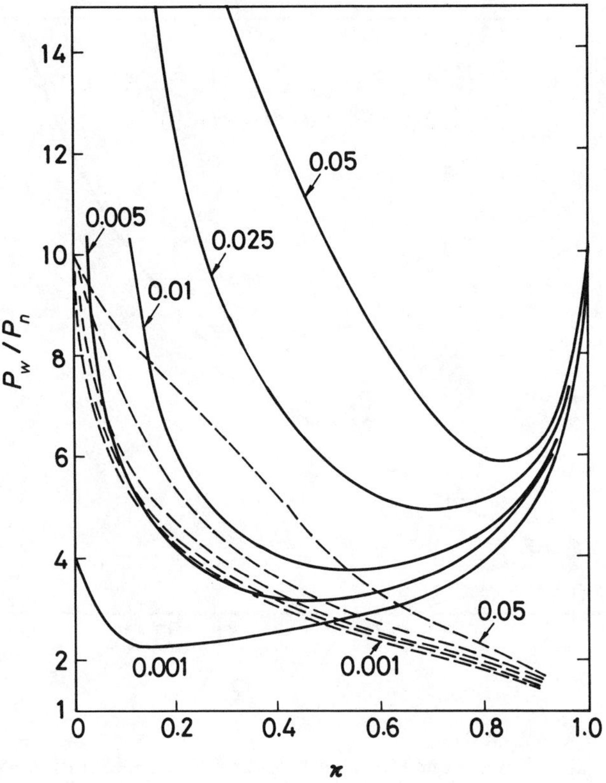

Fig. 2.22. Polydispersity indices of fractions derived from the concentrated phase (solid lines) and the dilute phase (broken lines) as functions of total precipitation ratio κ at various volume fractions of polymer in the original solution. The ordinate P_w/P_n should be read P_w''/P_n'' for the solid lines and P_w'/P_n' for the broken lines.

answer when fractionation efficiency is compared at a fixed ϕ_s, as has been done in Fig. 2.22. In general, in order to diminish R when ϕ_s is smaller than ϕ_{sc}, we may carry out phase separation at a temperature very close to the cloud-point curve. In this case, as R is lowered, the amount of polymer precipitated in the concentrated phase, i.e., the value of κ, becomes smaller. In other words, the guiding principle "make R smaller" is here equivalent to saying "make fraction size finer." The results in Fig. 2.22 show, however, that this simple principle is not always a guide to enhancing fractionation efficiency.

It will be noticed in examining Tung's numerical method that σ and R are not mutually independent when ϕ_s is fixed. Figure 2.23 illustrates an example

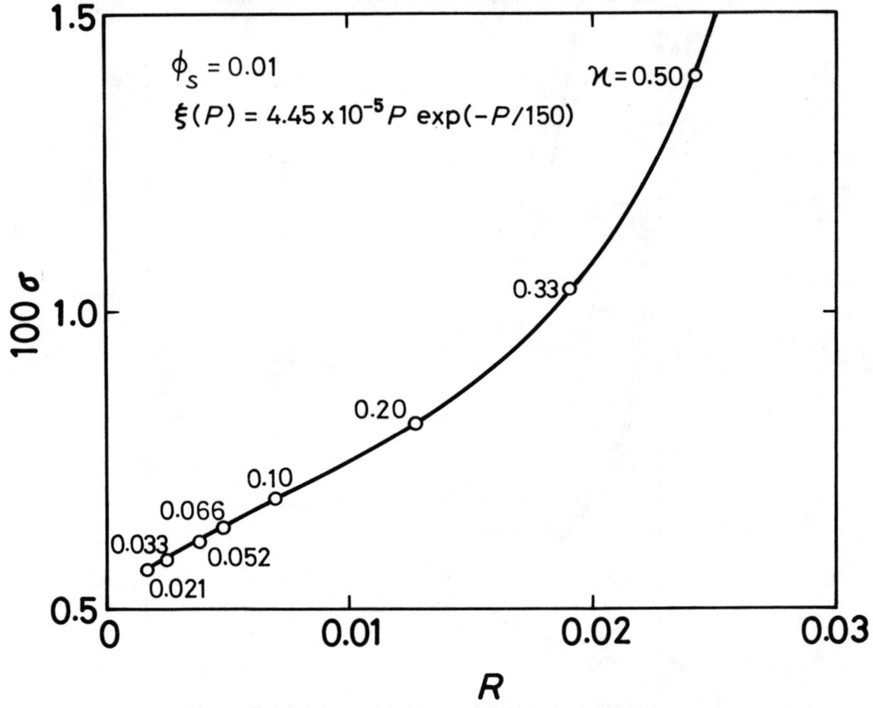

Fig. 2.23. Dependence of σ on R and on κ at fixed initial polymer concentration in fractional precipitation; $\xi(P)$ is the chain length distribution in the original sample.

given by Kamide et al. [29] of the interdependence of these parameters. In this case, if R is made smaller σ also becomes smaller and, as a result, the ratio ϕ_i''/ϕ_i' becomes less dependent on chain length. Therefore, as Kamide et al. suggested, it is reasonable to compare fractionation efficiencies at fixed ϕ_s in terms of a measure defined by

$$(dK/dP)_{P=P*} = \sigma/4 \tag{2.2.143}$$

Table 2.5 also indicates that when κ is fixed R decreases while σ increases as ϕ_s is lowered. The new measure defined here is useful for this case too.

2.3. Solutions of Chemically Different Polymers

2.3.1. Spinodal and Critical Point

The discussion developed in the preceding sections has been exclusively concerned with solutions of homologous polymers in a single solvent. In the following sections, we deal with the more general case of a solution which contains chemically different polymer solutes. We also treat mixed-solvent systems.

Consider an $(r+1)$-component system, in which the amount and volume fraction of component i are n_i and ϕ_i, respectively, and let the molecule of component i consist of P_i segments. The increase in local free energy accompanying the exchange of nearest-neighbor segments, which is schematically represented by

$$(S_i - S_i) + (S_j - S_j) \rightarrow (S_i - S_j) + (S_j - S_i) \tag{2.3.1}$$

is denoted by $2\,\Delta w_{ij}$. Here S_i designates a segment of species i. By analogy with the Flory-Huggins treatment of binary polymer solutions, we put

$$\chi_{ij} \equiv z\,\Delta w_{ij}/(kT) \tag{2.3.2}$$

and call χ_{ij} the interaction coefficient between components i and j. For simplicity, we assume that the χ_{ij} depends only on temperature. With these assumptions we obtain the expressions

$$\Delta_m G = RT\left[\sum_{i=0}^{r} n_i \ln \phi_i + \left(\sum_{i=0}^{r} P_i n_i\right) \sum_{i=0}^{r-1} \sum_{j=i+1}^{r} \chi_{ij}\phi_i\phi_j\right] \tag{2.3.3}$$

$$\Delta_m \mu_i/RT = \ln \phi_i + \sum_{j\neq i} (1 - P_i P_j^{-1})\,\phi_j$$

$$+ P_i \left[(1-\phi_i)\sum_{j\neq i} \chi_{ij}\phi_j - \tfrac{1}{2}\sum_{j\neq i}\sum_{k\neq i} \chi_{jk}\phi_j\phi_k\right], \tag{2.3.4}$$

$$(i = 0, 1, \ldots, r)$$

which correspond to Eqs. (2.2.26), (2.2.27a), and (2.2.27b). Also, if the volume per mole of lattice cells is denoted by V_τ and the Gibbs free energy per unit volume by G_ϕ, we obtain

$$\frac{V_\tau}{RT}\left(\frac{\partial^2 G_\phi}{\partial\phi_i\,\partial\phi_j}\right) = \frac{1}{P_0\,\phi_0} - \chi_{0i} - \chi_{0j} + \chi_{ij} \equiv J_{ij} \tag{2.3.5a}$$

$$\frac{V_T}{RT}\left(\frac{\partial^2 G_\phi}{\partial \phi_i^2}\right) = \frac{1}{P_0 \phi_0} + \frac{1}{P_i \phi_i} - 2\chi_{0i} \equiv J_{ii} \tag{2.3.5b}$$

which correspond to Eqs. (2.2.61a) and (2.2.61b). The spinodal is determined by

$$D \equiv |J| = 0 \tag{2.3.6}$$

To find the critical solution point this is combined with the condition

$$\begin{vmatrix} D_1 & D_2 & \cdots & D_r \\ J_{21} & J_{22} & \cdots & J_{2r} \\ \cdots & \cdots & \cdots & \cdots \\ J_{r1} & J_{r2} & \cdots & J_{rr} \end{vmatrix} = 0 \tag{2.3.7}$$

where

$$D_i \equiv (\partial D/\partial \phi_i)_{\phi_j} \tag{2.3.8}$$

Now, following Scott [30] and Tompa [31], we seek solutions to these equations for a ternary system. For simplicity, we put

$$2\chi_0 \equiv \chi_{01} + \chi_{02} - \chi_{12}, \quad 2\chi_1 \equiv \chi_{01} + \chi_{12} - \chi_{02}$$
$$2\chi_2 \equiv \chi_{02} + \chi_{12} - \chi_{01} \tag{2.3.9}$$

Then

$$\chi_{ij} = \chi_i + \chi_j, \quad (i, j = 0, 1, 2;\ i \neq j) \tag{2.3.10}$$

Further, in terms of J_i defined by

$$J_i \equiv (P_i \phi_i)^{-1} - 2\chi_i, \quad (i = 0, 1, 2) \tag{2.3.11}$$

J_{ij} may be written

$$J_{12} = J_0, \quad J_{11} = J_0 + J_1, \quad J_{22} = J_0 + J_2 \tag{2.3.12}$$

By these substitutions Eq. (2.3.6) is brought to exactly the same form as Eq. (2.2.62). Hence, the spinodal is expressed by Eq. (2.2.65) with J simply replaced by J_0. For convenience, we introduce Ψ_i $(i = 0, 1, 2)$ defined by

$$\Psi_i \equiv 1/J_i = P_i \phi_i/(1 - 2\chi_i P_i \phi_i) \tag{2.3.13}$$

and write the equation for the spinodal in the form

$$\Psi_0 + \Psi_1 + \Psi_2 = 0 \tag{2.3.14}$$

In terms of ϕ_i, this equation is

$$\sum_{i=0}^{2} P_i \phi_i - 2 \sum_{i=0}^{1} \sum_{j=i+1}^{2} P_i P_j \chi_{ij} \phi_i \phi_j + P_0 P_1 P_2 q \phi_0 \phi_1 \phi_2 = 0 \qquad (2.3.15)$$

with q defined by

$$\begin{aligned} q &\equiv 4 (\chi_0 \chi_1 + \chi_1 \chi_2 + \chi_2 \chi_0) \\ &= 2\chi_{01} \chi_{02} + 2\chi_{01} \chi_{12} + 2\chi_{02} \chi_{12} - \chi_{01}^2 - \chi_{02}^2 - \chi_{12}^2 \\ &= [(\chi_{01})^{1/2} + (\chi_{02})^{1/2} + (\chi_{12})^{1/2}] [-(\chi_{01})^{1/2} + (\chi_{02})^{1/2} + (\chi_{12})^{1/2}] \\ &\quad \times [(\chi_{01})^{1/2} - (\chi_{02})^{1/2} + (\chi_{12})^{1/2}] [(\chi_{01})^{1/2} + (\chi_{02})^{1/2} - (\chi_{12})^{1/2}] \end{aligned} \qquad (2.3.16)$$

Of course, ϕ_0, ϕ_1, and ϕ_2 are not mutually independent, since $\phi_0 + \phi_1 + \phi_2 = 1$. Hence, Ψ_0, Ψ_1, and Ψ_2 are constrained by the relation

$$\frac{\Psi_0}{P_0 (1 + 2 \chi_0 \Psi_0)} + \frac{\Psi_1}{P_1 (1 + 2 \chi_1 \Psi_1)} + \frac{\Psi_2}{P_2 (1 + 2 \chi_2 \Psi_2)} = 1 \qquad (2.3.17)$$

Next, we insert Eq. (2.3.6) into Eq. (2.3.8) to obtain

$$\begin{aligned} D_1 &= (J_1 + J_2)/(P_0 \phi_0^2) - (J_0 + J_2)/(P_1 \phi_1^2) \\ &= P_0 (J_0 + 2\chi_0)^2 (J_1 + J_2) - P_1 (J_1 + 2\chi_1)^2 (J_0 + J_2) \end{aligned} \qquad (2.3.18a)$$

$$D_2 = P_0 (J_0 + 2\chi_0)^2 (J_1 + J_2) - P_2 (J_2 + 2\chi_2)^2 (J_0 + J_1) \qquad (2.3.18b)$$

It then follows from Eq. (2.3.7) that

$$P_0 \Psi_0 (1 + 2\chi_0 \Psi_0)^2 + P_1 \Psi_1 (1 + 2\chi_1 \Psi_1)^2 + P_2 \Psi_2 (1 + 2\chi_2 \Psi_2)^2 = 0 \qquad (2.3.19)$$

which may be expressed in terms of ϕ_i as

$$\sum_{i=0}^{2} P_i^2 \phi_i/(1 - 2\chi_i P_i \phi_i)^3 = 0 \qquad (2.3.20)$$

When ϕ_1 and ϕ_2 are used to describe the state of the system, the spinodal is given by Eq. (2.3.15) and the critical point is determined by solving the set of Eqs. (2.3.15) and (2.3.20). However, if Ψ_1 and Ψ_2 are taken as the state variables, the spinodal is determined by the set of Eqs. (2.3.14) and (2.3.17), and the critical point by the set of Eqs. (2.3.14), (2.3.17), and (2.3.19), subject to the restriction

$$\Psi_i^{-1} = J_i \geqslant P_i^{-1} - 2\chi_i \qquad (2.3.21)$$

which is equivalent to the requirement $0 \leqslant \phi_i \leqslant 1$ $(i = 1, 2)$.

Given P_i and χ_i for $i = 0, 1, 2$, the above sets of equations may be solved by the following procedure. First, a trial value of Ψ_2 is assigned and $\Psi_0 = -\Psi_1 - \Psi_2$ is substituted into Eq. (2.3.17) to obtain a quadratic equation for Ψ_1. A real root that satisfies the condition in Eq. (2.3.21) gives one point on the spinodal, which corresponds to the initially assigned Ψ_2 value. Next, this Ψ_1 value is substituted into $\Psi_0 + \Psi_1 + \Psi_2 = 0$, together with the original Ψ_2. The calculated Ψ_0 and the Ψ_1 value are used to compute Ψ_2 from Eq. (2.3.19). If the result does not coincide with the initial trial value of Ψ_2, the latter is altered and the calculation is reiterated. This process is repeated until a convergent value of Ψ_2 is obtained. The final values of Ψ_0, Ψ_1, and Ψ_2 determine the desired critical point.

We may solve Eqs. (1.3.61a,b,c) in order to calculate the binodal. This problem is discussed below for different combinations of the three components.

2.3.2. Two Polymer Species in a Single Solvent

For a binary mixture of pure polymer 1 and pure polymer 2, Eq. (2.3.4) can be written

$$\Delta_m\mu_1/RT = \ln \phi_1 + [1 - (P_1/P_2)]\,\phi_2 + P_1\,\chi_{12}\,\phi_2^2 \tag{2.3.22a}$$

$$\Delta_m\mu_2/RT = \ln \phi_2 + [1 - (P_2/P_1)]\,\phi_1 + P_2\,\chi_{12}\,\phi_1^2 \tag{2.3.22b}$$

The critical point of this polymer mixture is given by

$$\phi_{1c} = [1 + (P_1/P_2)^{1/2}]^{-1} \tag{2.3.23}$$

$$\chi_{12c} = (1/2)\,[(P_1)^{-1/2} + (P_2)^{-1/2}]^2 \tag{2.3.24}$$

Thus when P_1 and P_2 are sufficiently large, χ_{12c} approaches zero. It has already been mentioned in § 1.5.2 that if the London dispersion forces dominate molecular interactions, the energy of creating molecular contacts between different species is positive. This indicates that $\Delta w_{12} \geqslant 0$, because each molecule may be replaced by a segment of a particular polymer chain. Therefore, except for the special case in which Δw_{12} happens to vanish, $\chi_{12c} \approx 0$ implies $T_c = \infty$, which means that different polymer species should be incompatible.

We proceed to a ternary solution made up by dissolving the above mixture of two polymers in a single solvent (component 0). For simplicity, we let P_0 be unity and consider the special case where $P_1 = P_2 = P$ and $\chi_{01} = \chi_{02} = \chi$. The conditions for phase equilibrium are derived from Eqs. (2.3.4) and (1.3.61):

$$\begin{aligned} &\ln \phi_0' + (1 - P^{-1})(1 - \phi_0') + \chi(1 - \phi_0')^2 - \chi_{12}\,\phi_1'\phi_2' \\ &= \ln \phi_0'' + (1 - P^{-1})(1 - \phi_0'') + \chi(1 - \phi_0'')^2 - \chi_{12}\,\phi_1''\phi_2'' \end{aligned} \tag{2.3.25a}$$

$$\begin{aligned}&\ln \phi_1' + (1-P)\phi_0' + P\chi\phi_0'^2 + P\chi_{12}(1-\phi_1')\phi_2' \\ &= \ln \phi_1'' + (1-P)\phi_0'' + P\chi\phi_0''^2 + P\chi_{12}(1-\phi_1'')\phi_2''\end{aligned} \tag{2.3.25b}$$

$$\begin{aligned}&\ln \phi_2' + (1-P)\phi_0' + P\chi\phi_0'^2 + P\chi_{12}(1-\phi_2')\phi_1' \\ &= \ln \phi_2'' + (1-P)\phi_0'' + P\chi\phi_0''^2 + P\chi_{12}(1-\phi_2'')\phi_1''\end{aligned} \tag{2.3.25c}$$

These three equations have a solution symmetric with respect to components 1 and 2:

$$\phi_0' = \phi_0'' = \phi_0, \quad \phi_1' = \phi_2'', \quad \phi_2' = \phi_1'' \tag{2.3.26a}$$

$$\ln \phi_1' + P\chi_{12}\phi_2' = \ln \phi_1'' + P\chi_{12}\phi_2'' \tag{2.3.26b}$$

If ξ_1 and ξ_2 are defined as

$$\xi_1 = \phi_1/(\phi_1+\phi_2) = \phi_1/(1-\phi_0), \quad \xi_2 = \phi_2/(1-\phi_0) \tag{2.3.27}$$

the above set of solutions is equivalent to

$$\xi_1' = 1 - \xi_1'' \tag{2.3.28a}$$

$$\begin{aligned}&\ln \xi_1' + P(1-\phi_0)\chi_{12}(1-\xi_1')^2 \\ &= \ln \xi_1'' + P(1-\phi_0)\chi_{12}(1-\xi_1'')^2\end{aligned} \tag{2.3.28b}$$

These expressions are identical with the conditions for phase equilibrium of a mixture of two polymers of equal chain length with an interaction parameter $(1-\phi_0)\chi_{12}$. Therefore, simply by referring to Eqs. (2.3.23) and (2.3.24), we can write

$$\xi_{1c} = \xi_{2c} = 1/2, \quad (1-\phi_{0c})\chi_{12} = 2/P \tag{2.3.29}$$

or

$$\phi_{0c} = 1 - 2/(P\chi_{12}), \quad \phi_{1c} = \phi_{2c} = 1/(P\chi_{12}) \tag{2.3.30}$$

Note that Eqs. (2.3.28a) and (2.3.28b) predict a binodal which is independent of the polymer-solvent interaction coefficients χ_{01} and χ_{02}. Values of $P\chi_{12}$ for pairs of conventional polymers are positive and large. Therefore, the critical point given by Eqs. (2.3.30) is located very near the apex for the pure solvent on the composition triangle, as in Fig. 2.24. If a homogeneous solution of polymer 1 (point A) is mixed with a homogeneous solution of polymer 2 (point B) in proportions indicated by the point Q in the figure, the mixture separates into a phase Q′ which contains little polymer 2 and a phase Q″ which contains little polymer 1. This phenomenon occurs even at very low polymer

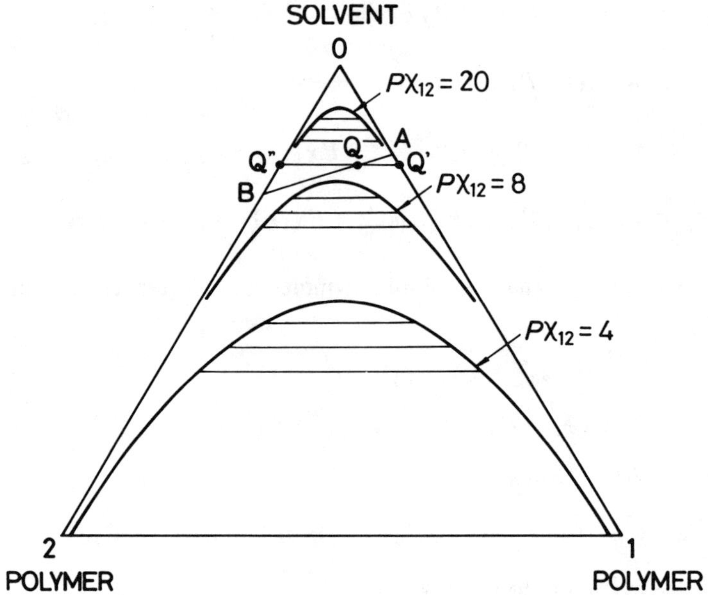

Fig. 2.24. Phase diagram showing demixing of polymer 1 and polymer 2.

concentrations, and has long drawn attention. It is referred to as **demixing**. When polymers undergo demixing in a certain solvent, changing the solvent cannot prevent its occurrence.

Figure 2.25 presents data of Sakurada and Seki [32] on benzene solutions of samples of poly(vinyl acetate) and polystyrene of approximately equal molecular weight. The open circles were determined from the compositions of equilibrated phases, while the filled circles were obtained from cloud-point determinations. This diagram bears a resemblance to the theoretical one shown in Fig. 2.24. The fact that the experimental tie lines are almost parallel to the triangle base is worthy of special note. Extensive studies of demixing were made by Dobry and Boyer-Kawenoki [33].

Although, as noted above, different polymer species are in general incompatible, a number of polymers are now known to exhibit compatibility. In many such pairs, one or both components possess strong polar groups, which make χ_{12} negative, i.e., there is a tendency for complex formation between the polymer species.

The conclusions derived above apply, with no great modification, to asymmetric systems in which $P_1 \neq P_2$ and $\chi_{01} \neq \chi_{02}$. In particular, when $|\chi_{01} - \chi_{02}| \ll 1$ and $(P_1)^{1/2} < P_2 < P_1^2$, the critical point is given, to a good approximation,

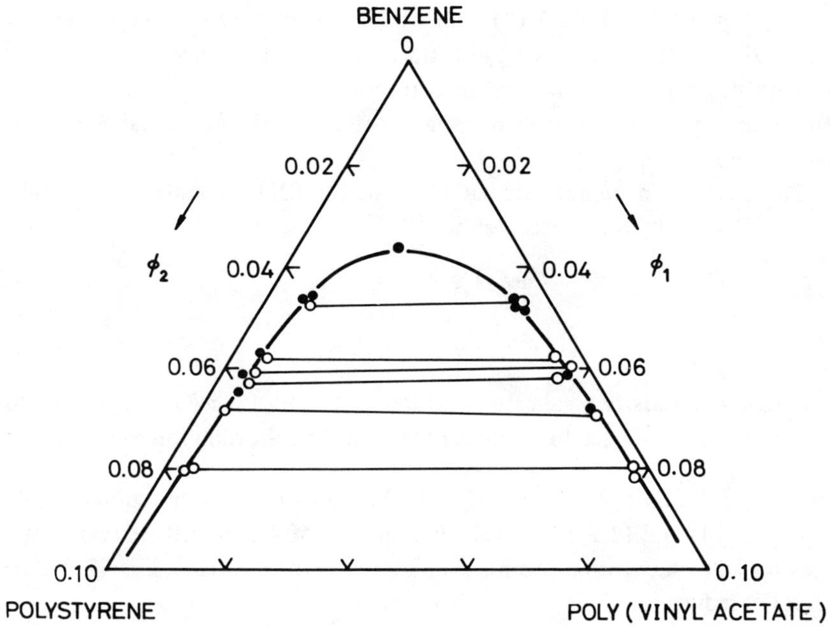

Fig. 2.25. Demixing of poly(vinyl acetate) and polystyrene in benzene. Data of Seki and Sakurada [32].

by

$$\xi_{1c} = (P_2)^{1/2}/[(P_1)^{1/2} + (P_2)^{1/2}], \quad \xi_{2c} = 1 - \xi_{1c} \tag{2.3.31a}$$

$$(1 - \phi_{0c})\chi_{12} = (1/2)\,[(P_1)^{-1/2} + (P_2)^{-1/2}]^2 \tag{2.3.31b}$$

These expressions can be rewritten

$$\phi_{0c} = 1 - [1/(2\chi_{12})]\;[(P_1)^{-1/2} + (P_2)^{-1/2}]^2 \tag{2.3.32a}$$

$$\begin{aligned}\phi_{1c} &= [1/(2\chi_{12})]\;[1 + (P_1/P_2)^{1/2}]^{-1}\;[(P_1)^{-1/2} + (P_2)^{-1/2}]^2 \\ &= (P_2/P_1)^{1/2}\,\phi_{2c}\end{aligned} \tag{2.3.32b}$$

Comparison of Eqs. (2.3.31a) and (2.3.31b) with Eqs. (2.3.23) and (2.3.24) for the mixture of two different polymers without diluent indicates that, in this case too, the presence of a solvent merely reduces the interaction coefficient χ_{12}.

2.3.3. One Polymer in a Mixture of Two Solvents

In this section, we consider a ternary system consisting of two solvents (components 0 and 1) and a monodisperse polymer (component 2). For convenience, we let $P_0 = P_1 = 1$ and $P_2 = P$. The critical point can be determined by solving Eqs. (2.3.14), (2.3.17), and (2.3.19) simultaneously. These equations are of first order, third order, and third order, respectively, so that there are, in general, nine roots; but only the real roots which are consistent with the condition in Eq. (2.3.21) have physical significance. We examine these for the special case $P = \infty$.

The third term on the left-hand side of Eq. (2.3.19) must remain finite as $P \to \infty$. Hence, Ψ_2 must satisfy either

$$\text{Case (1)},\ \Psi_2 = 0 \tag{2.3.33}$$

or

$$\text{Case (2)},\ 1 + 2\chi_2 \Psi_2 = 0 \tag{2.3.34}$$

Case (2) comprises two subcases with $P(1 + 2\chi_2\Psi_2)^2$ or $P(1 + 2\chi_2\Psi_2)$ finite for $P = \infty$. Thus we have to be concerned with the following three cases.

(A) $\Psi_2 = 0$ and $P\Psi_2$ is finite or zero. In this case, the terms containing Ψ_2 in Eqs. (2.3.14) and (2.3.17) vanish. Elimination of Ψ_0 from these two equations then leads to an equation quadratic in Ψ_1. In the present case, Eqs. (2.3.14) and (2.3.17) reduce to

$$1 - 2\chi_{01}\phi_0\phi_1 = 0, \quad \phi_0 + \phi_1 = 1 \tag{2.3.35}$$

which may be solved to obtain

$$\phi_0 = 1 - \phi_1,\ \phi_1 = (1/2)\left\{1 \pm [1 - (2/\chi_{01})]^{1/2}\right\},\ \phi_2 = 0 \tag{2.3.36}$$

Thus we have two roots. When P is finite, we choose these roots as first approximations and find the second approximation to Ψ_2 from Eq. (2.3.19). This is inserted into Eqs. (2.3.17) and (2.3.19), and the second-approximation values of Ψ_0 and Ψ_1 are calculated. Without giving details, we note that the second approximation to Ψ_2 is of the form $\Psi_2 = b/P$ and those to Ψ_0 and Ψ_1 are of the form $\Psi_i = k_i' + k_i''/P^{1/2}$ $(i = 0, 1)$.

(B) $1 + 2\chi_2\Psi_2 = 0$ and $P(1 + 2\chi_2\Psi_2)^2$ is finite. The latter condition may be written $1 + 2\chi_2\Psi_2 = \pm(c/P)^{1/2}$ where c is a constant. With Eq. (2.3.13), this gives

$$\phi_2 = [1/(2\chi_2)]\,[P^{-1} \mp (cP)^{-1/2}] \tag{2.3.37}$$

Hence we see that $\phi_2 \to 0, P\phi_2 \to \infty$, and $\Psi_2 \to -1/(2\chi_2)$ as $P \to \infty$.

In this case too, the third term in Eq. (2.3.17) tends to zero as $P \to \infty$. If Ψ_0 is eliminated from this limiting form of Eq. (2.3.17) and Eq. (2.3.14) with $\Psi_2 = -1/(2\chi_2)$, a quadratic equation for Ψ_1 is obtained. Alternatively, in this case, Eqs. (2.3.14) and (2.3.17) reduce to

$$1 - 2\chi_{02}\phi_0 - 2\chi_{12}\phi_1 + q\phi_0\phi_1 = 0 \tag{2.3.38a}$$

$$\phi_0 + \phi_1 = 1 \tag{2.3.38b}$$

where q is a parameter defined by Eq. (2.3.16). Each of the two sets of roots of Eqs. (2.3.38a) and (2.3.38b) consists of a double root. For finite P, it follows that $\phi_0 + \phi_1 = 1 \pm O(P^{-1/2})$ in correspondence with the plus and minus signs in Eq. (2.3.37). Thus the roots become distinct, and one of them gives a critical point.

In particular, when χ_{ij} is written in the same form as b_{ij} in Eq. (1.5.21), one of the last three factors in Eq. (2.3.16) vanishes. Then Eq. (2.3.38a) becomes linear in ϕ_1, and we get a set of double roots:

$$\phi_0 = \frac{1 - 2\chi_{12}}{2(\chi_{02} - \chi_{12})}, \quad \phi_1 = \frac{1 - 2\chi_{02}}{2(\chi_{12} - \chi_{02})}, \quad \phi_2 = 0 \tag{2.3.39}$$

(C) $1 + 2\chi_2\Psi_2 = 0$ and $P(1 + 2\chi_2\Psi_2)$ is finite. As $P \to \infty$, the third term in Eq. (2.3.19) vanishes. Solving Eqs. (2.3.14) and (2.3.19) with Ψ_2 set equal to $-1/(2\chi_2)$, we obtain three sets of distinct roots for Ψ_0 and Ψ_1. Substitution of these in Eq. (2.3.17) allows Ψ_2 to be determined. When $q = 0$, the following two sets of roots have physical significance:

$$\phi_0 = \frac{1}{2(\chi_{01}\chi_{02})^{1/2}}, \quad \phi_1 = \frac{1}{2(\chi_{01}\chi_{12})^{1/2}}, \quad \phi_2 = 1 - \frac{1}{2(\chi_{02}\chi_{12})^{1/2}} \tag{2.3.40a}$$

$$\phi_0 = \frac{(\chi_{12})^{1/2}}{2\chi_{02}(\chi_{01})^{1/2}}, \quad \phi_1 = \frac{(\chi_{02})^{1/2}}{2\chi_{12}(\chi_{01})^{1/2}},$$
$$\phi_2 = 1 - \frac{\chi_{02} - (\chi_{02}\chi_{12})^{1/2} + \chi_{12}}{2\chi_{02}\chi_{12}} \tag{2.3.40b}$$

For finite P, the first-order corrections to these ϕ_i are of order P^{-1}.

To sum up, there are nine roots altogether: two from (A), four from (B), and three from (C).

a. Precipitant + Good Solvent + Polymer. Except for exothermic systems, the best solvent for a polymer is the monomer unit itself. Therefore, the ternary solution of a binary mixture of polymers with different chain lengths, treated in § 2.2.3, becomes a typical precipitant + good solvent + polymer system when $P_1 = 1$ and $P_2 \gg 1$. In this case, we may assume that

$$\chi_{01} = \chi_{02} \equiv \chi, \quad \chi_{12} = 0 \tag{2.3.41}$$

A schematic phase diagram of this system has already been presented in Fig. 2.9. Three-phase separation occurs only in the vicinity of the critical point for the binary mixture of components 0 and 1. Therefore, at temperatures somewhat away from the critical temperature for the mixture of good solvent and precipitant, the phase diagram of the ternary system may look like Fig. 2.26. Here the thick solid line is a binodal and the open circle is the critical point. Further, the dot-dash lines are lines of constant θ_1, the volume fraction of component 1 in the mixed solvent (solvent + precipitant):

$$\theta_1 \equiv \frac{\phi_1}{\phi_0 + \phi_1} = \frac{\phi_1}{1 - \phi_2}, \quad \theta_0 = 1 - \theta_1 \tag{2.3.42}$$

These lines of constant solvent composition make large angles with the tie lines P′ P″, Q′ Q″, etc. This implies that $\theta_1{}' < \theta_1{}''$, the single and double primes indicating, as before, dilute and concentrated in the polymer. In other words, component 1 (good solvent), which has a greater affinity for the polymer than does component 0 (precipitant), is selectively taken into the concentrated phase. This phenomenon is called **selective adsorption** of good solvent.

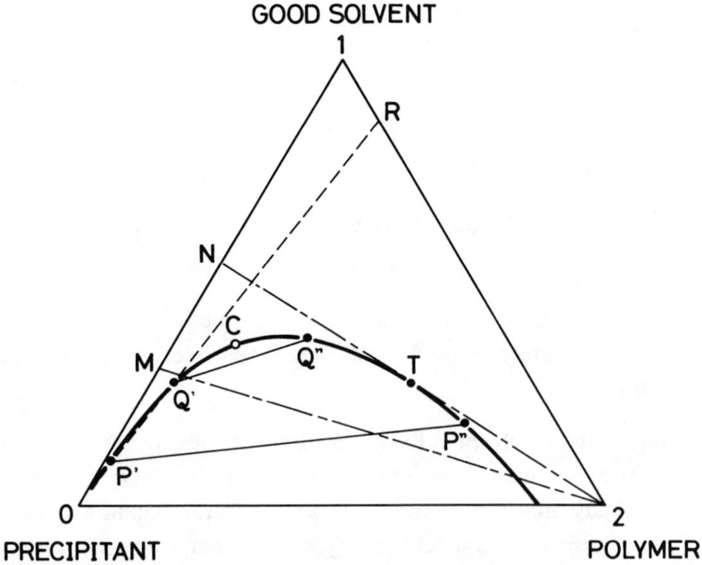

Fig. 2.26. Phase diagram of polymer 2 in mixtures of good solvent 1 and precipitant 0.

Suppose that a polymer (component 2) is dissolved in a good solvent to make a solution represented by point R in Fig. 2.26 and then a precipitant is added to the solution. The composition of the solution follows the line R0, and when it reaches point Q′, phase separation occurs. As more precipitant is added, the composition of the dilute phase, the principal phase in this case, changes from point Q′ to point P′ along the binodal, while that of the concentrated phase changes from point Q″ to point P″ along the binodal. The usual molecular weight fractionation takes advantage of this type of composition change accompanying the addition of a precipitant. In order to enhance fractionation efficiency it is necessary to adjust the initial composition so that line R0 intersects the binodal to the left of the critical point C. Note also that point T in the figure represents the threshold cloud-point of the system. No phase separation takes place at any polymer concentration if the representative point is in the region above the constant solvent composition line N2.

b. Precipitant + Precipitant + Polymer. In the limit $P_2 \to \infty$, components 0 and 1 are precipitants when $\chi_{02} > 1/2$ and $\chi_{12} > 1/2$. Scott [30] has shown that if $q > 0$ [see Eq. (2.3.16)] and χ_{01} is in the range

$$2 > \chi_{01} > [\chi_{02} + \chi_{12} - 1 + (2\chi_{01} - 1)^{1/2}(2\chi_{12} - 1)^{1/2}] \qquad (2.3.43)$$

the phase diagram of the system becomes like that shown in Fig. 2.27. The first inequality is the condition for complete miscibility of the two precipitants, while the second is the condition for Eq. (2.3.38) to have two real roots. When $P_2 = \infty$, the two critical points corresponding to these roots appear on side $\overline{01}$ of the triangle, and the branch of each binodal at polymer concentrations below the critical value merges with $\overline{01}$. For finite P_2 the critical point and binodals move away from $\overline{01}$.

If the solubility parameter δ_2 of the polymer component is intermediate between the solubility parameters of the precipitants, the mean value of δ_0 and δ_1 approximates δ_2. Then, according to the solubility parameter theory, the mixture of precipitants can be a solvent for the polymer. However, as has been noted above, this theory gives $q = 0$. Thus it is impossible to explain fully the solubility of a polymer in a mixture of precipitants by the solubility-parameter idea. This solubility phenomenon is observed, for example, in the system chloroform + ethanol + cellulose acetate [34].

c. Solvent + Solvent + Polymer. No phase separation occurs on $\overline{01}$, $\overline{02}$, and $\overline{12}$, but phase separation may occur inside the composition triangle. In this case, the binodal describes a loop and has two critical points. The reader is referred to Tompa's book [35] for details.

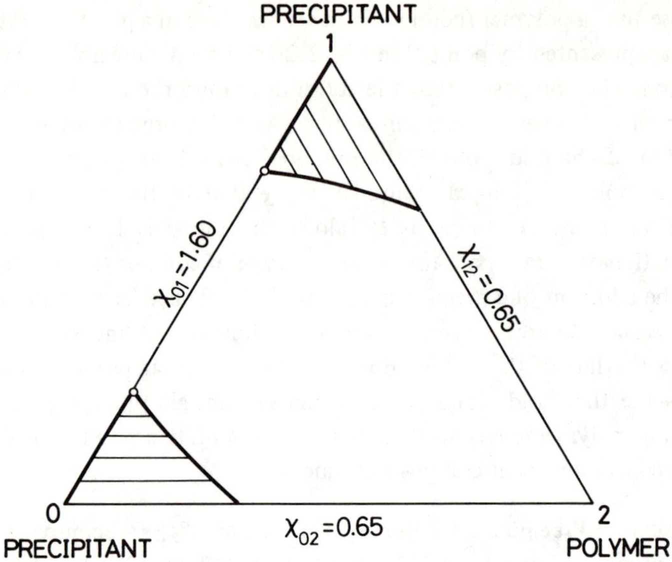

Fig. 2.27. Phase diagram of polymer 2 in mixtures of precipitant 0 and precipitant 1.

2.3.4. Osmotic Equilibrium in Mixed-Solvent Systems – 1. Constant-μ_D System

Osmotic equilibrium in mixed-solvent systems is obviously important in connection with molecular weight determinations of polyelectrolytes in aqueous salt solutions. It also has to do with the proper definition of the refractive index increment in light scattering analysis and of the density increment in sedimentation-equilibrium analysis. A general treatment is described in this section and the next.

As has been mentioned in Chapter 1, there are two ways of formulating osmotic equilibrium, one for the system at constant solvent chemical potential μ_0 and the other for the system at constant pressure. In this section, we extend the first formulation to the case of constant chemical potentials of all diffusible components.

The components which can permeate a semipermeable membrane are called **diffusible**, while those which do not permeate the same membrane are called **nondiffusible**. Of the $r + 1$ components in a solution, the diffusible ones are numbered $0, 1, \ldots, d$, and the nondiffusible ones are numbered $d + 1, d + 2, \ldots, d + s$ ($r = d + s$). We place a solution composed of $d + 1$ diffusible components only (i.e., a mixed solvent) on one side (the outer phase) of a semipermeable membrane and fill the phase on the other side (the inner phase) of the membrane with the $r+1$ component solution. We let thermodynamic quantities associated with the outer phase be distinguished by a single prime from those associated with the inner phase. Osmotic equilibrium between the two phases at fixed temperature is described by $d+1$ relations

$$\mu_i'(T, p', c_1', \ldots, c_d') = \mu_i(T, p, c_1, \ldots, c_d, c_{d+1}, \ldots, c_{d+s}), \quad (i = 0, 1, \ldots, d) \tag{2.3.44}$$

where c_i is the molar concentration of component i. The osmotic pressure π between the two phases is

$$\pi = p - p' \tag{2.3.45}$$

When all the nondiffusible components in the inner phase are infinitely dilute, the solution is said to be infinitely diluted. If the pressure, the molar concentration of diffusible component i, and its chemical potential in the inner phase at infinite dilution are denoted by p^o, c_i^o, and μ_i^o, we have $p^o = p'$, $c_i^o = c_i'$, and $\mu_i^o = \mu_i'$.

When the outer phase is at fixed pressure, temperature, and composition, values of μ_i' are constant. Hence, by Eq. (2.3.44), all μ_i ($i = 0, 1, \ldots, d$) are also constant. In this book, for convenience, the osmotic equilibrium under such conditions is termed constant-μ_D equilibrium, μ_D standing for the set of potentials $\mu_0, \mu_1, \ldots, \mu_d$. The osmotic pressure at constant μ_D is denoted by π^* in order to distinguish it from the osmotic pressure π at constant solution pressure. At the stage when Eq. (2.3.45) was written for π, we made no distinction between these two pressures. Since p' is constant for constant μ_D, Eq. (2.3.45) gives

$$d\pi^* = (dp)_{T, \mu_D} \tag{2.3.46}$$

A solution which is at equilibrium at constant μ_D may be called a constant-μ_D system. Since a systematic treatment of constant-μ_D systems is not given in existing textbooks, we summarize the essential points here. First, consider a ternary solution consisting of two diffusible components (0 and 1) and one nondiffusible component (2). If T, V, n_0, n_1, and n_2 are chosen as the set of state variables, the characteristic function for the solution is the Helmholtz free energy $A(T, V, n)$, where n means the set of moles n_0, n_1, and n_2. The term "characteristic function" implies that for any virtual displacement under constraints of fixed T, V, and n, the following relations must hold:

$$(\delta A)_{T, V, n} = 0, \quad (\delta^2 A)_{T, V, n} > 0 \tag{2.3.47}$$

The inequality is the necessary condition for stable equilibrium.

The total differential of A is given by Eq. (1.2.36), i.e.,

$$dA = -SdT - pdV + \mu_0\, dn_0 + \mu_1\, dn_1 + \mu_2\, dn_2 \tag{2.3.48}$$

If we introduce a new function $\mathcal{A}$ defined by

$$\mathcal{A} \equiv A - \mu_0 n_0 - \mu_1 n_1 \tag{2.3.49}$$

the total differential of $\mathcal{A}$ is written by use of Eq. (2.3.48) as

$$d\mathcal{A} = -SdT - pdV - n_0\, d\mu_0 - n_1\, d\mu_1 + \mu_2\, dn_2 \tag{2.3.50}$$

From this it follows that

$$S = -(\partial \mathcal{A}/\partial T)_{V, \mu_0, \mu_1, n_2} \tag{2.3.51}$$

$$p = -(\partial \mathcal{A}/\partial V)_{T, \mu_0, \mu_1, n_2} \tag{2.3.52}$$

$$n_0 = -(\partial \mathcal{A}/\partial \mu_0)_{T, V, \mu_1, n_2}, \quad n_1 = -(\partial \mathcal{A}/\partial \mu_1)_{T, V, \mu_0, n_2} \tag{2.3.53}$$

$$\mu_2 = (\partial \mathcal{A}/\partial n_2)_{T, V, \mu_0, \mu_1} \tag{2.3.54}$$

The quantity $\mathcal{A}$ is the characteristic function for T, V, μ_0, μ_1, n_2 as the set of state variables to describe the ternary solution. In other words, it is the characteristic function for a constant-μ_D ternary solution.

It follows that the condition for stable equilibrium of a constant-μ_D ternary solution (at constant T, V, and n_2) is

$$(\delta^2 \mathcal{A})_{T, V, \mu_D, n_2} > 0 \tag{2.3.55}$$

We take a small region α of fixed volume V^α in such a solution, calling the rest of the solution region β. Denoting the uniform molar concentration of component 2 by c_2, we have

$$c_2 = n_2/V = n_2{}^\alpha/V^\alpha = n_2{}^\beta/V^\beta, \quad (V^\alpha \ll V^\beta) \tag{2.3.56}$$

where the other notations are self-explanatory. For a virtual displacement $\delta n_2{}^\alpha$ from region β to region α, Eq. (2.3.55) gives

$$\begin{aligned}(\delta^2 \mathcal{A})_{T, V, \mu_D, n_2} &\approx (\delta^2 \mathcal{A})_{T, V^\alpha, \mu_D} \\ &= (1/2)\,(\partial^2 \mathcal{A}/\partial n_2{}^2)^\alpha_{T, V^\alpha, \mu_D}\,(\delta n_2{}^\alpha)^2 > 0\end{aligned} \tag{2.3.57}$$

The contribution of region β to $\delta^2 \mathcal{A}$, i.e., $(\delta^2 \mathcal{A})_{T, V^\beta, \mu_D}$, has been neglected by an argument similar to that in § 1.3.3. With Eqs. (2.3.54) and (2.3.56), the above inequality leads to

$$\mu_{22}{}^* \equiv (\partial \mu_2/\partial c_2)_{T, \mu_D} > 0 \tag{2.3.58}$$

which is the necessary condition for stable equilibrium of a uniform ternary solution at constant T and μ_D. The spinodal of such a solution is therefore given by

$$(\partial \mu_2/\partial c_2)_{T,\mu_D} = 0 \tag{2.3.59}$$

and the critical point is determined by the conditions

$$(\partial \mu_2/\partial c_2)_{T,\mu_D} = 0, \quad (\partial^2 \mu_2/\partial c_2{}^2)_{T,\mu_D} = 0 \tag{2.3.60}$$

which correspond to Eqs. (1.3.46) and (1.3.47) for ternary solutions at constant T and p.

The Gibbs-Duhem relation for ternary solutions at constant T and μ_D gives

$$(dp)_{T,\mu_D} - c_2\,(d\mu_2)_{T,\mu_D} = 0 \tag{2.3.61}$$

Combination of this with Eq. (2.3.46) leads to

$$d\pi^*/dc_2 = (\partial p/\partial c_2)_{T,\mu_D} = c_2\,(\partial \mu_2/\partial c_2)_{T,\mu_D} \tag{2.3.62}$$

This allows Eqs. (2.3.60) to be written

$$d\pi^*/dc_2 = 0, \quad d^2\pi^*/dc_2{}^2 = 0 \tag{2.3.63}$$

Thus we see that, even in solutions in mixed solvents, the following correspondences hold exactly at constant-μ_D equilibrium: $\pi^* \longleftrightarrow$ gas pressure and solute concentration $\longleftrightarrow$ gas density. Hence π^* obeys van't Hoff's law at the limit of infinite dilution.

We can express μ_2 at constant T and μ_D in the form

$$\begin{aligned}\mu_2\,(T, \mu_D, C_2) &= \mu_2{}^{\theta}\,(T, \mu_D) + RT\ln(C_2/C^{\theta}) \\ &\quad + RTM_2\,[B_2\,(T, \mu_D)\,C_2 + (1/2)\,B_3\,(T, \mu_D)\,C_2{}^2 + \ldots]\end{aligned} \tag{2.3.64}$$

by expanding the activity coefficient $y_2\,(T, \mu_D, C_2)$ in powers of mass concentration C_2. Introduction of this into Eq. (2.3.62), followed by integration, yields

$$\pi^*/(RT) = C_2/M_2 + (1/2)\,B_2\,(T, \mu_D)\,C_2{}^2 + (1/3)\,B_3\,(T, \mu_D)\,C_2{}^3 + \ldots \tag{2.3.65}$$

Neglecting terms higher than $C_2{}^3$ and using Eq. (2.3.63), we obtain

$$M_2{}^{-1} + [B_2\,(T_c, \mu_{Dc}) + B_3\,(T_c, \mu_{Dc})\,C_{2c}]\,C_{2c} = 0 \tag{2.3.66a}$$

$$B_2\,(T_c, \mu_{Dc}) + 2\,B_3\,(T_c, \mu_{Dc})\,C_{2c} = 0 \tag{2.3.66b}$$

where the subscript c indicates the critical point. These equations can be solved to give

$$B_{2,c}C_{2c} = -2/M_2, \quad B_{3,c}C_{2c}^2 = 1/M_2 \tag{2.3.67}$$

where $B_{2,c} \equiv B_2(T_c, \mu_{Dc})$ and $B_{3,c} \equiv B_3(T_c, \mu_{Dc})$, with μ_{Dc} denoting the value of μ_D for temperature T_c and composition m_1' of the mixed solvent in the outer phase at osmotic equilibrium. In order to have a critical point, i.e., to have positive values of T_c and C_{2c} as solutions to Eq. (2.3.67) it is required that B_2 and B_3 satisfy the conditions $B_{2,c} < 0$ and $B_{3,c} > 0$. Under these conditions Eq. (2.3.67) gives

$$C_{2c} = (M_2 B_{3,c})^{-1/2}, \quad B_{2,c} = -2(B_{3,c}/M_2)^{1/2} \tag{2.3.68}$$

We note that the condition $B_{3,c} > 0$ is not fully definitive because the second relation in Eq. (2.3.67) has been derived by neglecting terms higher than C_2^3 in Eq. (2.3.65). The critical osmotic pressure $\pi_c{}^*$ is obtained by inserting Eq. (2.3.67) in Eq. (2.3.65):

$$\pi_c{}^*/(RT_c C_{2c}) = 1/(3M_2) \tag{2.3.69}$$

The ratio on the left-hand side corresponds to the **critical ratio** in the theory of real gases.

Now consider the situation in which as the temperature is varied at fixed m_1', the solution becomes turbid at a certain temperature. By repeating similar experiments at different C_2 we may determine a cloud-point curve, and T_c and C_{2c} can then be evaluated from its maximum (or minimum). The determination of cloud-point curves for a series of molecular weights M_2 gives critical points T_c and C_{2c} as functions of M_2. If we let the limiting values of T_c and C_{2c} at $M_2 = \infty$ be denoted by $T_c{}^\infty$ and $C_{2c}{}^\infty$, we have, according to Eq. (2.3.68),

$$C_{2c}{}^\infty = 0 \tag{2.3.70}$$

$$B_{2,c}{}^\infty = B_2(T_c{}^\infty, \mu_{Dc}{}^\infty) = 0 \tag{2.3.71}$$

where $\mu_{Dc}{}^\infty \equiv \mu_D(T_c{}^\infty, m_1')$. Equation (2.3.70) shows that the limiting critical concentration appears at infinite dilution. Equation (2.3.71) indicates that the second virial coefficient vanishes at the limiting critical temperature and that this temperature depends on m_1' (which is equal to the solvent composition of the solution at infinite dilution). Thus we find that the theta temperature of a ternary solution of a polymer in a mixed solvent may be determined by finding $T_c{}^\infty$ from phase separation experiments done with solutions osmotically equilibrated against the mixed solvent.

When the ternary solution is at constant-μ_D equilibrium, C_1 may vary as C_2 is changed. If the solution is dilute with respect to the polymer component, this dependence of C_1 on C_2 can be represented by

$$C_1 = C_1{}^o + (\partial C_1/\partial C_2)^o_{T,\mu_D}\, C_2 + \ldots \tag{2.3.72}$$

where $C_1{}^o$ is the mass concentration of component 1 in the outer phase. The coefficient $(\partial C_1/\partial C_2)^o_{T,\mu_D}$ is a measure of selective adsorption of component 1 by component 2 at specified T and μ_D (and at infinite dilution). Using Eqs. (2.3.53) and (2.3.54), we obtain

$$\left(\frac{\partial C_1}{\partial C_2}\right)^o_{T,\mu_D} = \frac{M_1}{M_2}\left(\frac{\partial c_1}{\partial c_2}\right)^o_{T,\mu_D} \tag{2.3.73}$$

$$\left(\frac{\partial c_1}{\partial c_2}\right)^o_{T,\mu_D} = \left(\frac{\partial n_1}{\partial n_2}\right)^o_{T,V,\mu_D} = -\left(\frac{\partial^2 \mathcal{A}}{\partial n_2\, \partial \mu_1}\right)^o = -\left(\frac{\partial \mu_2}{\partial \mu_1}\right)^o_{T,\mu_0} \tag{2.3.74}$$

The thermodynamic interactions between components 0 and 2 and between components 1 and 2 in a ternary solution are hidden in the virial coefficients B_2, B_3, etc. when T, V, μ_0, μ_1, and C_2 are chosen as the state variables, i.e., when the solution is placed at constant-μ_D equilibrium. If we want to consider explicitly effects of interaction between components 1 and 2, it is necessary to change to the set of state variables T, V, μ_0, C_1, and C_2, i.e., to replace μ_1 by C_1. This new set of variables is appropriate for a solution osmotically equilibrated against pure liquid component 0. For such a solution, μ_1 and μ_2 become functions of C_1 when T, μ_0, and C_2 are fixed, and hence the derivative $(\partial \mu_2/\partial \mu_1)_{T,\mu_0,C_2}$ may have a nonzero value. In fact, we find from Eqs. (2.3.73) and (2.3.74) that this derivative is related to $(\partial C_1/\partial C_2)^o_{T,\mu_D}$ by

$$\begin{aligned}\left(\frac{\partial C_1}{\partial C_2}\right)^o_{T,\mu_D} &= -\frac{M_1}{M_2}\left(\frac{\partial \mu_2}{\partial \mu_1}\right)^o_{T,\mu_0} \\ &= -\frac{M_1}{M_2}\,\frac{(\partial \mu_2/\partial C_1)^o_{T,\mu_0}}{(\partial \mu_1/\partial C_1)^o_{T,\mu_0}}\end{aligned} \tag{2.3.75}$$

Although proof is omitted here, the condition for stable equilibrium requires $(\partial \mu_1/\partial C_1)_{T,\mu_0,C_2} > 0$. Therefore, when there is thermodynamic affinity between components 1 and 2 at fixed T, μ_0, and C_2, i.e., $(\partial \mu_2/\partial C_1)_{T,\mu_0,C_2} < 0$, we find from Eq. (2.3.75) that $(\partial C_1/\partial C_2)^o_{T,\mu_D} > 0$, which means that as C_2 increases, component 1 is preferentially taken up into the solution from the outer phase.

The theory of ternary solutions described above can easily be extended to the general multicomponent case. If components 0 to d are diffusible, thus constituting the mixed solvent, and components $d+1$ to $d+s$ are nondiffusible polymer solutes, the appropriate characteristic function for constant-μ_D equilibrium is

$$\mathcal{A}\,(T, V, \mu_0, \ldots, \mu_d, n_{d+1}, \ldots, n_{d+s}) \equiv A - \sum_{i=0}^{d} \mu_i n_i \tag{2.3.76}$$

Equation (2.3.62) for the osmotic pressure is replaced by

$$\left(\frac{\partial \pi^*}{\partial c_j}\right)_{c_k} = \left(\frac{\partial p}{\partial c_j}\right)_{T,\mu_D,c_k} = \sum_{i=d+1}^{d+s} c_i \left(\frac{\partial \mu_i}{\partial c_j}\right)_{T,\mu_D,c_k} \tag{2.3.77a}$$

or

$$\left(\frac{\partial \pi^*}{\partial C_j}\right)_{C_k} = \sum_{i=d+1}^{d+s} \frac{C_i}{M_i} \left(\frac{\partial \mu_i}{\partial C_j}\right)_{T,\mu_D,C_k}, \quad (j, k = d+1, \ldots, d+s) \tag{2.3.77b}$$

These relations are of the same form as Eq. (1.4.66) for solutions with a single solvent. For a polydisperse solute of nondiffusible components mixed in fixed proportions, C_i can be written

$$C_i = \xi_i C_s, \quad (i = d+1, \ldots, d+s) \tag{2.3.78}$$

where C_s is the total mass concentration of the solute mixture and ξ_i is the weight fraction of component i in the mixture. Defining an average chemical potential of the solute by

$$\mu_s \equiv \frac{\sum_{i=d+1}^{d+s} (\xi_i/M_i)\,\mu_i}{\sum_{i=d+1}^{d+s} \xi_i/M_i} = M_n \sum_{i=d+1}^{d+s} (\xi_i/M_i)\,\mu_i \tag{2.3.79}$$

we have

$$\begin{aligned} \frac{d\pi^*}{dC_s} &= \sum_{i=d+1}^{d+s} \xi_j \left(\frac{\partial \pi^*}{\partial C_j}\right)_{C_k} = \Sigma_j \xi_j \Sigma_i \frac{\xi_i C_s}{M_i} \left(\frac{\partial \mu_i}{\partial C_j}\right)_{T,\mu_D,C_k} \\ &= \frac{C_s}{M_n} \Sigma_j \xi_j \left(\frac{\partial \mu_s}{\partial C_j}\right)_{T,\mu_D,C_k} = \frac{C_s}{M_n} \left(\frac{\partial \mu_s}{\partial C_s}\right)_{T,\mu_D} \end{aligned} \tag{2.3.80}$$

where M_n denotes the number-average molar mass of the nondiffusible components. This expression is formally the same as that for a solution containing a single nondiffusible solute.

2.3.5. Osmotic Equilibrium in Mixed-Solvent Systems – 2. Constant-Pressure System

When the pressure of the inner phase p is held constant, we refer to constant-pressure osmotic equilibrium. We give a treatment of multicomponent solutions under this constraint following Scatchard [36] and Casassa and Eisenberg [37]. The appropriate composition variables are now molalities m_i defined with reference to the principal solvent (component 0). We consider a simple system which contains only a single nondiffusible component r. The externally assignable state variables are the temperature T, the composition of the outer phase m_i' $(i = 1, 2, \ldots, d)$, the pressure of the inner phase p, and the solute molality m_r. The pressure of the outer phase p' and the molalities of diffusible components in the inner phase m_i $(i = 1, 2, \ldots, d)$ must be treated as unknown quantities which are to be determined as the solutions to $d + 1$ simultaneous relations, Eq. (2.3.44). For the osmotic pressure, which should be written π in the present case, we have from Eq. (2.3.45)

$$d\pi = -(dp')_{T,p} \tag{2.3.81}$$

The inner phase is a system at constant temperature and pressure, so that μ_i can be considered a function of m_i $(i = 1, 2, \ldots, d, r)$ only. It follows that

$$(d\mu_i)_{T,p} = \sum_{j=1}^{r} \left(\frac{\partial \mu_i}{\partial m_j}\right)_{T,p,m_k} dm_j = \sum_{j=1}^{r} \mu_{ij}\, dm_j, \qquad (i = 1, 2, \ldots, d, r) \tag{2.3.82}$$

These linear equations can be solved for $r = d + 1$ unknowns dm_i

$$|\mu|\, dm_j = \begin{vmatrix} \mu_{11} & \cdots & \mu_{1,j-1} & (d\mu_1)_{T,p} & \mu_{1,j+1} & \cdots & \mu_{1r} \\ \mu_{12} & \cdots & \mu_{2,j-1} & (d\mu_2)_{T,p} & \mu_{2,j+1} & \cdots & \mu_{2r} \\ \cdots & \cdots & \cdots & \cdots & \cdots & \cdots & \cdots \\ \cdots & \cdots & \cdots & \cdots & \cdots & \cdots & \cdots \\ \cdots & \cdots & \cdots & \cdots & \cdots & \cdots & \cdots \\ \mu_{r1} & \cdots & \mu_{r,j-1} & (d\mu_r)_{T,p} & \mu_{r,j+1} & \cdots & \mu_{rr} \end{vmatrix}$$

$$= \sum_{k=1}^{r} (d\mu_k)_{T,p}\, \Delta_{kj}, \quad (j = 1, 2, \ldots, d, r) \tag{2.3.83}$$

where Δ_{kj} is the cofactor of element μ_{kj} in the determinant $|\mu|$.

If we change m_r by a small amount, keeping the other external variables T, p, and m_i' $(i = 1, 2, \ldots, d)$ constant, Eq. (2.3.83) can be written

$$|\mu| \frac{dm_j}{dm_r} = \sum_{k=1}^{r} \Delta_{kj} \frac{d\mu_k}{dm_r} \tag{2.3.84}$$

Here, for simplicity, the ordinary differential notation d is used to express changes in thermodynamic quantities dependent on dm_r. According to the Gibbs-Duhem relation, we have

$$\frac{1}{M_0} \frac{d\mu_0}{dm_r} + \sum_{i=1}^{d} m_i \frac{d\mu_i}{dm_r} + m_r \frac{d\mu_r}{dm_r} = 0 \tag{2.3.85}$$

By use of Eqs. (2.3.44) and (2.3.81), the $d\mu_i/dm_r$ in this equation may be written as

$$\frac{d\mu_i}{dm_r} = \frac{d\mu_i'}{dm_r} = \left(\frac{\partial \mu_i'}{\partial p'}\right)_{T,m'} \frac{dp'}{dm_r} = -V_i' \frac{d\pi}{dm_r}, \qquad (i = 0, 1, \ldots, d) \tag{2.3.86}$$

Introduction of this into Eq. (2.3.85) leads to

$$m_r \frac{d\mu_r}{dm_r} = \left(\frac{V_0'}{M_0} + \sum_{i=1}^{d} m_i V_i'\right) \frac{d\pi}{dm_r} \tag{2.3.87}$$

If we insert Eqs. (2.3.86) and (2.3.87) into Eq. (2.3.84), we obtain

$$|\mu| \frac{dm_j}{dm_r} = \frac{1}{m_r} \frac{d\pi}{dm_r} \left[\left(v_0' + \sum_{i=1}^{d} m_i V_i'\right) \Delta_{rj} - m_r \sum_{k=1}^{d} V_k' \Delta_{kj}\right] \tag{2.3.88}$$

For $j = r$, this gives the general expression for π:

$$\frac{d\pi}{dm_r} = \frac{m_r |\mu|}{(v_0' + \Sigma_i m_i V_i') \Delta_{rr} - m_r \Sigma_k V_k' \Delta_{kr}} \tag{2.3.89}$$

Using this to eliminate $d\pi/dm_r$ from Eq. (2.3.88), we get

$$\frac{dm_j}{dm_r} = \frac{(v_0' + \Sigma_i m_i V_i') \Delta_{rj} - m_r \Sigma_k V_k' \Delta_{kj}}{(v_0' + \Sigma_i m_i V_i') \Delta_{rr} - m_r \Sigma_k V_k' \Delta_{kr}} \tag{2.3.90}$$

which is the general expression for the selective adsorption of component j by component r, at constant T and p.

At infinite dilution ($m_r \to 0$), it follows from Eq. (1.4.60) that

$$(m_r \mu_{rr})^o = RT$$
$$(m_r \mu_{ri})^o = (m_r \mu_{ij})^o = 0, \quad (i, j \neq r) \tag{2.3.91a}$$

$$\left(m_r \frac{|\mu|}{\Delta_{rr}}\right)^o = \left[\frac{m_r}{\Delta_{rr}}\left(\mu_{rr}\Delta_{rr} + \sum_{i=1}^{d} \mu_{ri}\Delta_{ri}\right)\right]^o = (m_r \mu_{rr})^o = RT \tag{2.3.91b}$$

$$(m_r \Delta_{ij})^o = [m_r \mu_{rr} (\Delta_D)_{ij}]^o = RT (\Delta_D)^o_{ij}, \quad (i, j \neq r) \tag{2.3.91c}$$

where $(\Delta_D)_{ij}$ is the cofactor of element μ_{ij} in the determinant of the $\mu_{k\ell}$ (k, $\ell = 1, 2, \ldots, d$) for the d diffusible components. Since $m_i^o = m_i'$, we have from Eq. (1.1.70)

$$\left(v_0' + \sum_{k=1}^{d} m_k V_k'\right)^o = v_M' = v_M^o \tag{2.3.92}$$

where v_M is the solution volume per unit mass of the principal solvent (component 0). With Eqs. (2.3.91a) through (2.3.92), Eq. (2.3.90) yields at infinite dilution

$$\left(\frac{dm_j}{dm_r}\right)^o = \left(\frac{\Delta_{rj}}{\Delta_{rr}}\right)^o - \sum_{k=1}^{d}\left[\frac{V_k RT (\Delta_D)_{kj}}{v_M \Delta_{rr}}\right]^o \qquad (j = 1, 2, \ldots, d) \tag{2.3.93}$$

If this is used in expanding m_i in the denominator of Eq. (2.3.89) in powers of m_r, we arrive at

$$\frac{d\pi}{dm_r} = \frac{m_r |\mu|}{v_M^o \Delta_{rr}} \left\{1 + m_r \sum_{i=1}^{d} \sum_{j=1}^{d} \left[\frac{V_i V_j RT (\Delta_D)_{ij}}{v_M^2 \Delta_{rr}}\right]^o + \ldots\right\} \tag{2.3.94}$$

and

$$\left(\frac{d\pi}{dm_r}\right)^o = \frac{RT}{v_M^o} \tag{2.3.95}$$

Casassa and Eisenberg treated constant-μ_D systems in a way similar to that described above, and derived

$$\frac{d\pi^*}{dm_r} = \frac{m_r\,|\mu|}{v_M{}^o\,\Delta_{rr}}\left\{1 + m_r \sum_{i=1}^{d}\sum_{j=1}^{d}\left[\frac{V_i V_j RT(\Delta_D)_{ij}}{v_M{}^2\,\Delta_{rr}} + \frac{\kappa RT}{v_M}\right]^o + \ldots\right\} \tag{2.3.96}$$

Here the term containing the isothermal compressibility κ is the contribution from higher terms ignored in deriving Eq. (1.4.24). This term is almost completely negligible in ordinary circumstances. Hence we find that, even in mixed-solvent systems, π and π^* expanded in powers of m_r agree up to the order of $m_r{}^2$.

References

1. H. Staudinger and E. Husemann, Justus Liebigs Ann. Chem. **530**, 1 (1937).
2. C. Tanford, "Physical Chemistry of Macromolecules," pp. 233-236, Wiley, New York and London, 1961.
3. W. A. Caspari, J. Chem. Soc. **105**, 2139 (1914).
4. H. Staudinger and K. Fisher, J. Prakt. Chem. **157**, 19 (1940).
5. P. J. Flory, J. Chem. Phys. **9**, 660 (1941); **10**, 51 (1942).
6. M. L. Huggins, J. Chem. Phys. **9**, 440 (1941); Ann. N.Y. Acad. Sci. **43**, 1 (1942).
7. A. R. Shultz and P. J. Flory, J. Am. Chem. Soc. **74**, 4760 (1952).
8. H. Tompa, Trans. Faraday Soc. **46**, 970 (1950).
9. M. L. Huggins, Ann. N.Y. Acad. Sci. **44**, 431 (1943).
10. H. Tompa, C. R. 2^e Reun. Annu. Soc. Chim. Phys. Paris, 163 (1952).
11. R. Koningsveld, J. Polym. Sci. A-2, **6**, 325 (1968).
12. Th. G. Scholte, J. Polym. Sci. A-2, **8**, 841 (1970).
13. W. H. Stockmayer, J. Chem. Phys. **17**, 588 (1949).
14. M. Gordon, H. A. G. Chermin, and R. Koningsveld, Macromolecules, **2**, 207 (1969).
15. R. Koningsveld, Thesis, Univ. Leiden, Leiden, 1967; Pure Appl. Chem. **20**, 271 (1969).
16. H. Tompa, Trans. Faraday Soc. **45**, 1142 (1949).
17. K. Šolc, private communication.
18. K. Šolc, Macromolecules, **3**, 665 (1970); **8**, 819 (1975); J. Polym. Sci. Polym. Phys. Ed. **12**, 555, 1865 (1974).
19. A. R. Schultz, J. Polym. Sci. **11**, 93 (1953).
20. G. Rehage, D. Möller, and O. Ernst, Makromol. Chem. **88**, 232 (1965).
21. R. Koningsveld, L. A. Kleintjens, and A. R. Shultz, J. Polym. Sci. A-2, **8**, 1261 (1970).
22. W. R. Krigbaum and D. O. Geymer, J. Am. Chem. Soc. **81**, 1859 (1959).
23. G. Gee and W. J. Orr, Trans. Faraday Soc. **42**, 507 (1946); G. Gee and L. R. G. Treloar, Trans. Faraday Soc. **38**, 147 (1942).

24. G. V. Schulz, Z. Phys. Chem. **B43**, 25 (1939); **B46**, 137 (1940); **B47**, 515 (1940).
25. See, for example, A. Kotera, in "Polymer Fractionation," M. J. R. Cantow, Ed., Chap. B1, p. 65, 66, Academic Press, New York, 1967.
26. P. J. Flory, J. Chem. Phys. **12**, 425 (1944).
27. L. H. Tung, J. Polym. Sci. **61**, 449 (1962).
28. R. Koningsveld and A. J. Staverman, J. Polym. Sci. A-2, **6**, 367 (1968).
29. K. Kamide and K. Sugamiya, Makromol. Chem. 139, 197 (1970); K. Kamide, T. Ogawa, M. Sanada, and M. Matsumoto, Chem. High Polym. Japan, **25**, 440 (1968); K. Kamide, T. Ogawa, and M. Matsumoto, Chem. High Polym. Japan, **25**, 788 (1968).
30. R. L. Scott, J. Chem. Phys. **17**, 279 (1949).
31. H. Tompa, J. Chem. Phys. **17**, 1006 (1949).
32. I. Sakurada and K. Seki, quoted in A. Nakajima and M. Hosono, "Kobunshi no Bunshi Tokusei (Molecular Properties of Polymers)," Part 2, p. 404, Kagaku-dojin, Kyoto, 1969.
33. A. Dobry and F. Boyer-Kawenoki, J. Polym. Sci. **2**, 90 (1947).
34. C. H. Bamford and H. Tompa, Trans. Faraday Soc. **46**, 310 (1950).
35. H. Tompa, "Polymer Solutions," Chap. 7, Butterworths, London, 1956.
36. G. Scatchard, J. Am. Chem. Soc. **68**, 2315 (1946).
37. E. F. Casassa and H. Eisenberg, Adv. Protein Chem. **11**, 287 (1964).

Light Scattering

CHAPTER 3

Light Scattering

3.1. Fluctuations of Thermodynamic Quantities

3.1.1. Bernoulli, Gauss, and Poisson Distributions

Light scattering (Rayleigh scattering) affords an absolute means of molecular weight determination. The phenomenon is directly related to fluctuations of solute concentrations. Knowledge of fluctuations is also essential for understanding the bridge between thermodynamics and statistical mechanics.

We begin our consideration of fluctuations with simple probability arguments. If the probability of occurrence of a particular event A in any one trial is p, the probability q that all events other than A, i.e., B, will take place in one trial is evidently given by

$$q = 1 - p \tag{3.1.1}$$

The probability of obtaining a particular sequence of events A and B, such as AABAB . . . , after n trials is given by $p^x q^{n-x}$ if the sequence contains x events A. However, without regard to the order of occurrence of events, the probability $w_n(x)$ of finding a sequence containing x events A after n trials is

$$w_n(x) = \frac{n!}{x!\,(n-x)!}\, p^x q^{n-x} \tag{3.1.2}$$

or, more compactly, in terms of the binomial coefficients ${}_nC_x$:

$$w_n(x) = {}_nC_x\, p^x q^{n-x} \tag{3.1.3}$$

The distribution of x is called the **Bernoulli distribution**. Since $w_n(x)$ is the coefficient of u^x in the expansion of $(pu + q)^n$, the function $(pu + q)^n$ is the **generating function** for $w_n(x)$. If we put u equal to unity after expanding this generating function in powers of u, the result is simply the sum of the coefficients of the series:

$$\sum_{x=0}^{n} w_n(x) = [(pu+q)^n]_{u=1} = 1 \tag{3.1.4}$$

The mean of x is denoted by

$$\langle x\rangle \equiv \sum_{x=0}^{n} x\, w_n(x) \tag{3.1.5}$$

but since $xw_n(x)$ is the coefficient of u^{x-1} in the expansion of $d(pu+q)^n/du$, we can write

$$\langle x\rangle = \left[\frac{d(pu+q)^n}{du}\right]_{u=1} = [np(pu+q)^{n-1}]_{u=1} = np \tag{3.1.6a}$$

The mean of x^2 is found in a similar way:

$$\langle x^2\rangle \equiv \sum_{x=0}^{n} x^2 w_n(x) = \left\{\frac{d}{du}\left[u\frac{d(pu+q)^n}{du}\right]\right\}_{u=1} = n^2p^2 + npq \tag{3.1.6b}$$

Therefore the mean square of the fluctuation $\Delta x \equiv x - \langle x\rangle$ is given by

$$\langle(\Delta x)^2\rangle = \langle(x-\langle x\rangle)^2\rangle = \langle x^2\rangle - \langle x\rangle^2 = npq \tag{3.1.7}$$

For very large values of n the Bernoulli distribution approaches two different asymptotic forms, depending on the magnitude of p. First, consider the case in which p is finite so that the mean value $\langle x\rangle = np$ increases indefinitely as $n\to\infty$. Then applying Stirling's formula, we get

$$\ell n\, x! = (x+\tfrac{1}{2})\,\ell n\, x - x + \tfrac{1}{2}\,\ell n\,(2\pi) \tag{3.1.8}$$

for x near $\langle x\rangle$. Also, in the expansion

$$\begin{aligned}\ell n\, x &= \ell n\,(\langle x\rangle + \Delta x)\\ &= \ell n\,(np) + \frac{\Delta x}{np} - \frac{1}{2}\left(\frac{\Delta x}{np}\right)^2 + \ldots\end{aligned} \tag{3.1.9}$$

terms of higher order than $(\Delta x/np)^2$ may be ignored. With these approximate expressions for $\ell n\, x!$ and $\ell n\, x$, Eq. (3.1.2) becomes

$$w_n(\Delta x) = \frac{1}{(2\pi npq)^{1/2}}\exp\left[-\frac{(\Delta x)^2}{2npq}\right] \tag{3.1.10}$$

where $w_n(\Delta x)$ is a shorthand notation for $w_n(\langle x\rangle + \Delta x)$. The distribution function for Δx defined by Eq. (3.1.10) is called **Gaussian**. The normalization condition is

$$\int_{-\infty}^{\infty} w_n(\Delta x)\, d(\Delta x) = 1 \tag{3.1.11}$$

and we have for $\langle(\Delta x)^2\rangle$

$$\langle(\Delta x)^2\rangle = \int_{-\infty}^{\infty} (\Delta x)^2\, w_n(\Delta x) d(\Delta x) = npq \tag{3.1.12}$$

which agrees with the exact result given by the Bernoulli distribution.

Next, consider the case in which np remains finite as $n \to \infty$. Now we have $n \gg x$ for $x \approx \langle x\rangle = np$. Hence if we rewrite Eq. (3.1.2) as

$$\begin{aligned} w_n(x) &= \frac{p^x}{x!}\,[n(n-1)\ldots(n-x+1)]\,(1-p)^{n-x} \\ &= \frac{\langle x\rangle^x}{x!}\left[1\cdot\left(1-\frac{1}{n}\right)\cdots\left(1-\frac{x-1}{n}\right)\right]\left(1-\frac{\langle x\rangle}{n}\right)^{n-x} \end{aligned} \tag{3.1.13}$$

and introduce the condition $\langle x\rangle/n \ll 1$, then, in the limit $n \to \infty$, we get

$$w_n(x) = \frac{\langle x\rangle^x}{x!} \lim_{n\to\infty}\left(1-\frac{\langle x\rangle}{n}\right)^n = \frac{\langle x\rangle^x}{x!}\exp(-\langle x\rangle) \tag{3.1.14}$$

This is called the **Poisson distribution**. The normalization condition is

$$\sum_{x=0}^{\infty} w_n(x) = e^{-\langle x\rangle}\sum_{x=0}^{\infty}\frac{1}{x!}\langle x\rangle^x = e^{-\langle x\rangle}e^{\langle x\rangle} = 1 \tag{3.1.15}$$

This distribution gives the following expressions for $\langle x^2\rangle$ and $\langle(\Delta x)^2\rangle$:

$$\langle x^2\rangle = np + n^2p^2 \tag{3.1.16}$$

$$\langle(\Delta x)^2\rangle = \langle x^2\rangle - \langle x\rangle^2 = np \tag{3.1.17}$$

Equation (3.1.17) conforms to Eq. (3.1.7) with $q = 1$.

For $\langle x\rangle \gg 1$ it follows from Eq. (3.1.14) that

$$\begin{aligned} \ln w_n(x) &\approx x\ln\langle x\rangle - \langle x\rangle - [(x+\tfrac{1}{2})\ln x - x + \tfrac{1}{2}\ln(2\pi)] \\ &= -(\langle x\rangle + \Delta x + \tfrac{1}{2})\ln\left(1+\frac{\Delta x}{\langle x\rangle}\right) + \Delta x - \tfrac{1}{2}\ln(2\pi\langle x\rangle) \\ &\simeq -\frac{(\Delta x)^2}{2\langle x\rangle} - \tfrac{1}{2}\ln(2\pi\langle x\rangle) \end{aligned}$$

so that

$$w_n(\Delta x) = \frac{1}{(2\pi\langle x\rangle)^{1/2}}\exp\left[-\frac{(\Delta x)^2}{2\langle x\rangle}\right] \tag{3.1.18}$$

Thus the Poisson distribution also asymptotically tends to the Gaussian distribution as $\langle x \rangle$ increases indefinitely. This is an example of the **central limit theorem** in statistical phenomena.

One-dimensional **random walks** are governed by the Gaussian distribution even at relatively small n, because p is large. Consider a particle which moves with equal step-lengths ℓ either to the right or to the left on a straight line. After N steps, the particle will reach one of the $N + 1$ points which are distant from the starting point by $-N\ell$, $(-N + 2)\,\ell, \ldots, (N - 2)\,\ell$, and $N\ell$. The probability that the particle will be found at a distance $m\ell$ after N steps is denoted by $W_N(m)$. This is given by Eq. (3.1.2), with n and x replaced by N and $(N + m)/2$, respectively. We note that m is equal to the difference between the numbers of steps to the right and to the left. If each step is taken at random, it follows that $p = q = 1/2$. Hence

$$W_N(m) = \frac{N!}{[(N+m)/2]!\,[(N-m)/2]!} \left(\tfrac{1}{2}\right)^N \tag{3.1.19}$$

As $N \to \infty$, this distribution approaches a Gaussian function represented by

$$W_N(m) = \left(\frac{2}{\pi N}\right)^{1/2} \exp\left(-\frac{m^2}{2N}\right) \tag{3.1.20}$$

In Table 3.1, Eqs. (3.1.19) and (3.1.20) are compared for $N = 10$ [1]. It is seen that even at this small N the Gaussian distribution gives a good approximation. For N of order 10^2 the difference between Eqs. (3.1.19) and (3.1.20) becomes negligible.

When N is very large, it is more convenient to express the distribution by a continuous function of x, the displacement from the starting point, rather than by a discrete function of m. We designate the probability of finding the particle in the interval between x and $x + dx$ after N steps by $P_N(x)dx$, and call $P_N(x)$

Table 3.1. Probability $W_N(m)$ for one-dimensional random walk ($N = 10$)

m	Bernoullian Eq. (3.1.19)	Gaussian Eq. (3.1.20)
0	0.24609	0.252
2	0.20503	0.207
4	0.11715	0.113
6	0.04374	0.042
8	0.00977	0.010
10	0.00098	0.002

the **probability density** or the **distribution function** for x. Since the linear density of possible end-points of a walk of N steps is $1/(2\ell)$ in the range $-N\ell < x < N\ell$, the probability $P_N(x)dx$ is equal to $W_N(m)\,dx/(2\ell)$. Hence Eq. (3.1.20) gives

$$P_N(x) = \frac{1}{(2\pi N\ell^2)^{1/2}} \exp\left(-\frac{x^2}{2N\ell^2}\right) \tag{3.1.21}$$

Next, consider a particle which jumps on a simple cubic lattice, taking $N/3$ steps in each of the mutually perpendicular x, y, and z directions. The distribution function $P_N(\underset{\sim}{R})$ of the end-point $\underset{\sim}{R}(x, y, z)$ is given by

$$P_N(\underset{\sim}{R}) = P_{N/3}(x)\,P_{N/3}(y)\,P_{N/3}(z) = \gamma^3\,(\pi)^{-3/2} \exp(-\gamma^2 R^2) \tag{3.1.22}$$

where

$$\gamma^2 \equiv 3/(2N\ell^2) \tag{3.1.23}$$

ℓ being the distance between neighboring lattice points. The normalization condition on $P_N(\underset{\sim}{R})$ is

$$\int P_N(\underset{\sim}{R})\,d\underset{\sim}{R} = \int_0^\infty P_N(\underset{\sim}{R})\,4\pi R^2\,dR = 1 \tag{3.1.24}$$

and $\langle R^2 \rangle$, the mean-square R, is given by

$$\langle R^2 \rangle = \int R^2 P_N(\underset{\sim}{R})\,d\underset{\sim}{R} = 3/(2\gamma^2) = N\ell^2 \tag{3.1.25}$$

If each step of the particle is visualized as the orientation of a skeletal bond of a polymer chain, $\underset{\sim}{R}$ after N steps can be compared to the end-to-end distance of a polymer molecule consisting of N skeletal bonds. The result in Eq. (3.1.25), which means that the mean-square end-to-end distance is proportional to the number of skeletal bonds, is an important property of flexible polymers in theta solvents.

It is easily shown that fluctuations of the number of molecules in a very small fixed volume chosen within a macroscopic ideal-gas system obey the Poisson distribution. The volume of the system and the number of molecules in it are denoted by V_t and N_t, respectively. The corresponding quantities associated with the small region are denoted by V and N. Since the gas is ideal, the probability p of finding a given molecule in volume V is given by

$$p = V/V_t \ll 1 \tag{3.1.26}$$

Therefore, the probability $W_{N_t}(N)$ that N molecules out of N_t will be found in volume V is given by the Poisson distribution. If the instantaneous molecular number density in V is η, then

$$\langle \eta \rangle = \frac{\langle N \rangle}{V} = \frac{pN_t}{V} = \frac{N_t}{V_t} \tag{3.1.27}$$

The mean-square fluctuation of the number of molecules in volume V is given, according to Eq. (3.1.17), by

$$\langle (\Delta N)^2 \rangle = \langle N \rangle = V\langle \eta \rangle \tag{3.1.28}$$

Hence

$$\langle (\Delta \eta)^2 \rangle / \langle \eta \rangle^2 = \langle (\Delta N)^2 \rangle / (V\langle \eta \rangle)^2 = 1/(V\langle \eta \rangle) = 1/\langle N \rangle \tag{3.1.29}$$

which indicates that the mean-square relative fluctuation of molecular number density increases as V becomes smaller.

3.1.2. Fluctuations of Thermodynamic Quantity in a Closed System [2]

Consider a small closed system which is in contact with a large surrounding medium kept at constant temperature T_0 and constant pressure p_0. By assuming that the combined system consisting of the small system and its surroundings is completely isolated, the volume V_t and the internal energy U_t of the combined system are held constant.

At equilibrium, the state x of the small system is not fixed, but fluctuates with time about a mean or "equilibrium" state x^*. Here x designates the set of state variables for the small system. We ask the probability $w(x)$ that the small system at equilibrium will be found in state x. Let the entropy of the combined system when the small system is in state x be denoted by $S_t(x)$. Since the combined system is isolated, then, according to the Boltzmann theorem in statistical mechanics, we have

$$w(x) \propto \exp\,[S_t(x)/k] \tag{3.1.30}$$

where k is the Boltzmann constant. This relation may be rewritten

$$w(x) = w(x^*) \exp\,[\Delta S_t/k] \tag{3.1.31}$$

with ΔS_t defined by

$$\Delta S_t = S_t(x) - S_t(x^*) \tag{3.1.32}$$

In Fig. 3.1, the solid line represents schematically a relation between mean values of S_t and U_t at a given V_t. Thus point A represents the mean value of S_t at the fixed U_t and V_t, while point B represents the value of S_t when the small system happens to be in state x at the same U_t and V_t. Note particularly that the small system associated with point B is in part of the distribution of states which determines its equilibrium state at the given U_t and V_t. Consider

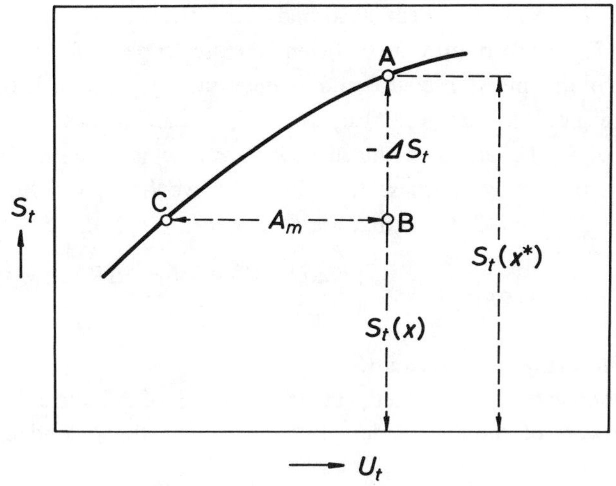

Fig. 3.1. S_t versus U_t at fixed V_t. Solid line: relation between mean S_t and mean U_t. Point B: one state in the equilibrium distribution of states at given U_t and V_t.

point C which has the same S_t as point B and denote the difference in U_t between these two points by A_m. If point C is sufficiently close to point A, i.e., the fluctuation $\Delta x = x - x^*$ is very small, the entropy difference ΔS_t between points A and B can be expressed by

$$\Delta S_t = -(\partial S_t/\partial U_t)^*_{V_t} A_m = -A_m/T^* \tag{3.1.33}$$

Here the derivative $(\partial S_t/\partial U_t)_{V_t}$ is the slope of the tangent to the solid line, and the asterisk indicates that it should be evaluated at point A, i.e., at the equilibrium state of the combined system at the given U_t and V_t. Hence, as indicated in the above equation, this derivative can be equated to $1/T^*$, T^* being the mean temperature of the small system at the given U_t and V_t.

We operate quasistatically a work reservoir thermally isolated from the combined system to put energy A_m into the small system associated with point C. Then the small system is reversibly brought to the state associated with point B. The increases ΔU, ΔV, and ΔS of internal energy, volume, and entropy of the small system which accompany this reversible change are related to A_m by

$$\Delta U = A_m - p_0 \Delta V + T_0 \Delta S \tag{3.1.34}$$

Here, during the change of the combined system from point C to point B, $p_0 \Delta V$ is the work done by the small system against the pressure p_0 of the surrounding medium and $T_0 \Delta S$ is the heat absorbed from the surrounding medium of temperature T_0. Since point C is sufficiently close to point A, p_0 and T_0 may be equated to the mean pressure and temperature, p^* and T^*, of the small system at the given U_t and V_t. Also, ΔU, ΔV, and ΔS may be considered the fluctuations of U, V, and S of the small system associated with the state x at the given U_t and V_t. With these facts in mind, we find from Eqs. (3.1.31), (3.1.32), (3.1.33), and (3.1.34) that the desired $w(x)$ is given by

$$w(x) = C \exp\left(-\frac{\Delta U - T^* \Delta S + P^* \Delta V}{kT^*}\right) \tag{3.1.35}$$

where C is a normalization constant.

Since the small system is closed, ΔU is a function of ΔS and ΔV. Expansion of ΔU in powers of these variables gives, correct to the second order in ΔS and ΔV,

$$\Delta U = \left(\frac{\partial U}{\partial S}\right)_V \Delta S + \left(\frac{\partial U}{\partial V}\right)_S \Delta V + \tfrac{1}{2}\left[\left(\frac{\partial^2 U}{\partial S^2}\right)(\Delta S)^2 + 2\left(\frac{\partial^2 U}{\partial S \partial V}\right)\Delta S \Delta V + \left(\frac{\partial^2 U}{\partial V^2}\right)(\Delta V)^2\right] \tag{3.1.36}$$

Although the temperature and pressure of the surroundings are kept constant, the temperature T and pressure p of the small system fluctuate with the time about the mean values T^* and p^*. These fluctuations ΔT and Δp are also functions of ΔS and ΔV, and, with Eqs. (1.2.18) and (1.2.19), can be expressed, correct to the first order in ΔS and ΔV, as

$$\Delta T = \left(\frac{\partial^2 U}{\partial S^2}\right)\Delta S + \left(\frac{\partial^2 U}{\partial S \partial V}\right)\Delta V \tag{3.1.37a}$$

$$\Delta p = -\left(\frac{\partial^2 U}{\partial S \partial V}\right)\Delta S - \left(\frac{\partial^2 U}{\partial V^2}\right)\Delta V \tag{3.1.37b}$$

The higher terms not explicitly written in these expressions are immaterial for subsequent treatments. By application of Eqs. (3.1.37a) and (3.1.37b), together with the general relations $(\partial U/\partial V)_S = -p$ and $(\partial U/\partial S)_V = T$, the right-hand side of Eq. (3.1.36) can be written $T^* \Delta S - p^* \Delta V + (1/2)(\Delta T \Delta S - \Delta p \Delta V)$. Hence, Eq. (3.1.35) becomes

$$w(x) = C \exp\left(\frac{-\Delta T \Delta S + \Delta p \Delta V}{2kT^*}\right) \tag{3.1.38}$$

This is the general expression for $w(x)$ which governs fluctuations of thermodynamic quantities in a closed system at equilibrium.

By changing from independent state variables S, V to T, V and using Eqs. (1.2.23), (1.2.25), and (1.2.30), we can express ΔS and Δp, correct to the first order in ΔT and ΔV, as

$$\Delta S = (C_V/T^*)\,\Delta T + (\alpha/\kappa)\,\Delta V \qquad (3.1.39a)$$

$$\Delta p = (\alpha/\kappa)\,\Delta T - (1/\kappa V^*)\,\Delta V \qquad (3.1.39b)$$

where the heat capacity at constant volume C_V, the isobaric thermal expansivity α, and the isothermal compressibility κ refer to the equilibrium state specified by the mean temperature T^* and the mean volume V^*. Introducing the above expressions into Eq. (3.1.38), we obtain

$$w(\Delta T, \Delta V) = C \exp\left[-\frac{C_V}{2k(T^*)^2}(\Delta T)^2 - \frac{1}{2k\kappa V^* T^*}(\Delta V)^2\right] \qquad (3.1.40)$$

This expression yields

$$\langle(\Delta T)^2\rangle = k(T^*)^2/C_V, \quad \langle(\Delta V)^2\rangle = \kappa k V^* T^*, \qquad \langle\Delta T\,\Delta V\rangle = 0 \qquad (3.1.41)$$

If the small system under consideration consists of $r + 1$ nonreacting components, its mass density ρ is expressed by

$$\rho = \sum_{i=0}^{r} n_i M_i / V \qquad (3.1.42)$$

where n_i and M_i are the amount and molar mass of component i. Since the system is closed, it follows from Eq. (3.1.42) that

$$\Delta V/V^* = -\Delta\rho/\rho^* \qquad (3.1.43)$$

where V^* and ρ^* represent the mean values of V and ρ, respectively. Substitution of this into Eq. (3.1.41) yields

$$\langle(\Delta\rho)^2\rangle = (\rho^*)^2\,\kappa k T^*/V^* \qquad (3.1.44a)$$

$$\langle\Delta\rho\,\Delta T\rangle = 0 \qquad (3.1.44b)$$

Equation (3.1.44a) reduces to Eq. (3.1.29) for one-component ideal gases ($pV = NkT$). Equation (3.1.44b) indicates that, in a closed system, fluctuations of temperature and mass density take place independently, i.e., they are uncorrelated. It will be shown later that these two fluctuations also are uncorrelated in open systems.

The mean-square pressure fluctuation $\langle(\Delta p)^2\rangle$ and the mean correlation $\langle \Delta p \, \Delta T\rangle$ between pressure and temperature fluctuations can be calculated by use of Eqs. (3.1.39b), (3.1.41), and (1.2.31), yielding

$$\langle(\Delta p)^2\rangle = C_p \, kT^* / (\kappa \, C_V \, V^*) \tag{3.1.45a}$$

$$\langle \Delta p \, \Delta T\rangle = \alpha k (T^*)^2 / (\kappa \, C_V) \tag{3.1.45b}$$

where C_p denotes the heat capacity at constant pressure (at T^* and V^*). Thus we see that, in closed systems, pressure fluctuations are correlated with temperature fluctuations.

3.1.3. Fluctuations of Concentrations in Multicomponent Solutions

We extend the treatment of closed systems presented above to multicomponent open systems. We consider a large medium which consists of $r + 1$ nonreacting components and which is kept at temperature T_0, pressure p_0, and chemical potentials $(\mu_i)_0$ $(i = 0, 1, \ldots, r)$. A small open system composed of the same $r + 1$ components is placed in contact with it. We denote fluctuations of entropy, volume, and amount of component i which occur in this system by ΔS, ΔV, and Δn_i. The accompanying fluctuation of internal energy, ΔU, is the sum of $A_m - p_0 \, \Delta V + T_0 \, \Delta S$ given by Eq. (3.1.34) and the contribution made by fluctuations Δn_i. The latter is given by $\Sigma_i (\mu_i)_0 \, \Delta n_i$. Thus, in place of Eq. (3.1.35), we obtain

$$w(x) = C \exp\left(-\frac{\Delta U - T^* \, \Delta S + p^* \, \Delta V - \Sigma_i \, \mu_i^* \, \Delta n_i}{kT^*}\right) \tag{3.1.46}$$

where we have replaced T_0, p_0, and $(\mu_i)_0$ by the mean or equilibrium temperature T^*, pressure p^*, and chemical potential μ_i^*.

If the contributions of ΔS, ΔV, and Δn_i to ΔU up to the second order are considered, we obtain, after an operation similar to that described in the preceding section,

$$\Delta U - T^* \, \Delta S + p^* \, \Delta V - \Sigma_i \mu_i^* \, \Delta n_i = \tfrac{1}{2} \left(\Delta T \, \Delta S - \Delta p \, \Delta V + \Sigma_i \, \Delta \mu_i \, \Delta n_i\right)$$

Therefore, Eq. (3.1.46) becomes

$$w(x) = C \exp\left(\frac{-\Delta T \, \Delta S + \Delta p \, \Delta V - \Sigma_i \, \Delta \mu_i \, \Delta n_i}{2kT^*}\right) \tag{3.1.47}$$

This general expression allows us to calculate the mean-square and mean correlations of fluctuations of thermodynamic quantities which appear in a small region of a multicomponent solution at equilibrium. The calculation varies depending on what set of state variables is chosen to describe the small system. However, the calculated results for an average, even though different in mathematical form, ought to be independent of the choice of state variables. In what follows, we present such calculations corresponding to three typical sets of variables.

a. Isochoric Systems-1: T, V, and n $(n_0, n_1, \ldots, n_r)$ as State Variables. Take the small system as a region of fixed volume V and choose T and n as the additional state variables to describe it. By definition, the term $\Delta p\,\Delta V$ disappears from Eq. (3.1.47). Fluctuations of S and μ_i $(i = 0, 1, \ldots, r)$ which occur in the small system are written

$$\begin{aligned}\Delta S &= \left(\frac{\partial S}{\partial T}\right)_{V,n} \Delta T + \sum_{j=0}^{r} \left(\frac{\partial S}{\partial n_j}\right)_{T,V,n_k} \Delta n_j \\ &= \frac{C_V}{T^*} \Delta T - \sum_{j=0}^{r} \left(\frac{\partial^2 A}{\partial n_j \partial T}\right)_{V,n_k} \Delta n_j\end{aligned} \tag{3.1.48a}$$

$$\begin{aligned}\Delta \mu_i &= \left(\frac{\partial \mu_i}{\partial T}\right)_{V,n} \Delta T + \sum_{j=0}^{r} \left(\frac{\partial \mu_i}{\partial n_j}\right)_{T,V,n_k} \Delta n_j \\ &= \left(\frac{\partial^2 A}{\partial n_i \partial T}\right)_{V,n_k} \Delta T + \sum_{j=0}^{r} \left(\frac{\partial^2 A}{\partial n_i \partial n_j}\right)_{T,V,n_k} \Delta n_j\end{aligned} \tag{3.1.48b}$$

correct to the first order in ΔT and Δn $(\Delta n_0, \Delta n_1, \ldots, \Delta n_r)$. If these relations are inserted into Eq. (3.1.47), then

$$w(\Delta T, \Delta n) = C \exp\left[-\frac{C_V}{2k(T^*)^2}(\Delta T)^2 - \frac{1}{2kT^*}\sum_{i=0}^{r}\sum_{j=0}^{r}\left(\frac{\partial^2 A}{\partial n_i \partial n_j}\right)_{T,V,n_k} \Delta n_i \Delta n_j\right] \tag{3.1.49}$$

If Eq. (3.1.49) is used to calculate $\langle(\Delta T)^2\rangle$, we recover Eq. (3.1.41). The second term in the brackets of Eq. (3.1.49) is a quadratic form of the $r + 1$ fluctuations Δn_i. By application of Eq. (A.72), the mean correlation between Δn_i and Δn_j is found to be

$$\langle \Delta n_i \Delta n_j \rangle = kT^* (\Delta_A)_{ij} / |A|, \quad (i, j = 0, 1, \ldots, r) \tag{3.1.50}$$

where $|A|$ is a determinant of order $r + 1$ with elements A_{ij} defined by

$$A_{ij} \equiv \left(\frac{\partial^2 A}{\partial n_i \partial n_j}\right)_{T,V,n_k} = \left(\frac{\partial \mu_i}{\partial n_j}\right)_{T,V,n_k} \tag{3.1.51}$$

and $(\Delta_A)_{ij}$ is the cofactor of element A_{ij}. Here, since V is fixed, Eq. (3.1.50) is equivalent to

$$\langle \Delta c_i \Delta c_j \rangle = (kT^*/V^2)(\Delta_A)_{ij}/|A|, \quad (i,j = 0,1,\ldots,r) \qquad (3.1.52)$$

where $c_i = n_i/V$ is the molar concentration of component i. This expression gives the mean correlation of concentrations of any pair of components in the solution.

b. Constant-n_0 Systems: T, p, n_0, and m $(m_1, m_2, \ldots, m_r)$ as State Variables. Take the small system as a region containing a fixed amount n_0 of component 0 (the principal solvent) and choose T, p, and molalities $m_1, m_2, \ldots, m_r$ of all other components as the additional variables. This choice is due to Stockmayer [3]. All quantities other than n_0 of the system fluctuate. Fluctuations of S, V, and μ_i $(i = 0, 1, \ldots, r)$ in the small system, correct to the first order in ΔT, Δp, and Δm_i $(\Delta m_1, \Delta m_2, \ldots, \Delta m_r)$, are expressed as

$$\Delta S = \left(\frac{\partial S}{\partial T}\right)_{p,n_0,m} \Delta T + \left(\frac{\partial S}{\partial p}\right)_{T,n_0,m} \Delta p + \sum_{j=1}^{r} \left(\frac{\partial S}{\partial m_j}\right)_{T,p,n_0,m_k} \Delta m_j \qquad (3.1.53a)$$

$$\Delta V = \left(\frac{\partial V}{\partial T}\right)_{p,n_0,m} \Delta T + \left(\frac{\partial V}{\partial p}\right)_{T,n_0,m} \Delta p + \sum_{j=1}^{r} \left(\frac{\partial V}{\partial m_j}\right)_{T,p,n_0,m_k} \Delta m_j \qquad (3.1.53b)$$

$$\Delta\mu_i = -S_i \Delta T + V_i \Delta p + \sum_{j=1}^{r} \mu_{ij} \Delta m_j, \quad (i = 0, 1, \ldots, r) \qquad (3.1.53c)$$

where S_i, V_i, and μ_{ij} are defined by Eqs. (1.2.43), (1.2.44), and (1.3.42), respectively. The term $\Delta\mu_0 \Delta n_0$ disappears from Eq. (3.1.46), since the small system chosen has n_0 fixed, and also Δn_i is equal to $n_0 M_0 \Delta m_i$ for $i = 1, 2, \ldots, r$. Thus, with Eqs. (3.1.53a) through (3.1.53c) and use of Eqs. (1.2.27) and (1.2.30), Eq. (3.1.47) becomes

$$w(\Delta T, \Delta p, \Delta m)$$

$$= C \exp\left\{ -\frac{1}{2kT^*} \left[\frac{C_p}{T^*} (\Delta T)^2 - 2\alpha V^* \Delta T \Delta p + \kappa V^* (\Delta p)^2 \right] - \frac{n_0 M_0}{2kT^*} \sum_{i=1}^{r} \sum_{j=1}^{r} \mu_{ij} \Delta m_i \Delta m_j \right\} \qquad (3.1.54)$$

With the aid of Eq. (A.72), this expression gives for the mean correlation between composition fluctuations Δm_i and Δm_j

$$\langle \Delta m_i \Delta m_j \rangle = \frac{kT^*}{n_0 M_0} \frac{\Delta_{ij}}{|\mu|} = \frac{v_M{}^* kT^*}{V^*} \frac{\Delta_{ij}}{|\mu|}, \qquad (i, j = 1, 2, \ldots, r) \tag{3.1.55}$$

where $|\mu|$ is a determinant of order r with elements μ_{ij}, Δ_{ij} is the cofactor of element μ_{ij}, and $v_M{}^*$ is the mean volume of the small system per unit mass of component 0.

Equation (3.1.54) also yields

$$\langle (\Delta T)^2 \rangle = \frac{\kappa k (T^*)^2}{\kappa C_p - \alpha^2 T^* V^*}, \quad \langle (\Delta p)^2 \rangle = \frac{k C_p T^*}{V^* (\kappa C_p - \alpha^2 T^* V^*)},$$
$$\langle \Delta p \, \Delta T \rangle = \frac{\alpha k (T^*)^2}{\kappa C_p - \alpha^2 T^* V^*} \tag{3.1.56}$$

Using Eq. (1.2.31), we confirm that these expressions agree with Eq. (3.1.41), (3.1.45a), and (3.1.45b) derived for closed systems.

Fluctuations of mass density ρ in the small system under consideration can be expressed as

$$\begin{aligned} \Delta \rho &= \left(\frac{\partial \rho}{\partial T}\right)_{p,m} \Delta T + \left(\frac{\partial \rho}{\partial p}\right)_{T,m} \Delta p + \sum_{j=1}^{r} \left(\frac{\partial \rho}{\partial m_j}\right)_{T,p,m_k} \Delta m_j \\ &= -\alpha \rho^* \Delta T + \kappa \rho^* \Delta p + \sum_{j=1}^{r} \Gamma_j \Delta m_j \end{aligned} \tag{3.1.57}$$

where use has been made of the relations

$$\alpha \equiv \frac{1}{V} \left(\frac{\partial V}{\partial T}\right)_{p,n} = -\frac{1}{\rho} \left(\frac{\partial \rho}{\partial T}\right)_{p,m} \tag{3.1.58a}$$

$$\kappa \equiv -\frac{1}{V} \left(\frac{\partial V}{\partial p}\right)_{T,n} = \frac{1}{\rho} \left(\frac{\partial \rho}{\partial p}\right)_{T,m} \tag{3.1.58b}$$

and Γ_j is defined by

$$\Gamma_j \equiv (\partial \rho / \partial m_j)_{T,p,m_k} \tag{3.1.58c}$$

The quantity Γ_j is called the density increment on the molality scale. It will prove to be important in the theory of sedimentation equilibrium discussed later (see Chapter 4). Since it follows from Eq. (3.1.54) that composition fluctuations are not correlated with temperature or pressure fluctuations, we find that

$$\langle(\Delta\rho)^2\rangle = (\rho^*)^2 \,[\alpha^2 \langle(\Delta T)^2\rangle - 2\alpha\kappa \langle\Delta T\Delta p\rangle + \kappa^2 \langle(\Delta p)^2\rangle] + \sum_{i=1}^{r} \sum_{j=1}^{r} \Gamma_i\Gamma_j \langle\Delta m_i \Delta m_j\rangle \tag{3.1.59a}$$

$$= \kappa(\rho^*)^2\, kT^*/V^* + (v_M{}^* kT^*/V^*) \sum_{i=1}^{r} \sum_{j=1}^{r} \Gamma_i\Gamma_j \frac{\Delta_{ij}}{|\mu|}$$

and

$$\langle\Delta\rho\Delta T\rangle = -\,\alpha\rho^* \langle(\Delta T)^2\rangle + \kappa\rho^* \langle\Delta p\Delta T\rangle = 0 \tag{3.1.59b}$$

where Eqs. (3.1.55) and (3.1.56) have been used. Equation (3.1.59b) agrees with Eq. (3.1.44b), indicating that density and temperature fluctuations occur independently in open systems as well as in closed systems. However, Eq. (3.1.59a) differs from Eq. (3.1.44a) for closed systems by the term associated with composition fluctuations. The two equations agree for one-component systems in which no composition fluctuation is possible.

c. **Isochoric Systems-2: T, V, μ_D (chemical potentials of diffusible components), and c' (molar concentrations of nondiffusible components) as State Variables.** In the most general case, a polymer solution may be described as one which consists of diffusible low-molecular-weight components (which constitute a mixed solvent) and nondiffusible polymer components. This classification of components, made in § 2.3.4, was not explicitly taken into account in the preceding sections a and b. The analysis described below treats diffusible and nondiffusible components on a different footing, as was first done by Shogenji and Ooi [4]. Thus we choose, besides T and V, the chemical potentials μ_D of diffusible components and the molar concentrations c' of nondiffusible components as the independent state variables to describe a small region within the solution. For this choice of state variables V must not fluctuate, i.e., the small region must be isochoric, since otherwise the size of the system and hence its extensive properties are left indeterminate.

For simplicity, we show the process of calculation for the case in which the solution is composed of one diffusible component 0 and r nondiffusible components. Thus our set of state variables is T, V, μ_0, and c' ($c_1, c_2, \ldots, c_r$). By definition, V does not fluctuate. Fluctuations of S, c_0, and μ_i ($i = 1, 2, \ldots, r$), correct to the first order in ΔT, $\Delta\mu_0$, and $\Delta c'$, are as follows:

$$\Delta S = \left(\frac{\partial S}{\partial T}\right)_{V,\mu_0,c'} \Delta T + \left(\frac{\partial S}{\partial \mu_0}\right)_{T,V,c'} \Delta\mu_0 + \sum_{j=1}^{r} \left(\frac{\partial S}{\partial c_j}\right)_{T,V,\mu_0,c_k} \Delta c_j \tag{3.1.60a}$$

$$\Delta c_0 = \left(\frac{\partial c_0}{\partial T}\right)_{V, \mu_0, c'} \Delta T + \left(\frac{\partial c_0}{\partial \mu_0}\right)_{T, V, c'} \Delta \mu_0 + \sum_{j=1}^{r} \left(\frac{\partial c_0}{\partial c_j}\right)_{T, V, \mu_0, c_k} \Delta c_j \tag{3.1.60b}$$

$$\Delta \mu_i = \left(\frac{\partial \mu_i}{\partial T}\right)_{V, \mu_0, c'} \Delta T + \left(\frac{\partial \mu_i}{\partial \mu_0}\right)_{T, V, c'} \Delta \mu_0 + \sum_{j=1}^{r} \left(\frac{\partial \mu_i}{\partial c_j}\right)_{T, V, \mu_0, c_k} \Delta c_j \tag{3.1.60c}$$

All the derivatives in these equations can be expressed in terms of the second derivatives of the characteristic function $\mathcal{A}(T, V, \mu_0, n')$ for constant-μ_0 systems. Here n' denotes the set of composition variables $n_1, n_2, \ldots, n_r$. Substitution of Eqs. (3.1.60a) through (3.1.60c) into Eq. (3.1.47) and use of Eqs. (2.3.51) and (2.3.54) yield

$$\begin{aligned} w(\Delta T, \Delta\mu_0, \Delta c') &= C \exp\left\{\frac{1}{2kT^*}\left[\mathcal{A}_{TT}(\Delta T)^2 + 2\mathcal{A}_{T0}\,\Delta T \Delta\mu_0 + \mathcal{A}_{00}(\Delta\mu_0)^2\right]\right. \\ &\quad \left. - \frac{V}{2kT^*}\sum_{i=1}^{r}\sum_{j=1}^{r}\left(\frac{\partial \mu_i}{\partial c_j}\right)_{T, V, \mu_0, c_k} \Delta c_i \Delta c_j\right\} \end{aligned} \tag{3.1.61}$$

where

$$\mathcal{A}_{TT} \equiv (\partial^2 \mathcal{A}/\partial T^2)_{V, \mu_0, n'}, \quad \mathcal{A}_{00} \equiv (\partial^2 \mathcal{A}/\partial \mu_0{}^2)_{T, V, n'},$$
$$\mathcal{A}_{T0} \equiv (\partial^2 \mathcal{A}/\partial T \partial \mu_0)_{V, n'} = \mathcal{A}_{0T} \tag{3.1.62}$$

With Eq. (3.1.61) the mean correlation between Δc_i and Δc_j can be calculated as

$$\langle \Delta c_i \Delta c_j \rangle = \frac{kT^*}{V} \frac{(\Delta_c)_{ij}}{|\mu_c|}, \quad (i, j = 1, 2, \ldots, r) \tag{3.1.63a}$$

where $|\mu_c|$ is a determinant of order r with elements $\mu_{ij}{}^{(c)}$ defined by

$$\mu_{ij}{}^{(c)} \equiv (\partial \mu_i/\partial c_j)_{T, \mu_0, c_k} = \mu_{ji}{}^{(c)} \tag{3.1.63b}$$

and $(\Delta_c)_{ij}$ is the cofactor of $\mu_{ij}{}^{(c)}$. The corresponding equations expressed in terms of mass concentrations C_i are

$$\langle \Delta C_i \Delta C_j \rangle = \frac{kT^*}{V} \frac{(\Delta_C)_{ij}}{|\mu_C|}, \quad (i, j = 1, 2, \ldots, r) \tag{3.1.64a}$$

and

$$\mu_{ij}{}^{(C)} \equiv (1/M_i)(\partial \mu_i/\partial C_j)_{T, \mu_0, C_k} \tag{3.1.64b}$$

Using these results for the mean correlation between concentration fluctuations of nondiffusible components, we can derive $\langle \Delta c_0 \, \Delta c_i \rangle$, but this will not be shown here.

Equation (3.1.63a) should be compared with Eq. (3.1.52). Though they are formally different, these equations represent the same physical quantity $\langle \Delta c_i \, \Delta c_j \rangle$. Note that the order of $|\mu_c|$ is lower by one than that of $|A|$ and that the elements of $|\mu_c|$ consist of derivatives involving the chemical potentials and concentrations of nondiffusible components only, which is not the case with the elements of $|A|$.

By use of Eq. (3.1.61) the following averages are also derived:

$$\langle (\Delta T)^2 \rangle = -\frac{kT^* \mathcal{A}_{00}}{\mathcal{A}_{TT}\mathcal{A}_{00} - \mathcal{A}_{T0}^{\ 2}} \tag{3.1.65a}$$

$$\langle \Delta T \, \Delta \mu_0 \rangle = \frac{kT^* \mathcal{A}_{T0}}{\mathcal{A}_{TT}\mathcal{A}_{00} - \mathcal{A}_{T0}^{\ 2}} \tag{3.1.65b}$$

$$\langle (\Delta \mu_0)^2 \rangle = -\frac{kT^* \mathcal{A}_{TT}}{\mathcal{A}_{TT}\mathcal{A}_{00} - \mathcal{A}_{T0}^{\ 2}} \tag{3.1.65c}$$

where the relation $\mathcal{A}_{T0} = \mathcal{A}_{0T}$ has been used. Application of Eqs. (1.2.25), (A.13), (2.3.51), (2.3.53), and (3.1.62) yields

$$\begin{aligned} C_V/T &= (\partial S/\partial T)_{V, n_0, n'} \\ &= \left(\frac{\partial S}{\partial T}\right)_{V, \mu_0, n'} - \left(\frac{\partial S}{\partial \mu_0}\right)_{T, V, n'} \frac{(\partial n_0/\partial T)_{V, \mu_0, n'}}{(\partial n_0/\partial \mu_0)_{T, V, n'}} \\ &= -\mathcal{A}_{TT} + (\mathcal{A}_{T0}^{\ 2}/\mathcal{A}_{00}) \end{aligned} \tag{3.1.66}$$

If this is substituted into Eq. (3.1.65a), we recover Eq. (3.1.41), which again confirms that the mean-square temperature fluctuation is the same for both closed and open systems.

An analysis similar to that described above can be carried out for solutions which contain more than one diffusible component. With components 0, 1, $\ldots, d$ diffusible and components $d+1, d+2, \ldots, d+s$ nondiffusible, the set of state variables to be chosen is $T, V, \mu_0, \mu_1, \ldots, \mu_d, c_{d+1}, c_{d+2}, \ldots, c_{d+s}$. Then the mean fluctuation between concentration fluctuations of nondiffusible components is expressed by the same relation as Eq. (3.1.63a), with $i, j = d+1, d+2, \ldots, d+s$ and with the subscript μ_0 in Eq. (3.1.63b) replaced by μ_D. Thus, in this general case, $|\mu_c|$ is a determinant of order s and, again, its elements are derivatives involving the chemical potentials and concentrations of nondiffusible components only.

3.1.4. Density Fluctuations Near the Critical Point

The isothermal compressibility κ of a liquid diverges at the gas-liquid critical point. Hence density fluctuations become extremely large as this point is approached and fluctuations at different points will tend to be correlated. Since the determinant $|\mu|$ vanishes at the critical point, concentration fluctuations also become very strongly correlated. In this section, we discuss the coupling effect for density fluctuations in a one-component system. Inasmuch as we wish to deal with density fluctuations at points separated by any distance, we consider a closed system of macroscopic fixed volume V. As has been shown, temperature and density fluctuations take place independently. Therefore, we assume in what follows that the temperature is uniform throughout the system and has the equilibrium value T^*.

We divide the system into a number of very small regions of constant volume and apply Eq. (3.1.47) to each of them (each small region is an isochoric open system). Thus for the α-th small region we have

$$w(x_\alpha) = C_\alpha \exp\left(-\frac{\Delta U_\alpha - T^* \Delta S_\alpha - \mu^* \Delta n_\alpha}{kT^*}\right) \qquad (3.1.67a)$$

The probability $w(x)$ that the entire system is in the state of density fluctuation specified by x $(x_1, x_2, \ldots)$ is obtained by multiplying together the $w(x_\alpha)$ for all the small regions and using the condition that the entire system is closed. The result is

$$w(x) = C \exp\left(-\frac{\Delta_\rho A}{kT^*}\right) \qquad (3.1.67b)$$

where $\Delta_\rho A$ represents the increase in the Helmholtz free energy of the system over the mean value which accompanies the density fluctuation x. If the corresponding local free energy increase per unit volume at point $\underset{\sim}{r}$ is designated by $\Delta_\rho A_\phi(\underset{\sim}{r})$, then $\Delta_\rho A$ can be written

$$\Delta_\rho A = \int_V \Delta_\rho A_\phi(\underset{\sim}{r})\, d\underset{\sim}{r} \qquad (3.1.68)$$

We may expand $\Delta_\rho A_\phi$ in powers of the density fluctuation $\Delta\rho$ as $\Delta_\rho A_\phi = (\partial A_\phi/\partial\rho)_T\ \Delta\rho + (1/2)\ (\partial^2 A_\phi/\partial\rho^2)_T\ (\Delta\rho)^2 + \ldots$. Substituting this into Eq. (3.1.68), considering again that the system is closed, and ignoring terms higher than the square of the density fluctuation, we obtain

$$\Delta_\rho A = \frac{1}{2\kappa(\rho^*)^2} \int_V (\Delta\rho)^2\, d\underset{\sim}{r} \qquad (3.1.69)$$

where ρ^* is the mean density of the system.

At the critical point, where κ is infinite, the right-hand side of Eq. (3.1.69) vanishes. Therefore, the contribution of higher terms in $\Delta\rho$ to $\Delta_\rho A_\phi$ must be taken into account to obtain $w(x)$ in the vicinity of this point. If density fluctuations were to take place uniformly in the system, the leading nonzero term would become proportional to $(\Delta\rho)^4$ [2]. However, such a mode of density fluctuation can be expected only in very small systems, and nonuniform density fluctuations may occur in systems of macroscopic size. In this case, the spatial gradient of $\Delta\rho$ becomes nonzero and the expansion of $\Delta_\rho A_\phi$ includes not only various powers of $\Delta\rho$ but also those of $\underset{\sim}{\nabla}(\Delta\rho)$, where $\underset{\sim}{\nabla}$ stands for the gradient operator [see Eq. (A.34)]. As will be illustrated in § 3.3.8, $[\underset{\sim}{\nabla}(\Delta\rho)]^2$ rather than $(\Delta\rho)^4$ dominates $\Delta_\rho A_\phi$. Thus, a first approximation to $\Delta_\rho A$ in the vicinity of the critical point would be [2]

$$\Delta_\rho A = \frac{1}{(\rho^*)^2} \int_V \left\{ \frac{1}{2\kappa} (\Delta\rho)^2 + \frac{b}{2} [\underset{\sim}{\nabla}(\Delta\rho)]^2 \right\} d\underset{\sim}{r} \tag{3.1.70}$$

where b is a positive constant.

To evaluate the integral in Eq. (3.1.70) we assume, without loss of generality, that the system under consideration is a parallelpiped with the sides X, Y, and Z. Then V is equal to XYZ. We expand $\Delta\rho$ as a function of the position vector $\underset{\sim}{r}$ in a Fourier series

$$\Delta\rho(\underset{\sim}{r}) = \Sigma_{\underset{\sim}{k}}\, \xi_{\underset{\sim}{k}} \exp(i\underset{\sim}{k} \cdot \underset{\sim}{r}) \tag{3.1.71a}$$

where

$$\xi_{\underset{\sim}{k}} = \frac{1}{V} \int_V \Delta\rho(\underset{\sim}{r}) \exp(-i\underset{\sim}{k} \cdot \underset{\sim}{r})\, d\underset{\sim}{r} \tag{3.1.71b}$$

and $\underset{\sim}{k}$ is a vector with x, y, and z components given by

$$k_x = 2\pi n_x/X, \quad k_y = 2\pi n_y/Y, \quad k_z = 2\pi n_z/Z \tag{3.1.71c}$$

n_x, n_y, and n_z being integers. The sum in Eq. (3.1.71a) is thus extended over all of $n_x, n_y, n_z = 0, \pm1, \pm2, \ldots$. From Eq. (3.1.71a) it follows that

$$\underset{\sim}{\nabla}(\Delta\rho) = \Sigma_{\underset{\sim}{k}}\, i\underset{\sim}{k}\, \xi_{\underset{\sim}{k}} \exp(i\underset{\sim}{k} \cdot \underset{\sim}{r}) \tag{3.1.72}$$

Substitution of this, together with Eq. (3.1.71a), into Eq. (3.1.70), followed by integration, yields

$$\Delta_\rho A = \frac{V}{2(\rho^*)^2} \Sigma_{\underset{\sim}{k}} \left(\frac{1}{\kappa} + bk^2 \right) \xi_{\underset{\sim}{k}}\, \xi_{\underset{\sim}{k}}^*, \quad (k = |\underset{\sim}{k}|) \tag{3.1.73}$$

where $\xi_{\underset{\sim}{k}}^*$ denotes the complex conjugate of $\xi_{\underset{\sim}{k}}$. With $w(x)$ obtained by inserting this into Eq. (3.1.67b), the following result can be derived:

$$\langle \xi_{\underset{\sim}{k}} \xi_{\underset{\sim}{k}}^* \rangle = \frac{(\rho^*)^2 k_B T^*}{V(\kappa^{-1} + bk^2)} \tag{3.1.74a}$$

$$\langle \xi_{\underset{\sim}{k}} \xi_{\underset{\sim}{k}'}^* \rangle = 0, \quad (\underset{\sim}{k} \neq \underset{\sim}{k}') \tag{3.1.74b}$$

where the Boltzmann constant has been designated k_B to distinguish it from the running index k.

Using these results, we can calculate the mean correlation of density fluctuations at two different points, $\underset{\sim}{r}^\alpha$ and $\underset{\sim}{r}^\beta$, in the system

$$\begin{aligned}
&\langle \Delta\rho(\underset{\sim}{r}^\alpha)\, \Delta\rho(\underset{\sim}{r}^\beta) \rangle \equiv \langle \Delta\rho^\alpha\, \Delta\rho^\beta \rangle \\
&\quad = \langle \Sigma_{\underset{\sim}{k}}\, \xi_{\underset{\sim}{k}}\, \xi_{\underset{\sim}{k}}^* \exp\,(i\,\underset{\sim}{k} \cdot \underset{\sim}{r}^{\alpha\beta}) \rangle \\
&\quad = \frac{V}{(2\pi)^3} \int \langle \xi_{\underset{\sim}{k}}\, \xi_{\underset{\sim}{k}}^* \rangle \exp\,(i\,\underset{\sim}{k} \cdot \underset{\sim}{r}^{\alpha\beta})\, d\underset{\sim}{k} \\
&\quad = \frac{(\rho^*)^2 k_B T^*}{8\pi^3 b} \int_0^\infty \int_0^\pi \frac{1}{(b\kappa)^{-1} + k^2} \exp\,(ikr^{\alpha\beta} \cos\theta)\, 2\pi k^2 \sin\theta\, dk\, d\theta \\
&\quad = \frac{(\rho^*)^2 k_B T^*}{2\pi^2 b} \int_0^\infty \frac{k^2}{(b\kappa)^{-1} + k^2} \frac{\sin(kr^{\alpha\beta})}{kr^{\alpha\beta}}\, dk \\
&\quad = \frac{(\rho^*)^2 k_B T^*}{4\pi b r^{\alpha\beta}} \exp\left[-\frac{r^{\alpha\beta}}{(b\kappa)^{1/2}}\right]
\end{aligned} \tag{3.1.75}$$

where $\underset{\sim}{r}^{\alpha\beta} = \underset{\sim}{r}^\beta - \underset{\sim}{r}^\alpha$, $r^{\alpha\beta} = |\underset{\sim}{r}^{\alpha\beta}|$, and θ is the angle between vectors $\underset{\sim}{k}$ and $\underset{\sim}{r}^{\alpha\beta}$. The replacement of the sum by an integral in the above derivation is permitted, because V is of macroscopic size.

A mean distance L over which the coupling of density fluctuations extends about a given point may be defined by

$$L^2 \equiv \int_V (r^{\alpha\beta})^2 \langle \Delta\rho^\alpha\, \Delta\rho^\beta \rangle\, d\underset{\sim}{r}^{\alpha\beta} \Big/ \int_V \langle \Delta\rho^\alpha\, \Delta\rho^\beta \rangle\, d\underset{\sim}{r}^{\alpha\beta} \tag{3.1.76}$$

Introduction of Eq. (3.1.75) yields

$$L^2 = 6b\kappa \tag{3.1.77}$$

which indicates that in the region near the critical point where κ diverges to infinity, the correlation of density fluctuations in liquid becomes so strong that the correlation length L becomes of macroscopic order.

3.1.5. Fluctuations of Molecular Number Density and Distribution Functions

We can introduce various statistical mechanical concepts through a consideration of fluctuations. To show this consider a one-component monatomic fluid. In anticipation of the extension to polymeric systems, some special notations are introduced to describe the geometry of the system. The arrangement of a monatomic molecule σ in space is determined only by the Cartesian coordinates of its center of mass $\underset{\sim}{r}_\sigma = x_\sigma, y_\sigma, z_\sigma$. For a polyatomic molecule, in addition to $\underset{\sim}{r}_\sigma$, the coordinates of its constituting atoms must be given in order to determine its configuration. The set of coordinates determining the position and configuration of molecule σ is denoted by the symbol (σ) and the associated volume element is denoted by $d(\sigma)$. Thus for a monatomic molecule, we have

$$(\sigma) = \underset{\sim}{r}_\sigma = x_\sigma, y_\sigma, z_\sigma, \quad d(\sigma) = d\underset{\sim}{r}_\sigma = dx_\sigma\, dy_\sigma\, dz_\sigma \tag{3.1.78}$$

However, the usual notations $\underset{\sim}{r}_\sigma$ and $d\underset{\sim}{r}_\sigma$ also will be used as needed to indicate explicitly that we are concerned with the center of mass of the molecule, either monatomic or polyatomic. Thus both (σ) and $\underset{\sim}{r}_\sigma$ may appear in the same equation.

Consider a very large system which has a volume V_t and contains N_t monatomic molecules of the same kind. These molecules are numbered $1, 2, \ldots, N_t$, and the probability of finding a molecule σ in a volume element $d\underset{\sim}{r}_\sigma$ at a point $\underset{\sim}{r}_\sigma$ is denoted by $V_t^{-1} F_1(\sigma) d(\sigma)$. Further, the probability of finding a molecule σ in $d\underset{\sim}{r}_\sigma$ at $\underset{\sim}{r}_\sigma$ and, at the same instant, a molecule τ in $d\underset{\sim}{r}_\tau$ at $\underset{\sim}{r}_\tau$ is denoted by $V_t^{-2} F_2(\sigma, \tau)\, d(\sigma) d(\tau)$. Then it follows from these definitions that

$$V_t^{-1} \int_{V_t} F_1(\sigma)\, d(\sigma) = 1 \tag{3.1.79}$$

$$V_t^{-2} \int_{V_t} F_2(\sigma, \tau)\, d(\sigma)\, d(\tau) = 1 \tag{3.1.80}$$

$$F_1(\sigma) = V_t^{-1} \int_{V_t} F_2(\sigma, \tau)\, d(\tau) \tag{3.1.81}$$

The functions $F_1(\sigma)$ and $F_2(\sigma, \tau)$ are termed the **one-body molecular distribution function** and the **two-body molecular distribution function**, respectively. For molecules σ and τ which are located far apart we have the relation $F_2(\sigma, \tau) = F_1(\sigma) F_1(\tau)$, because the two molecules have no influence on each other. Thus we introduce a new function called the **molecular pair correlation function** $g_2(\sigma, \tau)$. The definition is

$$g_2(\sigma, \tau) \equiv F_2(\sigma, \tau) - F_1(\sigma) F_1(\tau) \tag{3.1.82}$$

This function vanishes for molecules σ and τ separated by a large distance. Since all the molecules in the system under consideration are identical, $F_1(\sigma)$, $F_2(\sigma, \tau)$, and $g_2(\sigma, \tau)$ have the same functional forms for every molecule and for every molecular pair. Furthermore, when a system of monatomic molecules is macroscopically homogeneous, then

$$F_1(\sigma) = 1 \tag{3.1.83}$$

Before proceeding further, note that in the literature the functions $f_1(\underset{\sim}{r}^{\alpha})$, $f_2(\underset{\sim}{r}^{\alpha}, \underset{\sim}{r}^{\beta})$, and $g_2(\underset{\sim}{r}^{\alpha}, \underset{\sim}{r}^{\beta})$ defined in the footnote† are also referred to as the one-body and two-body molecular distribution functions and the molecular pair correlation function, respectively.

Next, we consider a small domain of volume V in the system and examine the fluctuation of the molecular number density η that takes place in it. The remaining large region acts as an environmental medium for the small domain (see § 3.1.2). For convenience, we introduce a position function $m(\underset{\sim}{r})$ which has the property [5]:

$$m(\underset{\sim}{r}) = \begin{cases} 1 & \text{for } \underset{\sim}{r} \text{ inside } V \\ 0 & \text{for } \underset{\sim}{r} \text{ outside } V \end{cases} \tag{3.1.84}$$

Given the positions of all molecules, $\underset{\sim}{r}_\sigma$ ($\sigma = 1, 2, \ldots, N_t$), at a given instant, the number of molecules N present in V is written in terms of $m(\underset{\sim}{r})$ as

$$N = \sum_{\sigma=1}^{N_t} m(\underset{\sim}{r}_\sigma) \tag{3.1.85}$$

†The function $f_1(\underset{\sim}{r}^{\alpha})$ is the probability density of finding any one molecule at position $\underset{\sim}{r}^{\alpha}$ in volume V_t containing N_t identical molecules. It is related to $F_1(\sigma)$ by

$$f_1(\underset{\sim}{r}^{\alpha}) = V_t^{-1} \sum_{\sigma=1}^{N_t} [F_1(\underset{\sim}{r}_\sigma)]_{\underset{\sim}{r}_\sigma = \underset{\sim}{r}^{\alpha}} = (N_t/V_t) F_1(\underset{\sim}{r}^{\alpha})$$

The function $f_2(\underset{\sim}{r}^{\alpha}, \underset{\sim}{r}^{\beta})$ is the probability density of finding any two molecules simultaneously at positions $\underset{\sim}{r}^{\alpha}$ and $\underset{\sim}{r}^{\beta}$, and can be written in terms of $F_2(\sigma, \tau)$ as

$$f_2(\underset{\sim}{r}^{\alpha}, \underset{\sim}{r}^{\beta}) = [N_t(N_t - 1)/V_t^2]\, F_2(\underset{\sim}{r}^{\alpha}, \underset{\sim}{r}^{\beta})$$

The function $g_2(\underset{\sim}{r}^{\alpha}, \underset{\sim}{r}^{\beta})$ is defined, with $N_t \gg 1$, by

$$(N_t/V_t)^2\, g_2(\underset{\sim}{r}^{\alpha}, \underset{\sim}{r}^{\beta}) \equiv f_2(\underset{\sim}{r}^{\alpha}, \underset{\sim}{r}^{\beta}) - f_1(\underset{\sim}{r}^{\alpha}) f_1(\underset{\sim}{r}^{\beta})$$

Since $[m(\underset{\sim}{r})]^2 = m(\underset{\sim}{r}), N^2$ is represented by

$$
\begin{aligned}
N^2 &= \sum_{\sigma=1}^{N_t} \sum_{\tau=1}^{N_t} m(\underset{\sim}{r}_\sigma)\, m(\underset{\sim}{r}_\tau) \\
&= \sum_{\sigma=1}^{N_t} m(\underset{\sim}{r}_\sigma) + \sum_{\sigma=1}^{N_t} \sum_{\substack{\tau=1 \\ (\sigma \neq \tau)}}^{N_t} m(\underset{\sim}{r}_\sigma)\, m(\underset{\sim}{r}_\tau)
\end{aligned}
\tag{3.1.86}
$$

The mean molecular-number density $\langle \eta \rangle$ in V is therefore given by

$$
\begin{aligned}
\langle \eta \rangle &= \frac{\langle N \rangle}{V} = \frac{N_t}{V V_t} \int_{V_t} m(\underset{\sim}{r}_\sigma) F_1(\underset{\sim}{r}_\sigma)\, d\underset{\sim}{r}_\sigma \\
&= \frac{N_t}{V V_t} \int_V F_1(\sigma)\, d\sigma = \frac{N_t}{V_t}
\end{aligned}
\tag{3.1.87}
$$

The mean-square $\langle \eta^2 \rangle$ is calculated as follows:

$$
\begin{aligned}
\langle \eta^2 \rangle &= \frac{\langle N^2 \rangle}{V} \\
&= \frac{\langle N \rangle}{V^2} + \frac{N_t (N_t - 1)}{V^2 V_t^2} \int_{V_t} m(\underset{\sim}{r}_\sigma)\, m(\underset{\sim}{r}_\tau)\, F_2(\underset{\sim}{r}_\sigma, \underset{\sim}{r}_\tau)\, d\underset{\sim}{r}_\sigma\, d\underset{\sim}{r}_\tau \\
&= \frac{\langle \eta \rangle}{V} + \frac{\langle \eta \rangle^2}{V^2} \int_V F_2(\sigma, \tau)\, d(\sigma)\, d(\tau)
\end{aligned}
\tag{3.1.88}
$$

Therefore, the mean square of the number-density fluctuation, $\Delta\eta \equiv \eta - \langle \eta \rangle$, is given by

$$
\langle (\Delta\eta)^2 \rangle = \langle \eta^2 \rangle - \langle \eta \rangle^2 = \frac{\langle \eta \rangle}{V} \left[1 + \frac{\langle \eta \rangle}{V} \int_V g_2(\sigma, \tau)\, d(\sigma)\, d(\tau) \right]
\tag{3.1.89}
$$

If this is compared with Eq. (3.1.44a), we find that

$$
\eta \kappa kT - 1 = \eta \left[\frac{1}{V} \int_V g_2(\sigma, \tau)\, d(\sigma)\, d(\tau) \right]
\tag{3.1.90}
$$

where η means $\langle \eta \rangle$. This relation between κ and $g_2(\sigma, \tau)$ is called the **Ornstein-Zernike equation** [6].

In very dilute gas, i.e., in the limit $\eta \to 0$, the presence of other molecules has no effect on the correlation between molecules σ and τ. Hence if the potential of intermolecular force is designated by $u(r)$, we may put

$$g_2{}^o(r) = \exp[-u(r)/kT] - 1 \tag{3.1.91}$$

where r is the distance between molecules σ and τ, and the superscript circle indicates an infinitely dilute gas. Assuming for $u(r)$ a form like that shown in Fig. 3.2 and putting

$$u(r) = \begin{cases} \infty \text{ for } r \leqslant d \\ -c/r^m \text{ for } r > d \text{ and } m > 3 \end{cases} \tag{3.1.92}$$

we obtain

$$\begin{aligned} G^o &\equiv \frac{1}{V} \int_V g_2{}^o(\sigma, \tau)\, d(\sigma)\, d(\tau) \\ &= -\frac{4\pi d^3}{3}\left[1 - \frac{3c}{(m-3)\, d^m\, kT}\right] \end{aligned} \tag{3.1.93}$$

provided that $u(r) \ll kT$.

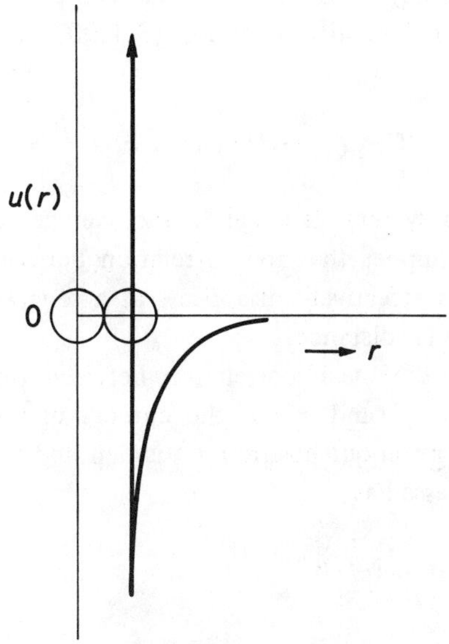

Fig. 3.2. Intermolecular interaction potential $u(r)$ defined by Eq. (3.1.92).

Now, Eq. (2.1.2) for gas pressure is used to calculate κ, and $g_2(\sigma, \tau)$ is expanded in powers of the density. Then it follows from the Ornstein-Zernike equation that the second virial coefficient of the gas, $A_2{}^g$, is expressed by

$$A_2{}^g = -\frac{N_A}{2M^2V} \int_V g_2{}^o(\sigma, \tau)\, d(\sigma)\, d(\tau) = -\frac{N_A}{2M^2} G^o \qquad (3.1.94)$$

Substitution of Eq. (3.1.93) gives

$$A_2{}^g = \frac{N_A}{2M^2}\left(\frac{4\pi d^3}{3}\right)\left(1 - \frac{T_B}{T}\right) \qquad (3.1.95)$$

with

$$T_B = \frac{3c}{(m-3)\, d^m k} \qquad (3.1.96)$$

Equation (3.1.95) agrees with the expression of $A_2{}^g$ for a spherical molecule of diameter d, which is given by Eq. (2.1.3) with b represented by Eq. (2.1.4). If $m \leqslant 3$, the case exemplified by the Coulomb force, G^o diverges, and the effect of other molecules on $g_2{}^o(\sigma, \tau)$ is no longer negligible, even at the limit of zero density. The intermolecular force with $m > 3$ and that with $m \leqslant 3$ are called **short-range** and **long-range**, respectively. Throughout this volume we tacitly confine our discussion to systems in which intermolecular and intramolecular forces are short-range. It is to be noted, however, that the Ornstein-Zernike equation is valid even for long-range intermolecular forces.

For condensed liquid systems the isothermal compressibility is so small that the first term on the left-hand side of Eq. (3.1.90) may be dropped, and we have

$$G \equiv \frac{1}{V} \int_V g_2(\sigma, \tau)\, d(\sigma)\, d(\tau) = -\frac{1}{\eta} = -\frac{V_t}{N_t} \qquad (3.1.97)$$

i.e., $-G$ for condensed systems is equal to the average volume occupied by a single molecule. This implies that the correlation between a pair of molecules in a condensed system effectively disappears at a separation comparable with the average intermolecular distance.

Next, we calculate the mean correlation between fluctuations of number density at two points $\underset{\sim}{r}^\alpha$ and $\underset{\sim}{r}^\beta$. If the number of molecules contained in the volume element $d\underset{\sim}{r}^\alpha$ about position $\underset{\sim}{r}^\alpha$ is denoted by $\eta^\alpha d\underset{\sim}{r}^\alpha$, the number density η^α can be expressed as

$$\begin{aligned} \eta^\alpha &= \lim_{V^\alpha \to 0} N^\alpha / V^\alpha \\ &= \sum_{\sigma=1}^{N_t} \left[\lim_{V^\alpha \to 0} \frac{m^\alpha(\underset{\sim}{r}_\sigma)}{V^\alpha} \right] = \sum_{\sigma=1}^{N_t} \delta(\underset{\sim}{r}_\sigma - \underset{\sim}{r}^\alpha) \end{aligned} \qquad (3.1.98)$$

where δ is the Dirac delta function, N^α/V^α is the molecular number density in a volume V^α around $\underset{\sim}{r}^\alpha$, and Eq. (3.1.85) has been used to express N^α. The product of number densities at two points $\underset{\sim}{r}^\alpha$ and $\underset{\sim}{r}^\beta$ are therefore given by

$$\eta^\alpha \eta^\beta = \sum_{\substack{\sigma=1 \\ (\sigma \neq \tau)}}^{N_t} \sum_{\tau=1}^{N_t} \delta(\underset{\sim}{r}_\sigma - \underset{\sim}{r}^\alpha)\,\delta(\underset{\sim}{r}_\tau - \underset{\sim}{r}^\beta) \tag{3.1.99}$$

Hence we get

$$\begin{aligned} \langle \eta^\alpha \eta^\beta \rangle &= \frac{N_t(N_t - 1)}{V_t^2} \int_{V_t} \delta(\underset{\sim}{r}_\sigma - \underset{\sim}{r}^\alpha)\,\delta(\underset{\sim}{r}_\tau - \underset{\sim}{r}^\beta) \\ &\times F_2(\underset{\sim}{r}_\sigma, \underset{\sim}{r}_\tau)\, d\underset{\sim}{r}_\sigma\, d\underset{\sim}{r}_\tau = \langle \eta \rangle^2 F_2(\underset{\sim}{r}^\alpha, \underset{\sim}{r}^\beta) \end{aligned} \tag{3.1.100}$$

where $N_t - 1$ has been approximated by N_t. With Eq. (3.1.100) we find that

$$\langle \Delta\eta^\alpha \Delta\eta^\beta \rangle = \langle \eta^\alpha \eta^\beta \rangle - \langle \eta \rangle^2 = \langle \eta \rangle^2 g_2(\underset{\sim}{r}^\alpha, \underset{\sim}{r}^\beta) \tag{3.1.101}$$

By comparison with Eq. (3.1.75) it follows that

$$g_2(r) = \frac{kT}{4\pi b r} \exp\left[-\frac{r}{(b\kappa)^{1/2}}\right] \tag{3.1.102}$$

In the vicinity of the critical point, κ is so large that the pair correlation function $g_2(r)$ decays very slowly with intermolecular distance r. In fact, the mean correlation persists up to a distance far beyond the average separation of molecules. In such a case, since κ is very large, the second term on the left-hand side of Eq. (3.1.90) may be neglected so that

$$\kappa k T \approx \int g_2(r)\, d\underset{\sim}{r} \tag{3.1.103}$$

3.1.6. The Kirkwood-Buff Theory of Solutions [7]

This section describes an extension of the theory presented in the preceding section to multicomponent systems. Consider a system of $r + 1$ nonreacting components in volume V_t. The molecules of component i are numbered 1, 2, $\ldots, \sigma_i, \ldots, N_{it}$. The probability of finding molecule σ_i in a volume element $d(\sigma_i)$ at position (σ_i) is $V_t^{-1} F_1(\sigma_i)\, d(\sigma_i)$, and the joint probability that molecule τ_j of component j is simultaneously in a volume element $d(\tau_j)$ at (τ_j) is denoted by $V_t^{-2} F_2(\sigma_i, \tau_j)\, d(\sigma_i)\, d(\tau_j)$. Relations like Eqs. (3.1.79) through (3.1.81) hold for the distribution functions $F_1(\sigma_i)$ and $F_2(\sigma_i, \tau_j)$. We also define the spatial correlation function for a pair of molecules σ_i and τ_j by

$$g_2(\sigma_i, \tau_j) \equiv F_2(\sigma_i, \tau_j) - F_1(\sigma_i) F_1(\tau_j) \tag{3.1.104}$$

Taking a volume in the system and denoting the number of molecules of component i in it by N_i, we have in place of Eq. (3.1.86)

$$N_i^2 = \sum_{\sigma_i=1}^{N_{it}} m(\underset{\sim}{r}_{\sigma_i}) + \sum_{\sigma_i=1}^{N_{it}} \sum_{\substack{\tau_i=1 \\ (\sigma_i \neq \tau_i)}}^{N_{it}} m(\underset{\sim}{r}_{\sigma_i}) m(\underset{\sim}{r}_{\tau_i}) \tag{3.1.105a}$$

$$N_i N_j = \sum_{\sigma_i=1}^{N_{it}} \sum_{\tau_j=1}^{N_{jt}} m(\underset{\sim}{r}_{\sigma_i}) m(\underset{\sim}{r}_{\tau_j}), \quad (i \neq j) \tag{3.1.105b}$$

Hence we get

$$\langle \eta_i \rangle = \frac{N_{it}}{V_t V} \int_V F_1(\sigma_i) \, d(\sigma_i) = \frac{N_{it}}{V_t} \tag{3.1.106a}$$

$$\langle \eta_i \eta_j \rangle = \frac{\langle \eta_i \rangle}{V} \delta_{ij} + \frac{\langle \eta_i \rangle \langle \eta_j \rangle}{V^2} \int_V F_2(\sigma_i, \tau_j) \, d(\sigma_i) \, d(\tau_j) \tag{3.1.106b}$$

where η_i is the number density of component i defined by $\eta_i = N_i/V$ and δ_{ij} is the Kronecker delta. Also for the fluctuation $\Delta\eta_i = \eta_i - \langle \eta_i \rangle$, we find

$$\langle \Delta\eta_i \, \Delta\eta_j \rangle = \frac{\langle \eta_i \rangle}{V} (\delta_{ij} + \langle \eta_j \rangle G_{ij}), \qquad (i, j = 0, 1, \ldots, r) \tag{3.1.107}$$

where

$$G_{ij} \equiv \frac{1}{V} \int_V g_2(\sigma_i, \tau_j) \, d(\sigma_i) \, d(\tau_j) \tag{3.1.108}$$

The functional forms of $F_2(\sigma_i, \tau_j)$ and $g_2(\sigma_i, \tau_j)$ depend on the kinds of components i and j, as is indicated by the subscript ij on G. When the interaction between these components is repulsive, G_{ij} is negative, while when the interaction is attractive, and thus promotes association, G_{ij} is positive. For this reason G_{ij} is called the **cluster integral** for the pair of components i and j. If Eq. (3.1.107) is compared with the correlations of concentration fluctuations derived in § 3.1.3, it is possible to relate various thermodynamic quantities with cluster integrals. For this purpose there are three methods which correspond to Eqs. (3.1.50), (3.1.55), and (3.1.63a).

a. Method Using Eq. (3.1.50) for Isochoric Systems. Introduction of $n_i = \eta_i V/N_A$ into Eq. (3.1.50) gives

$$\langle \Delta\eta_i \, \Delta\eta_j \rangle = \frac{N_A RT}{V^2} \frac{(\Delta_A)_{ij}}{|A|}, \qquad (i, j = 0, 1, \ldots, r) \tag{3.1.109}$$

Equating this with Eq. (3.1.107), we find

$$J_{ij} = J_{ji} \equiv \eta_i \delta_{ij} + \eta_i \eta_j G_{ij} = \frac{N_A RT}{V} \frac{(\Delta_A)_{ij}}{|A|} \tag{3.1.110}$$

This relation shows that a symmetric matrix with $VJ_{ij}/(N_A RT)$ as its i, j element is the inverse of a symmetric matrix $\underset{\sim}{A}$ whose i, j element is A_{ij} defined by Eq. (3.1.51). Therefore, $\underset{\sim}{A}$ is the inverse of the former matrix, so that

$$A_{ij} \equiv \left(\frac{\partial \mu_i}{\partial n_j}\right)_{T, V, n_k} = \frac{N_A RT}{V} \frac{(\Delta_J)_{ij}}{|J|}, \qquad (i, j = 0, 1, \ldots, r) \tag{3.1.111}$$

where $|J|$ is a determinant of order $r + 1$ whose i, j element is J_{ij} and $(\Delta_J)_{ij}$ is the cofactor of element J_{ij}. Since $(\partial \mu_i/\partial n_j)_{T, V, n_k} = (\partial \mu_i/\partial n_j)_{T, p, n_k} + V_i V_j/(\kappa V)$, Eq. (3.1.111) can be rewritten

$$\left(\frac{\partial \mu_i}{\partial n_j}\right)_{T, p, n_k} + \frac{V_i V_j}{\kappa V} = \frac{N_A RT}{V} \frac{(\Delta_J)_{ij}}{|J|} \tag{3.1.112}$$

Both sides are multiplied by n_i and summed over i from 0 to r to give

$$V_j = \frac{\kappa RT}{|J|} \sum_{i=0}^{r} \eta_i (\Delta_J)_{ij} \tag{3.1.113}$$

where the Gibbs-Duhem relation and $V = \Sigma_i n_i V_i$ have been used. With further multiplication by η_j and summation over j from 0 to r, it follows that

$$\kappa RT = N_A |J| \Big/ \sum_{i=0}^{r} \sum_{j=0}^{r} \eta_i \eta_j (\Delta_J)_{ij} \tag{3.1.114}$$

Elimination of κRT from Eqs. (3.1.113) and (3.1.114) yields

$$V_i/N_A = \sum_{j=0}^{r} \eta_j (\Delta_J)_{ij} \Big/ \sum_{j=0}^{r} \sum_{k=0}^{r} \eta_j \eta_k (\Delta_J)_{jk} \tag{3.1.115}$$

If this is introduced into Eq. (3.1.112), we obtain an equation which relates μ_i to the cluster integrals G_{ij}.

b. Method Using Eq. (3.1.55) for Constant-n_0 Systems. The molality m_i and number density η_i are related by

$$m_i = \frac{\eta_i}{\eta_0 M_0} = \frac{v_M{}^2}{N_A} \eta_i \tag{3.1.116a}$$

so that

$$\frac{\Delta m_i}{m_i} = \frac{\Delta \eta_i}{\eta_i} - \frac{\Delta \eta_0}{\eta_0} \tag{3.1.116b}$$

where, as before, v_M denotes the volume of the system containing unit mass of the principal solvent (component 0). Thus we have

$$\langle \Delta m_i \Delta m_j \rangle = \frac{v_M{}^2 \eta_i \eta_j}{N_A{}^2} \left[\frac{\langle \Delta\eta_i \Delta\eta_j \rangle}{\eta_i \eta_j} - \frac{\langle \Delta\eta_i \Delta\eta_0 \rangle}{\eta_i \eta_0} - \frac{\langle \Delta\eta_j \Delta\eta_0 \rangle}{\eta_j \eta_0} + \frac{\langle (\Delta\eta_0)^2 \rangle}{\eta_0{}^2} \right] \tag{3.1.117}$$

Substituting Eqs. (3.1.107) and using Eq. (3.1.55), we obtain

$$\eta_i \eta_j Q_{ij} = \frac{N_A RT}{v_M} \frac{\Delta_{ij}}{|\mu|} \qquad (i, j = 1, 2, \ldots, r) \tag{3.1.118}$$

with Q_{ij} defined by

$$Q_{ij} \equiv (\delta_{ij}/\eta_j) + (1/\eta_0) + G_{ij} + G_{00} - G_{i0} - G_{j0} \tag{3.1.119}$$

Equations (3.1.118) can be solved for $\mu_{ij} = (\partial \mu_i / \partial m_j)_{T,p,m_k}$ to obtain

$$\mu_{ij} = \frac{N_A RT}{v_M} \frac{(\Delta_Q)_{ij}}{\eta_i \eta_j |Q|} \tag{3.1.120}$$

where $(\Delta_Q)_{ij}$ is the cofactor of element Q_{ij} of the r-order determinant $|Q|$.

For two-component systems Eq. (3.1.115) gives

$$V_0 = \frac{1 + c_1 N_A (G_{11} - G_{10})}{c_0 + c_1 + c_0 c_1 N_A (G_{00} + G_{11} - 2G_{10})} \tag{3.1.121}$$

since $c_i = \eta_i / N_A$. With the relation $(\partial\mu_1/\partial m_1)_{T,p} = (c_0 V_0 / v_M)\,(\partial\mu_1/\partial c_1)_{T,p}$ and Eq. (3.1.121), Eq. (3.1.120) yields

$$\left(\frac{\partial \mu_1}{\partial c_1}\right)_{T,p} = \frac{RT}{c_1 [1 + c_1 N_A (G_{11} - G_{10})]} \tag{3.1.122}$$

The corresponding expressions for V_1 and $(\partial\mu_0/\partial c_0)_{T,p}$ are obtained by exchanging subscripts 0 and 1.

c. **Method Using Eq. (3.1.63a) for Isochoric Systems.** If Eq. (3.1.63a) is equated to Eq. (3.1.107) after the molar concentration c_i is replaced by the molecular number density η_i, we obtain

$$J_{ij} = N_A RT\,(\Delta_c)_{ij}/|\mu_c|,\qquad (i,j = 1,2,\ldots,r) \tag{3.1.123}$$

This set of equations may be solved for $\mu_{ij}^{(c)} \equiv (\partial\mu_i/\partial c_j)_{T,\mu_0,c_k}$ to obtain

$$\mu_{ij}^{(c)} = N_A RT\,\frac{(\Delta_{J'})_{ij}}{|J'|} \tag{3.1.124}$$

Here $|J'|$ is a determinant of order r obtained by deleting the first column and first row of $|J|$, and $(\Delta_{J'})_{ij}$ is the cofactor of element J_{ij} in $|J'|$.

For a mixed-solvent system in which components 0, 1, . . . , d are diffusible and components $d+1, d+2, \ldots, d+s$ are nondiffusible, we obtain

$$\mu_{ij}^{(c)} \equiv \left(\frac{\partial\mu_i}{\partial c_j}\right)_{T,\mu_D,c_k} = N_A RT\,\frac{(\Delta_s)_{ij}}{|J_s|},\qquad (i,j = d+1,d+2,\ldots,d+s) \tag{3.1.125}$$

where $|J_s|$ is a determinant of order s whose elements are the J_{ij} for the s nondiffusible components and $(\Delta_s)_{ij}$ is the cofactor of element J_{ij}. The osmotic pressure π^* measured at constant T and μ_D obeys Eq. (2.3.77a). Substitution of Eq. (3.1.125) into this equation gives

$$\left(\frac{\partial\pi^*}{\partial c_j}\right)_{T,\mu_D,c_k} = \frac{N_A RT}{|J_s|}\sum_{i=d+1}^{d+s} c_i\,(\Delta_s)_{ij} \tag{3.1.126}$$

For two-component systems Eqs. (3.1.125) and (3.1.126) reduce to

$$\left(\frac{\partial\mu_1}{\partial c_1}\right)_{T,\mu_0} = \frac{RT}{c_1\,(1 + c_1 N_A G_{11})} \tag{3.1.127}$$

$$\left(\frac{\partial\pi^*}{\partial c_1}\right)_{T,\mu_0} = \frac{RT}{1 + c_1 N_A G_{11}} \tag{3.1.128}$$

From the latter we get

$$\frac{\pi^*}{RT} = c_1 - \tfrac{1}{2} N_A G_{11}{}^o c_1{}^2 + \ldots \tag{3.1.129}$$

where $G_{11}{}^o$ denotes the infinite-dilution value of G_{11}. If c_1 is replaced by mass concentration $C_1 = M_1 c_1$, this series gives for the osmotic second virial coefficient

$$A_2 = -\frac{N_A}{2M_1{}^2 V} \int_V [F_2{}^o(\sigma_1, \tau_1) - F_1{}^o(\sigma_1) F_1{}^o(\tau_1)]\, d(\sigma_1)\, d(\tau_1) \tag{3.1.130}$$

The theory outlined above is due to Kirkwood and Buff [7].

Zimm [8] has shown that for two-component systems with very small κ a simple expression for $(\partial\mu_1/\partial c_1)_{T,p}$ can be derived from Eq. (3.1.128). If c_1 is regarded as a function of T, p, and μ_0 in a two-component system, we have

$$(\partial c_1/\partial p)_{T,m_1} = (\partial c_1/\partial p)_{T,\mu_0} + (\partial c_1/\partial\mu_0)_{T,p}(\partial\mu_0/\partial p)_{T,m_1} \tag{3.1.131}$$

But the following relations can be derived:

$$\left(\frac{\partial c_1}{\partial p}\right)_{T,m_1} = \left[\frac{\partial(n_1/V)}{\partial p}\right]_{T,n_0,n_1} = -\frac{n_1}{V^2}\left(\frac{\partial V}{\partial p}\right)_{T,n_0,n_1} = c_1\kappa \tag{3.1.132a}$$

$$\left(\frac{\partial c_1}{\partial\mu_0}\right)_{T,p} = \left(\frac{\partial c_1}{\partial\mu_1}\right)_{T,p}\left(\frac{\partial\mu_1}{\partial\mu_0}\right)_{T,p} = -\frac{c_0}{c_1}\left(\frac{\partial c_1}{\partial\mu_1}\right)_{T,p} \tag{3.1.132b}$$

and with these, Eq. (3.1.131) becomes

$$\left(\frac{\partial c_1}{\partial p}\right)_{T,\mu_0} = c_1\kappa + \frac{c_0 V_0}{c_1}\left(\frac{\partial c_1}{\partial\mu_1}\right)_{T,p} \approx \frac{c_0 V_0}{c_1}\left(\frac{\partial c_1}{\partial\mu_1}\right)_{T,p} \tag{3.1.133}$$

where the condition that κ be very small has been taken into account. Since $(\partial\pi^*/\partial c_1)_{T,\mu_0} = (\partial p/\partial c_1)_{T,\mu_0}$, it follows from Eqs. (3.1.128) and (3.1.133) that

$$\left(\frac{\partial\mu_1}{\partial c_1}\right)_{T,p} = \frac{c_0 V_0 RT}{c_1(1 + c_1 N_A G_{11})} \tag{3.1.134}$$

which is the desired Zimm expression. While Eq. (3.1.122) contains two cluster integrals G_{11} and G_{10}, this expression for $(\partial\mu_1/\partial c_1)_{T,p}$ contains only G_{11}. This fact simplifies molecular theoretical interpretation of the thermodynamic quantity $(\partial \mu_1/\partial c_1)_{T,p}$. In passing, we note that the right-hand sides of Eqs. (3.1.122) and (3.1.134) converge to the same limit RT/c_1 as c_1 approaches zero.

Exchanging the subscripts 0 and 1 in the Zimm expression, we get

$$\left(\frac{\partial \mu_0}{\partial c_0}\right)_{T,p} = \frac{c_1 V_1 RT}{c_0 (1 + c_0 N_A G_{00})} \tag{3.1.135}$$

which gives, when solved for G_{00},

$$\frac{N_A G_{00}}{V_0} = RT \frac{c_1 V_1}{c_0^2 V_0} \left(\frac{\partial c_0}{\partial \mu_0}\right)_{T,p} - \frac{1}{c_0 V_0} \tag{3.1.136}$$

If c_i is replaced by the volume fraction $\phi_i = V_i c_i$, with the concentration dependence of V_0 and V_1 neglected, Eq. (3.1.136) is written

$$\begin{aligned} \frac{N_A G_{00}}{V_0} &= RT \left(\frac{\phi_1}{\phi_0}\right) \left(\frac{\partial \ln \phi_0}{\partial \mu_0}\right)_{T,p} - \frac{1}{\phi_0} \\ &= - \phi_1 \left[\frac{\partial}{\partial a_0} \left(\frac{a_0}{\phi_0}\right)\right]_{T,p} - 1 \end{aligned} \tag{3.1.137}$$

where a_0 is the activity of component 0.

Figure 3.3 illustrates relations between the solvent activity a_0 and the solvent volume fraction ϕ_0 for three polymer + solvent systems. The values of G_{00}/V_0 computed from these data by Zimm and Lundberg [9] using Eq. (3.1.137) are plotted against ϕ_0 in Fig. 3.4. From the definition of G_{00} we see that G_{00}/V_0 vanishes if there is no interaction between solvent molecules. For a hypothetical ideal solution which consists of solvent molecules and polymer segments, ϕ_0 becomes the mole fraction of the solvent, so that a_0 is equal to ϕ_0, and it follows from Eq. (3.1.137) that $G_{00} = -V_0/N_A$, which is just the molecular volume of pure solvent. According to Flory's expression for μ_0 [see Eq. (2.1.23)], $\ln (a_0/\phi_0)$ is equal to ϕ_1 if $\chi_H = 0$ and $P \gg 1$. In this case, Eq. (3.1.137) yields

$$G_{00}/V_0 = 0 \tag{3.1.138}$$

Since this is larger than G_{00}/V_0 for the hypothetical ideal solution, we see that chain connectivity of polymer molecules increases the probability for solvent clustering. The observed values of G_{00}/V_0 for rubber in benzene and for polystyrene in toluene, shown in Fig. 3.4, are about 1 to 2, which implies that the clustering tendency of solvent molecules in these systems is greater than is predicted from Flory's theory. It has been shown that such a difference can

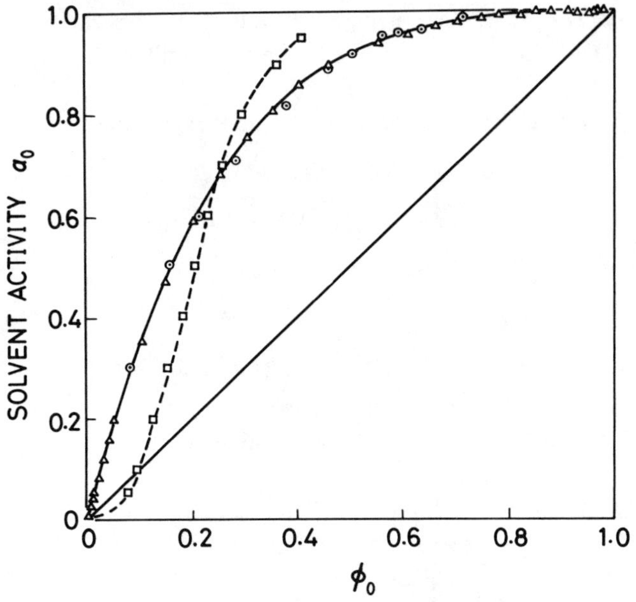

Fig. 3.3. Solvent activity a_0 as a function of solvent volume fraction ϕ_0 at 25°C for three systems: Δ, natural rubber + benzene [10]; ⊙, polystyrene + toluene [9]; □, collagen + water [11].

be explained theoretically by a thorough consideration of chain connectivity [12]. In Fig. 3.4, the behavior of G_{00}/V_0 for the system collagen + water at low water concentrations is shown to be quite different from that of the other systems shown. This difference may be ascribed, in part if not all, to selective adsorption of water molecules at discrete sites on the collagen molecule, because such an effect should appreciably lower the probability of water-water contacts below the value expected for collagen segments and water molecules mixed at random. Useful information of this sort will become available if thermodynamic data are properly related to cluster integrals.

3.2. Light Scattering from Gas and Liquid

3.2.1. Light Scattering from an Ideal Gas – The Theory of Rayleigh

When light impinges on a molecule, a dipole is induced by the action of the electric field $\underset{\sim}{E}_0$ of the incident beam. The dipole moment $\underset{\sim}{p}$ is proportional to $\underset{\sim}{E}_0$, so that we may write

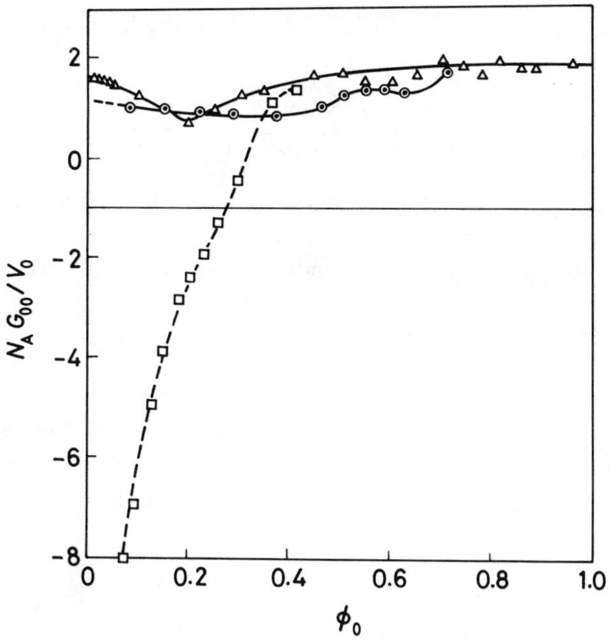

Fig. 3.4. Solvent cluster integral G_{00} in the three systems of Fig. 3.3.

$$\underset{\sim}{p} = \alpha \underset{\sim}{E}_0 \tag{3.2.1}$$

where α is called the **polarizability** of the molecule. If the molecule is optically isotropic, $\underset{\sim}{p}$ and $\underset{\sim}{E}_0$ have the same orientation, and α is a scalar quantity. The dipole oscillates at the same frequency as $\underset{\sim}{E}_0$ and radiates an electromagnetic wave. This is the basic mechanism of the light scattering phenomenon.

Suppose that monochromatic plane-polarized light traveling in the z direction impinges on an optically isotropic molecule in a gas. Its electric field is

$$\underset{\sim}{E}_0 = \underset{\sim}{E}_0{}^o \exp\,(i\omega t - ikz) \tag{3.2.2}$$

with

$$k \equiv 2\pi/\lambda \tag{3.2.3}$$

Here $\underset{\sim}{E}_0{}^o$ is the amplitude vector of $\underset{\sim}{E}_0$, ω is the angular frequency, and λ is the wavelength of the light in the gas. With the origin of the Cartesian coordinates x, y, z taken at the molecule, the dipole moment $\underset{\sim}{p}$ induced in the molecule is represented by

$$\underset{\sim}{p} = \underset{\sim}{p}^o \exp\,(i\omega t) \quad \text{with} \quad \underset{\sim}{p}^o = \alpha \underset{\sim}{E}_0{}^o \tag{3.2.4}$$

According to Eq. (A.108), the electric field $\underset{\sim}{E}$ of the light scattered from this oscillating dipole is given by

$$\underset{\sim}{E} = -\frac{1}{4\pi\epsilon_0}\left(\frac{2\pi}{\lambda_0}\right)^2 \frac{\underset{\sim}{r} \times (\underset{\sim}{r} \times \underset{\sim}{p}^o)}{r^3} \exp(i\omega t - ikr) \qquad (3.2.5)$$

where ϵ_0 is the permittivity of vacuum, λ_0 is the wavelength of incident light in vacuum, and $\underset{\sim}{r}$ is the position vector of the point of observation P. With the angle between the vectors $\underset{\sim}{r}$ and $\underset{\sim}{p}^o$ denoted by γ, Eq. (3.2.5) gives

$$E \equiv |\underset{\sim}{E}| = \frac{1}{4\pi\epsilon_0}\left(\frac{2\pi}{\lambda_0}\right)^2 \left(\frac{p^o \sin\gamma}{r}\right) \cos(\omega t - kr) \qquad (3.2.6)$$

where $p^o = |\underset{\sim}{p}^o|$. As is illustrated in Fig. 3.5, the vector product $\underset{\sim}{r} \times \underset{\sim}{p}^o$ is normal to the shaded plane containing $\underset{\sim}{r}$ and $\underset{\sim}{p}^o$. Hence the vector $\underset{\sim}{E}^o = -\underset{\sim}{r} \times (\underset{\sim}{r} \times \underset{\sim}{p}^o) = (\underset{\sim}{r} \times \underset{\sim}{p}^o) \times \underset{\sim}{r}$ lies in this shaded plane and points forward from

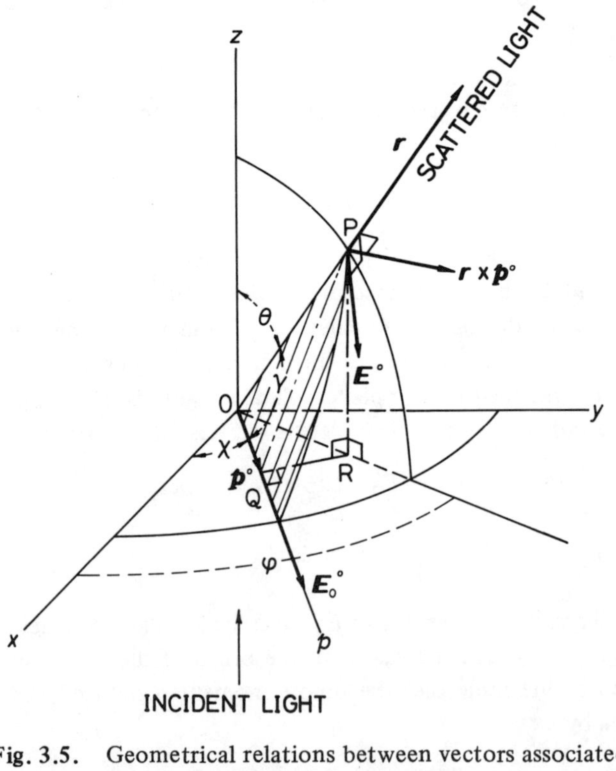

Fig. 3.5. Geometrical relations between vectors associated with incident and scattered light.

the sheet. We set the p axis in the direction of $\underset{\sim}{p}^o$ and drop perpendiculars from point P to the p axis and to the x, y plane to determine points Q and R. Then the angle $\angle$ OQR is 90°. Therefore, if the polar angles of $\underset{\sim}{r}$ are denoted by θ and φ and the angle between the p and x axes by χ, the angle γ is related to these angles by

$$\cos\gamma = \sin\theta\cos(\varphi - \chi) \tag{3.2.7}$$

For unpolarized incident light any orientation of $\underset{\sim}{E}_0{}^o$ in the x, y plane, i.e., the angle χ, is equally probable, so that

$$\langle\sin^2\chi\rangle = \langle\cos^2\chi\rangle = \tfrac{1}{2}, \quad \langle\sin\chi\cos\chi\rangle = 0 \tag{3.2.8a}$$

$$\langle\sin^2\gamma\rangle = \tfrac{1}{2}(1 + \cos^2\theta) \tag{3.2.8b}$$

where $\langle\cdot\cdot\cdot\rangle$ means the isotropic average over χ from 0 to 360°. Indicating the time average by a bar, we have

$$\overline{\cos^2(\omega t - kr)} = 1/2 \tag{3.2.8c}$$

Thus it follows from Eq. (3.2.6) that the intensity i of scattered light received at point P is given by

$$i = \epsilon\langle\overline{E^2}\rangle = \frac{\pi^2}{2r^2\epsilon_0{}^2\lambda_0{}^4}\alpha^2(1 + \cos^2\theta)I_0 \tag{3.2.9}$$

with

$$I_0 = \epsilon\overline{E_0{}^2} = \tfrac{1}{2}\epsilon(E_0{}^o)^2 \tag{3.2.10}$$

when the incident beam is unpolarized light. Here ϵ is the permittivity of the gas, $E_0{}^o = |\underset{\sim}{E}_0{}^o|$, and I_0 is the intensity of the incident beam. The angle θ, that the scattered light makes with the direction of the incident ray, is called the **scattering angle**.

Now, the gas is assumed to be ideal. Each molecule behaves independently of other molecules. Therefore, the intensity I of light scattered per unit volume at the coordinate origin is given by the product of i and the molecular number density (or concentration) η:

$$I = \frac{\pi^2}{2r^2\epsilon_0{}^2\lambda_0{}^4}\alpha^2(1 + \cos^2\theta)\eta I_0 \tag{3.2.11}$$

We introduce a quantity called **Rayleigh's ratio** $R(\theta)$ defined by

$$R(\theta) \equiv \frac{I(r, \theta)r^2}{I_0(1 + \cos^2\theta)} \tag{3.2.12}$$

Equation (3.2.11) gives for $R(\theta)$

$$R(\theta) = \frac{\pi^2}{2\lambda_0^4}\left(\frac{\alpha}{\epsilon_0}\right)^2 \eta \qquad (3.2.13)$$

The intensity of light transmitting a gas is attenuated as a result of scattering. The decrement τ in transmitted intensity per unit path length is expressed by

$$\tau = \frac{1}{I_0}\int_0^{\pi} I(r, \theta)\, 2\pi r^2 \sin\theta\, d\theta \qquad (3.2.14)$$

Introduction of Eq. (3.2.11) yields

$$\tau = \frac{8\pi^3}{3\lambda_0^4}\left(\frac{\alpha}{\epsilon_0}\right)^2 \eta \qquad (3.2.15)$$

The quantity τ defined by Eq. (3.2.14) is called the **turbidity** of the medium.

The important feature of Rayleigh's formula (3.2.13) or (3.2.15) is that $R(\theta)$ (or τ) is inversely proportional to λ_0^4. This dependence on wavelength explains why when sunlight penetrates the atmosphere, light scattered by gas molecules is blue, while transmitted light is reddish. Although light may also be scattered by particles suspended in the air, such as water droplets and dust, this kind of scattering has a minor effect, as can be seen from the fact that we can look up at a sky of deeper blue at high altitudes where the air is cleaner. In 1869 before Rayleigh's theory appeared, Tyndall had found that light scattered from colloid solutions assumed a blue color and that it was almost completely plane-polarized at a scattering angle of 90°. For this reason the name **Tyndall-Rayleigh scattering** is given to light scattering characterized by the λ_0^{-4} dependence. The polarization of scattered light is discussed in the next section.

In ideal gases, the polarizability α and the refractive index $\tilde{n}$ are related by

$$\tilde{n}^2 - 1 = \left(\frac{\alpha}{\epsilon_0}\right)\frac{N_A}{M}\rho \qquad (3.2.16)$$

so that

$$2\tilde{n}\frac{d\tilde{n}}{d\rho} = \left(\frac{\alpha}{\epsilon_0}\right)\frac{N_A}{M} \qquad (3.2.17)$$

where $\rho = M\eta/N_A$ is the gas density, with M the molar mass of the gas. Elimination of α/ϵ_0 from Eq. (3.2.13) with Eq. (3.2.17) yields

$$R(\theta) = K\left(\frac{d\tilde{n}}{d\rho}\right)^2 M\rho \qquad (3.2.18a)$$

with

$$K \equiv \frac{2\pi^2\tilde{n}^2}{N_A\lambda_0^4} \qquad (3.2.18b)$$

which indicates that M may be determined from measurements of $\rho, \tilde{n}, d\tilde{n}/d\rho$, and $R(\theta)$.

3.2.2. Depolarization and the Cabannes Corrections [13]

In the preceding section, it was assumed that the molecule is isotropic, so that the induced dipole moment vector $\underset{\sim}{p}$ has the same direction as the electric field of the incident beam. If this assumption does not hold, we must start from a general relation

$$\underset{\sim}{p} = \underset{\sim}{\alpha} \cdot \underset{\sim}{E}_0 \tag{3.2.19}$$

where $\underset{\sim}{\alpha}$ is the **polarizability tensor** of the gas molecule. This is a tensor of second rank, and hence may be represented by a dyadic. The dot between $\underset{\sim}{\alpha}$ and $\underset{\sim}{E}_0$ has the meaning explained in § A.4. If the molecule does not absorb light (all the ensuing discussion on light scattering from gas, liquid, and solutions is confined to situations in which this condition holds), $\underset{\sim}{\alpha}$ is a symmetric tensor and can be expressed as

$$\underset{\sim}{\alpha} = \underset{\sim}{i}'\underset{\sim}{i}'\alpha_{x'} + \underset{\sim}{j}'\underset{\sim}{j}'\alpha_{y'} + \underset{\sim}{k}'\underset{\sim}{k}'\alpha_{z'} \tag{3.2.20}$$

in a Cartesian coordinate system (x', y', z') such that the unit vectors $\underset{\sim}{i}'$, $\underset{\sim}{j}'$, and $\underset{\sim}{k}'$ are in the directions of the principal axes of the tensor $\underset{\sim}{\alpha}$. The symbols $\underset{\sim}{i}'\underset{\sim}{i}'$, etc. denote unit dyads (see § A.4). The quantities $\alpha_{x'}$, $\alpha_{y'}$, and $\alpha_{z'}$ are called the eigenvalues of $\underset{\sim}{\alpha}$ or the **principal polarizabilities** of the molecule. The coordinate system (x', y', z') is a molecule-fixed system, and rotates as the molecule rotates. However, the principal polarizabilities are not affected by molecular rotation. We define two parameters α and γ by

$$\alpha \equiv \tfrac{1}{3}(\alpha_{x'} + \alpha_{y'} + \alpha_{z'}) \tag{3.2.21}$$

$$\gamma^2 \equiv \tfrac{1}{18}\left[(\alpha_{x'} - \alpha_{y'})^2 + (\alpha_{y'} - \alpha_{z'})^2 + (\alpha_{z'} - \alpha_{x'})^2\right] \tag{3.2.22}$$

The former may be regarded as the magnitude of the polarizability and the latter as a measure of optical anisotropy of the molecule.

Experimental devices are always set relative to a laboratory-fixed coordinate system (x, y, z). If we denote the orthogonal unit vectors of this system by $\underset{\sim}{i}$, $\underset{\sim}{j}$, and $\underset{\sim}{k}$, we can write $\underset{\sim}{E}_0$ and $\underset{\sim}{p}$ in two ways:

$$\underset{\sim}{E}_0 = \underset{\sim}{i}E_{0x} + \underset{\sim}{j}E_{0y} + \underset{\sim}{k}E_{0z} = \underset{\sim}{i}'E_{0x'} + \underset{\sim}{j}'E_{0y'} + \underset{\sim}{k}'E_{0z'} \tag{3.2.23}$$

$$\underset{\sim}{p} = \underset{\sim}{i}p_x + \underset{\sim}{j}p_y + \underset{\sim}{k}p_z = \underset{\sim}{i}'p_{x'} + \underset{\sim}{j}'p_{y'} + \underset{\sim}{k}'p_{z'} \tag{3.2.24}$$

As before, we choose the direction of incident light as the z axis, and put $E_{0z} = 0$. Then according to Eq. (A.31), the following relations can be written:

$$\left.\begin{aligned} E_{0x'} &= \ell_{xx'} E_{0x} + \ell_{yx'} E_{0y} \\ E_{0y'} &= \ell_{xy'} E_{0x} + \ell_{yy'} E_{0y} \\ E_{0z'} &= \ell_{xz'} E_{0x} + \ell_{yz'} E_{0y} \end{aligned}\right\} \tag{3.2.25}$$

Substituting these, together with Eqs. (3.2.20) and (3.2.23), into Eq. (3.2.19) and using Eq. (A.49), we obtain an expression for the vector $\underset{\sim}{p}$ referred to the molecule-fixed coordinate system:

$$\begin{aligned} \underset{\sim}{p} = {} & \alpha_{x'} (\ell_{xx'} E_{0x} + \ell_{yx'} E_{0y}) \underset{\sim}{i}' + \alpha_{y'} (\ell_{xy'} E_{0x} + \ell_{yy'} E_{0y}) \underset{\sim}{j}' \\ & + \alpha_{z'} (\ell_{xz'} E_{0x} + \ell_{yz'} E_{0y}) \underset{\sim}{k}' \end{aligned} \tag{3.2.26}$$

If Eq. (A.32) is applied, we then get for the x, y, z components of vector $\underset{\sim}{p}$

$$\begin{aligned} p_x = {} & (\alpha_{x'} \ell_{xx'}{}^2 + \alpha_{y'} \ell_{xy'}{}^2 + \alpha_{z'} \ell_{xz'}{}^2) E_{0x} \\ & + (\alpha_{x'} \ell_{xx'} \ell_{yx'} + \alpha_{y'} \ell_{xy'} \ell_{yy'} + \alpha_{z'} \ell_{xz'} \ell_{yz'}) E_{0y} \end{aligned} \tag{3.2.27}$$

$$\begin{aligned} p_y = {} & (\alpha_{x'} \ell_{xx'} \ell_{yx'} + \alpha_{y'} \ell_{xy'} \ell_{yy'} + \alpha_{z'} \ell_{xz'} \ell_{yz'}) E_{0x} \\ & + (\alpha_{x'} \ell_{yx'}{}^2 + \alpha_{y'} \ell_{yy'}{}^2 + \alpha_{z'} \ell_{yz'}{}^2) E_{0y} \end{aligned} \tag{3.2.28}$$

$$\begin{aligned} p_z = {} & (\alpha_{x'} \ell_{xx'} \ell_{zx'} + \alpha_{y'} \ell_{xy'} \ell_{zy'} + \alpha_{z'} \ell_{xz'} \ell_{zz'}) E_{0x} \\ & + (\alpha_{x'} \ell_{yx'} \ell_{zx'} + \alpha_{y'} \ell_{yy'} \ell_{zy'} + \alpha_{z'} \ell_{yz'} \ell_{zz'}) E_{0y} \end{aligned} \tag{3.2.29}$$

Since the molecule under consideration is not fixed in space, any of its properties may be averaged uniformly over all molecular orientations. Thus by applying Eqs. (A.42), (A.44), (A.45), and (A.46) and by expressing the results in terms of α and γ defined by Eqs. (3.2.21) and (3.2.22), we derive

$$\langle p_x \rangle = \alpha E_{0x}, \quad \langle p_y \rangle = \alpha E_{0y}, \quad \langle p_z \rangle = 0 \tag{3.2.30}$$

$$\langle p_x{}^2 \rangle = (\alpha^2 + \tfrac{4}{5} \gamma^2) E_{0x}{}^2 + \tfrac{3}{5} \gamma^2 E_{0y}{}^2 \tag{3.2.31a}$$

$$\langle p_y{}^2 \rangle = \tfrac{3}{5} \gamma^2 E_{0x}{}^2 + (\alpha^2 + \tfrac{4}{5} \gamma^2) E_{0y}{}^2 \tag{3.2.31b}$$

$$\langle p_z{}^2 \rangle = \tfrac{3}{5} \gamma^2 (E_{0x}{}^2 + E_{0y}{}^2) \tag{3.2.31c}$$

and

$$\langle p_x p_y \rangle = \langle p_y p_z \rangle = \langle p_z p_x \rangle = 0 \tag{3.2.32}$$

In order to examine the relation between the polarizations of incident and scattered light we may place a **polarizer** in the incident beam before the scattering molecule, and an **analyzer** before the detector for scattered intensity. When the analyzer is set in such a way that its polarization axis, designated by a unit vector $\underset{\sim}{e}_a$, is perpendicular to the vector $\underset{\sim}{r}$, the electric field strength $\underset{\sim}{E}_a$ of the scattered light at point P in Fig. 3.6 is given by

$$\underset{\sim}{E}_a = \underset{\sim}{e}_a \cdot \underset{\sim}{E} = \frac{\pi}{\epsilon_0 r \lambda_0^2} (\underset{\sim}{e}_a \cdot \underset{\sim}{p}^o) \cos(\omega t - kr) \qquad (3.2.33)$$

where Eq. (A.27) has been used. Again we assume that each molecule in the gas scatters light independently. The scattered intensity I_a per unit volume is then given by

$$I_a = \epsilon \langle \overline{E_a^2} \rangle \eta = \frac{\pi^2 \epsilon}{2r^2 \epsilon_0^2 \lambda_0^4} \langle (\underset{\sim}{e}_a \cdot \underset{\sim}{p}^o)^2 \rangle \eta \qquad (3.2.34)$$

Here, in passing from the first to the second equality, we have averaged the exponential term in Eq. (3.2.5) with respect to time. Hence the symbol $\langle \cdots \rangle$ in the last expression includes both the average over molecular orientation and the average over the orientation of the vector $\underset{\sim}{E}_0$ when the incident beam is unpolarized.

For simplicity, assume that the point of observation P lies on the y, z plane, as illustrated in Fig. 3.6, and consider two cases: in one $\underset{\sim}{e}_a$ points in the direction of the x axis and in the other $\underset{\sim}{e}_a$ lies in the y, z plane and is perpendicular to $\underset{\sim}{r}$.

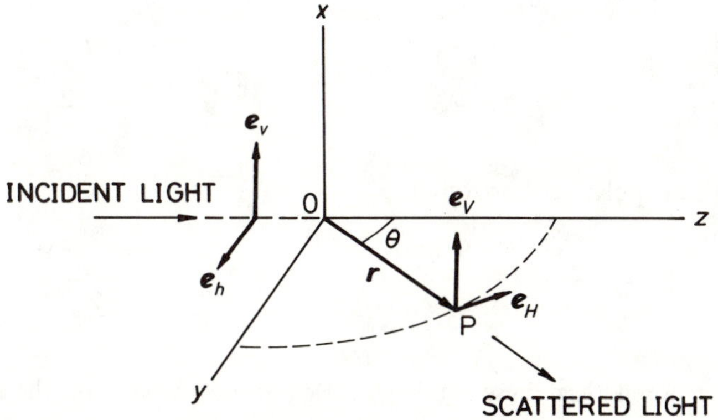

Fig. 3.6. Vertical (v and V) and horizontal (h and H) directions of the polarization axes of polarizer and analyzer, respectively.

The former arrangement is called the vertical direction of the analyzer and designated by subscript V, while the latter is called the horizontal direction of the analyzer and is distinguished by subscript H. The vertical and horizontal directions of the polarizer may be defined analogously with $\underset{\sim}{r}$ replaced by the z axis. They are designated by subscripts v and h, respectively. From these definitions it follows that

$$\underset{\sim}{e}_V = \underset{\sim}{e}_v = \underset{\sim}{e}_x, \quad \underset{\sim}{e}_H = -(\cos\theta)\underset{\sim}{e}_y + (\sin\theta)\underset{\sim}{e}_z, \quad \underset{\sim}{e}_h = \underset{\sim}{e}_y \tag{3.2.35}$$

where $\underset{\sim}{e}$ denotes a unit vector.

First, consider the case with the analyzer vertical. Using Eqs. (3.2.32b) and (3.2.34), we get

$$I_V = \frac{\pi^2 \epsilon}{2r^2 \epsilon_0^2 \lambda_0^4} \left[(\alpha^2 + \tfrac{4}{5}\gamma^2)\langle(E_{0x}{}^o)^2\rangle + \tfrac{3}{5}\gamma^2 \langle(E_{0y}{}^o)^2\rangle\right]\eta \tag{3.2.36}$$

where the superscript zero, as before, denotes an amplitude vector. For unpolarized incident light $E_{0x}{}^o = E_0{}^o \cos\chi$ and $E_{0y}{}^o = E_0{}^o \sin\chi$, where χ is the angle shown in Fig. 3.5. Substituting these relations in Eq. (3.2.36) and averaging uniformly over χ, we get

$$R_{Vu} \equiv \frac{I_{Vu} r^2}{I_{0u}} = \frac{\pi^2}{2\epsilon_0^2 \lambda_0^4}(\alpha^2 + \tfrac{7}{5}\gamma^2)\eta; \quad I_{0u} = \frac{\epsilon (E_0{}^o)^2}{2} \tag{3.2.37}$$

where the subscript u indicates unpolarized incident light. For vertically polarized incident light, $E_{0y}{}^o = 0$ and

$$R_{Vv} \equiv \frac{I_{Vv} r^2}{I_{0v}} = \frac{\pi^2}{\epsilon_0^2 \lambda_0^4}(\alpha^2 + \tfrac{4}{5}\gamma^2)\eta; \quad I_{0v} = \frac{\epsilon (E_{0x}{}^o)^2}{2} \tag{3.2.38}$$

Finally, with the polarizer horizontal

$$R_{Vh} \equiv \frac{I_{Vh} r^2}{I_{0h}} = \frac{\pi^2}{\epsilon_0^2 \lambda_0^4}(\tfrac{3}{5}\gamma^2)\eta; \quad I_{0h} = \frac{\epsilon (E_{0y}{}^o)^2}{2} \tag{3.2.39}$$

It should be noted that these Rayleigh ratios do not depend on the scattering angle θ.

When the analyzer has the $\underset{\sim}{e}_H$ orientation, Eq. (3.2.34) along with Eqs. (3.2.35), (3.2.31c), and (3.2.31b) leads to

$$I_H = \frac{\pi^2 \epsilon}{2r^2 \epsilon_0^2 \lambda_0^4} [\langle(p_y^o)^2\rangle \cos^2\theta + \langle(p_z^o)^2\rangle \sin^2\theta]\, \eta$$

$$= \frac{\pi^2 \epsilon}{2r^2 \epsilon_0^2 \lambda_0^4} \left\{ \tfrac{3}{5}\gamma^2 \langle(E_{0x})^2\rangle + \left[\alpha^2 \cos^2\theta + \frac{\gamma^2}{5}\sin^2\theta\right] \langle(E_{0y})^2\rangle \right\} \eta \tag{3.2.40}$$

Hence

$$R_{Hu} = \frac{\pi^2}{2\epsilon_0^2 \lambda_0^4} \left[\alpha^2 + \frac{\gamma^2}{5}(6 + \cos^2\theta)\right] \eta \tag{3.2.41}$$

$$R_{Hv} = \frac{\pi^2}{\epsilon_0^2 \lambda_0^4} (\tfrac{3}{5}\gamma^2)\eta \tag{3.2.42}$$

$$R_{Hh} = \frac{\pi^2}{\epsilon_0^2 \lambda_0^4} \left[\alpha^2 + \frac{\gamma^2}{5}(3 + \cos^2\theta)\right] \eta \tag{3.2.43}$$

It follows from Eq. (3.2.40) that if the molecule is isotropic ($\gamma = 0$), I_H vanishes at $\theta = \pi/2$, regardless of the state of polarization of the incident light. That is, light scattered in this direction is completely vertically polarized. In gases of anisotropic molecules, I_H at $\theta = \pi/2$ has a nonzero value. This is the phenomenon of **depolarization**. The **degree of depolarization** is expressed by either of the quantities Δ_u or Δ_v defined as

$$\Delta_u \equiv \left(\frac{I_{Hu}}{I_{Vu}}\right)_{\theta=\pi/2} = \frac{6\gamma^2}{5\alpha^2 + 7\gamma^2} \tag{3.2.44}$$

$$\Delta_v \equiv \frac{(I_{Hv})_{\theta=\pi/2}}{(I_{Vv})_{\theta=0}} = \frac{3\gamma^2}{5\alpha^2 + 4\gamma^2} \tag{3.2.45}$$

Since for ideal gases or more precisely for point-mass molecules I_{Vv} has no angular dependence, Δ_v is equal to $(I_{Hv}/I_{Vv})_{\theta=\pi/2}$. But in polymer solutions both I_{Vu} and I_{Vv} may have angular dependences, so that it is more general to define Δ_v by Eq. (3.2.45). The above two equations give

$$\gamma^2/\alpha^2 = 5\Delta_u/(6 - 7\Delta_u) = 5\Delta_v/(3 - 4\Delta_v) \tag{3.2.46}$$

Finally, when incident light is unpolarized and no analyzer is used, I_U is given by the sum of I_V and I_H. Hence

$$R_{Uu} = \frac{\pi^2}{2\epsilon_0^2 \lambda_0^4} \left[\alpha^2(1 + \cos^2\theta) + \frac{\gamma^2}{5}(13 + \cos^2\theta)\right] \eta \tag{3.2.47}$$

which may be used to determine γ^2/α^2 from measurement of R_{Uu} as a function of θ. The $\underset{\sim}{E}_0$ of unpolarized light is invariant in the x, y plane of Fig. 3.6. Therefore, R_{Uu} does not change when point P moves away from the y, z plane. Thus, substituting Eq. (3.2.47) into Eq. (3.2.14) and eliminating α/ϵ_0 by use of Eq. (3.2.17), we get

$$\tau = \int_0^{\pi} R_{Uu}\, 2\pi \sin\theta \, d\theta = \frac{32\pi^3}{3\lambda_0^4 N_A} \left(\tilde{n} \frac{d\tilde{n}}{d\rho} \right)^2 M\rho F_\tau \quad (3.2.48)$$

where

$$F_\tau = 1 + 2 \left(\frac{\gamma}{\alpha} \right)^2 = \frac{6 + 3\Delta_u}{6 - 7\Delta_u} = \frac{3 + 6\Delta_v}{3 - 4\Delta_v} \quad (3.2.49)$$

It also follows from Eqs. (3.2.47) and (3.2.17) that

$$R_{Uu}(\pi/2) = K \left(\frac{d\tilde{n}}{d\rho} \right)^2 M\rho F_{\pi/2} \quad (3.2.50)$$

where

$$F_{\pi/2} = 1 + \frac{13}{5} \left(\frac{\gamma}{\alpha} \right)^2 = \frac{6 + 6\Delta_u}{6 - 7\Delta_u} = \frac{3 + 9\Delta_v}{3 - 4\Delta_v} \quad (3.2.51)$$

Both F_τ and $F_{\pi/2}$ are measures of the optical anisotropy of the molecule, and are referred to as **Cabannes factors** [13].

3.2.3. Density Scattering from Real Gases and Liquids

The theory of light scattering outlined in the preceding sections is concerned with gases of noninteracting molecules, i.e., ideal gases. In real systems, the local electric field at each molecule cannot be simply equated to the electric field of the incident light. The electric fields created by the dipoles of other molecules must be taken into account. Equation (3.2.1) or (3.2.19) is no longer applicable to such systems. We also must consider interference of scattering from different molecules. This effect does not allow us to sum the intensities of scattered light from individual molecules. An appropriate approach to light scattering of actual systems, either gas or liquid, is to regard the system as a continuous dielectric medium and, using electromagnetic theory, to relate fluctuations of local permittivity to light scattering.

Consider a dielectric medium which is macroscopically homogeneous and isotropic, and which does not absorb light. If its permittivity $\underset{\sim}{\epsilon}$ is perfectly uniform, the medium does not scatter light; light travels through it with no attenuation. However, if the medium contains a local fluctuation of $\underset{\sim}{\epsilon}$, light is scattered at that point. The electric field $\underset{\sim}{E}(\underset{\sim}{r}, t)$ of light scattered from a small region of volume V is given by a relation of the same form as Eq. (3.2.5), but now with $\underset{\sim}{p}^o$ defined by

$$\underset{\sim}{p}^o = \int_V (\Delta\underset{\sim}{\epsilon}^\alpha \cdot \underset{\sim}{E}_0{}^o) \exp\,(i\underset{\sim}{s} \cdot \underset{\sim}{r}^\alpha)\, d\underset{\sim}{r}^\alpha \qquad (3.2.52)$$

which is the polarization of the volume V. Here it is assumed that the point of observation P is far from the scattering volume, so that $r \gg V^{1/3}$ $(r = |\underset{\sim}{r}|)$. In Eq. (3.2.52), $\underset{\sim}{s}$ is the scattering vector defined by

$$\underset{\sim}{s} = \underset{\sim}{k} - \underset{\sim}{k}_0 \qquad (3.2.53)$$

with $\underset{\sim}{k}_0$ and $\underset{\sim}{k}$ the wave vectors of the incident light and the scattered light, respectively, and $\Delta\underset{\sim}{\epsilon}^\alpha$ is the fluctuation of $\underset{\sim}{\epsilon}$ at point α labelled by the vector $\underset{\sim}{r}^\alpha$ in the volume V.

Even in a macroscopically isotropic medium, fluctuations of the local permittivity can be anisotropic, so that, in general, $\Delta\underset{\sim}{\epsilon}$ must be regarded as a tensor. However, our analysis begins by assuming that $\Delta\underset{\sim}{\epsilon}$ is isotropic, i.e., it is a scalar quantity $\Delta\epsilon$. In this case, p^o and $\underset{\sim}{E}_0{}^o$ have the same direction, and the calculation indicated in Eqs. (3.2.5) to (3.2.9) can be applied without modification. Thus

$$I = \frac{\pi^2}{2r^2\,\epsilon_0{}^2\,\lambda_0{}^4}\,(1 + \cos^2\theta)\, G_\epsilon(\theta)\, I_0 \qquad (3.2.54)$$

for the scattered intensity I per unit volume. Here $G_\epsilon(\theta)$ is defined by

$$G_\epsilon(\theta) \equiv \frac{1}{V}\,\langle \int_V (\Delta\epsilon^\alpha\, \Delta\epsilon^\beta)\, \exp\,(i\underset{\sim}{s} \cdot \underset{\sim}{r}^{\alpha\beta})\, d\underset{\sim}{r}^\alpha\, d\underset{\sim}{r}^\beta \rangle \qquad (3.2.55)$$

with $\underset{\sim}{r}^{\alpha\beta} = \underset{\sim}{r}^\beta - \underset{\sim}{r}^\alpha$. Using Eq. (A.110), we get

$$\Delta\epsilon = 2\,\epsilon_0\,\tilde{n}\,\Delta\tilde{n} \qquad (3.2.56)$$

where $\tilde{n}$ is the refractive index. From these equations we obtain

$$R(\theta) = \frac{2\pi^2\,\tilde{n}^2}{N_A\,\lambda_0{}^4}\; G_n(\theta) \qquad (3.2.57)$$

where

$$G_n(\theta) = \frac{N_A}{V}\,\langle \int_V (\Delta\tilde{n}^\alpha\, \Delta\tilde{n}^\beta)\, \exp\,(i\underset{\sim}{s} \cdot \underset{\sim}{r}^{\alpha\beta})\, d\underset{\sim}{r}^\alpha\, d\underset{\sim}{r}^\beta \rangle \qquad (3.2.58)$$

In an isotropic medium, the mean correlation of fluctuations of refractive index at different points $\underset{\sim}{r}^\alpha$ and $\underset{\sim}{r}^\beta$, i.e., $\langle\Delta\tilde{n}^\alpha\, \Delta\tilde{n}^\beta\rangle$ depends only on the distance

$r^{\alpha\beta} = |\underset{\sim}{r}^{\alpha\beta}|$. Therefore, if the angle between $\underset{\sim}{r}^{\alpha\beta}$ and $\underset{\sim}{s}$ is denoted by ψ, $G_n(\theta)$ becomes

$$G_n(\theta) = \frac{N_A}{V} \int_V d\underset{\sim}{r}^\alpha \int_0^\infty \int_0^\pi \langle \Delta\tilde{n}^\alpha \Delta\tilde{n}^\beta \rangle \exp(isr^{\alpha\beta} \cos\psi) \times 2\pi (r^{\alpha\beta})^2 \sin\psi \, dr^{\alpha\beta} d\psi \qquad (3.2.59)$$

$$= 4\pi N_A \int_0^\infty \langle \Delta\tilde{n}^\alpha \Delta\tilde{n}^\beta \rangle (r^{\alpha\beta})^2 \frac{\sin(sr^{\alpha\beta})}{sr^{\alpha\beta}} dr^{\alpha\beta}$$

Except in the region near the critical point, the mean correlation length L of $\Delta\tilde{n}$ is of the same order as the effective range ℓ of molecular interactions, i.e., a few times the molecular diameter d. Thus

$$sL = \frac{4\pi L}{\lambda} \sin\left(\frac{\theta}{2}\right) \ll 1 \qquad (3.2.60)$$

If we choose a volume σ so as to satisfy the conditions $L^3 < \sigma$ and $s\sigma^{1/3} \ll 1$ simultaneously, and subdivide the entire system into cells of volume σ, then by the condition $L^3 < \sigma$

$$\langle \Delta\tilde{n}^\alpha \Delta\tilde{n}^\beta \rangle = \begin{cases} \langle (\Delta\tilde{n})^2 \rangle & \text{for } \underset{\sim}{r}^\alpha \text{ and } \underset{\sim}{r}^\beta \text{ in the same cell} \\ 0 & \text{otherwise} \end{cases} \qquad (3.2.61)$$

Also by the condition $s\sigma^{1/3} \ll 1$ we may set $\sin(sr^{\alpha\beta})/(sr^{\alpha\beta})$ equal to unity for $\underset{\sim}{r}^\alpha$ and $\underset{\sim}{r}^\beta$ in the same cell. Then, Eq. (3.2.59) can be simplified to

$$G_n(\theta) = N_A \langle (\Delta\tilde{n})^2 \rangle \sigma \qquad (3.2.62)$$

For a one-component system, if we choose T and density ρ as the state variables, then we have

$$\Delta\tilde{n} = \left(\frac{\partial\tilde{n}}{\partial\rho}\right)_T \Delta\rho \qquad (3.2.63)$$

Substituting this into Eq. (3.2.62) and applying Eq. (3.1.44a), with $V^* = \sigma$, we obtain

$$G_n(\theta) = \left(\frac{\partial\tilde{n}}{\partial\rho}\right)_T^2 \rho^2 \kappa RT \qquad (3.2.64)$$

Note that σ has disappeared. Thus Eq. (3.2.57) becomes

$$R(\theta) = K\kappa RT \left(\rho \frac{\partial\tilde{n}}{\partial\rho}\right)_T^2 \qquad (3.2.65a)$$

with

$$K \equiv \frac{2\pi^2 \tilde{n}^2}{N_A \lambda_0^4} \qquad (3.2.65b)$$

Equation (3.2.65a) is the **Smoluchowski-Einstein formula for density scattering** [14].

The theory developed above is based on the assumption that $\Delta \underset{\sim}{\epsilon}$ is a scalar. In general, this is not valid, and anisotropies of molecular shape and molecular interactions bring about an anisotropic fluctuation of local permittivity. Molecular-theoretical treatments of $\Delta \underset{\sim}{\epsilon}$ are exceedingly difficult. However, if $\Delta \underset{\sim}{\epsilon}$ is a symmetric tensor and correlation of anisotropic contributions to $\Delta \underset{\sim}{\epsilon}$ extends only over a distance far less than L, anisotropic effects of $\Delta \underset{\sim}{\epsilon}$ can be treated phenomenologically in a very simple way. That is, we may perform a theoretical calculation similar to that described in § 3.2.2, paying attention to the point that $\underset{\sim}{p}^o$ in Eq. (3.2.5) is not parallel to $\underset{\sim}{E}_0{}^o$. Actually, Eqs. (3.2.27) through (3.2.51) for polarized scattered light apply as they stand. Thus from Eq. (3.2.47) for R_{Uu} and Eq. (3.2.46) for γ^2/α^2 we find that the Smoluchowski-Einstein formula corrected for molecular anisotropy is

$$R(\theta) = K\kappa RT \left(\rho \frac{\partial \tilde{n}}{\partial \rho}\right)_T^2 \left[1 + \frac{\Delta_v}{3 - 4\Delta_v}\left(\frac{13 + \cos^2\theta}{1 + \cos^2\theta}\right)\right] \quad (3.2.66)$$

For example, experimental data [15] on benzene (23°C, λ_0 = 436 nm) give $R(\pi/2) = 45.6 \times 10^{-6}$ cm^{-1}, $\Delta_u = 0.42$, and $\Delta_v = 0.265$. Hence, the second term in the brackets, i.e., the Cabannes correction, amounts to 2.78 at $\theta = 90°$. In general, the Cabannes corrections for liquids of small molecules are of this magnitude.

For dilute gases, elimination of κ from Eq. (3.2.65a) with the aid of Eq. (3.1.90), followed by use of Eq. (3.1.94) yields

$$R(\theta) = K\left(\frac{\partial \tilde{n}}{\partial \rho}\right)_T^2 M\rho\,[1 - 2A_2{}^g M\rho + O(\rho^2)] \quad (3.2.67)$$

This reduces to Eq. (3.2.18a) in the limit of vanishingly small density, i.e., as the ideal gas is approached.

In the vicinity of the critical point, at which κ becomes indefinitely large, very strong scattering of light is observed. This phenomenon is called **critical opalescence**. As the critical point is approached, the mean correlation length L for density fluctuations increases so that the condition in Eq. (3.2.60) is no longer obeyed except at scattering angles very close to zero and Eq. (3.2.64) ceases to be applicable.

If Eq. (3.2.63) is used in Eq. (3.2.59), we obtain

$$G_n(\theta) = 4\pi N_A \left(\frac{\partial \tilde{n}}{\partial \rho}\right)_T^2 \int_0^\infty \langle \Delta\rho^\alpha \Delta\rho^\beta \rangle \frac{\sin(sr^{\alpha\beta})}{sr^{\alpha\beta}} (r^{\alpha\beta})^2\, dr^{\alpha\beta} \quad (3.2.68)$$

In the vicinity of the critical point, $\langle \Delta\rho^\alpha \Delta\rho^\beta \rangle$ is given by Eq. (3.1.75), i.e.,

$$\langle \Delta\rho^\alpha \Delta\rho^\beta \rangle = \frac{\rho^2 kT}{4\pi b r^{\alpha\beta}} \exp\left[-\frac{r^{\alpha\beta}}{(b\kappa)^{1/2}}\right] \tag{3.2.69}$$

Substituting this into Eq. (3.2.68), carrying out the integration, and using Eq. (3.1.77), we obtain

$$G_n(\theta) = \left(\frac{\partial \tilde{n}}{\partial \rho}\right)_T^2 \frac{\rho^2 \kappa kT}{1 + (1/6) s^2 L^2} \tag{3.2.70}$$

In the limit $\theta \to 0$, i.e., $s \to 0$, this expression agrees with Eq. (3.2.64). For nonzero θ, sL becomes indefinitely large as the critical point is approached, and hence G_n tends to proportionality with λ^2. The wavelength dependence of critical opalescence thus changes from λ^{-4} type to the λ^{-2} type as the scattering angle increases.

3.2.4. Critical Exponents

The van der Waals equation (2.1.1) is the most famous of the equations of state which predict a gas-liquid phase transition. It may be put in the form

$$p = \frac{RT}{V_m - b} - \frac{a}{V_m{}^2} \tag{3.2.71}$$

where $V_m = V/n$ is the molar volume of the gas. The critical point of a gas is determined by imposing the two conditions:

$$(\partial p/\partial V_m)_T = 0, \quad (\partial^2 p/\partial V_m{}^2)_T = 0 \tag{3.2.72}$$

For the van der Waals gas, these conditions give

$$V_{mc} = 3b, \quad RT_c = 8a/(27b), \quad p_c = a/(27b^2) \tag{3.2.73}$$

where the subscript c denotes the critical point. From these relations we get

$$a = 3p_c(V_{mc})^2, \quad b = \frac{1}{3} V_{mc}, \quad \frac{p_c V_{mc}}{RT_c} = \frac{3}{8} \tag{3.2.74}$$

The dimensionless parameter $p_c V_{mc}/RT_c$ is called the **critical ratio** of the gas. Measured values of T_c, p_c, V_{mc}, and $p_c V_{mc}/RT_c$ for three monatomic gases are shown in Table 3.2. Substitution of the expressions for a and b given above into Eq. (3.2.71) gives

$$\frac{p}{p_c} = \frac{8}{3} \frac{T/T_c}{(V_m/V_{mc}) - (1/3)} - \frac{3}{(V_m/V_{mc})^2} \tag{3.2.75}$$

Table 3.2. Critical constants for monatomic gases [16]

Substance	T_c/K	p_c /atm	$V_{mc}/\text{cm}^3\ \text{mol}^{-1}$	$p_c V_{mc}/RT_c$
Argon Ar	150.7	48.0	75.2	0.292
Krypton Kr	209.4	54.3	92.2	0.291
Xenon Xe	289.8	58.0	118.8	0.290

which indicates that if p, T, and V_m are expressed as dimensionless quantities relative to their critical values, the van der Waals equation of state can be cast into a form that contains no parameter of the individual substance. When an equation of state has this property, it is said to satisfy the **corresponding-states principle**. Any "two-parameter" (e.g., a and b in the van der Waals relation) equation of state can be expressed in a reduced form that obeys this principle.

Equation (3.2.71) is cubic in V_m and for $T < T_c$ gives p versus V_m isotherms like the bottom curve in Fig. 3.7. The molar volumes V_{mg} and $V_{m\ell}$ of the vapor and liquid phases coexisting at a given temperature can be calculated

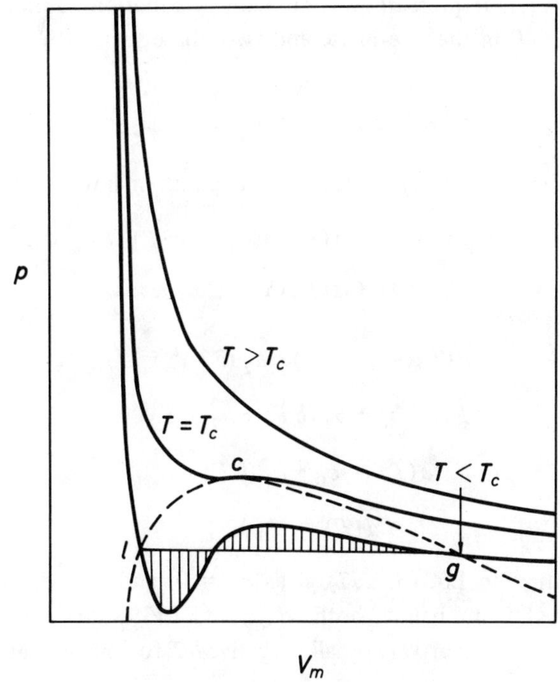

Fig. 3.7. Isotherms of the van der Waals type.

from the equilibrium conditions $\mu_g = \mu_\ell$ and $p_g = p_\ell = p_0$ (the external pressure). Since $V_m = (\partial\mu/\partial p)_T$, the first condition is written

$$\mu_g - \mu_\ell = \int_{p_\ell}^{p_g} V_m \, dp = 0 \tag{3.2.76}$$

where the integration is carried out along the isotherm at the given T. By integration by parts, Eq. (3.2.76) is rewritten

$$\int_{V_{m\ell}}^{V_{mg}} (p_0 - p) \, d V_m = 0 \tag{3.2.77}$$

where the condition $p_g = p_\ell = p_0$ has been used. This relation indicates that $V_{m\ell}$ and V_{mg} can be determined by drawing a horizontal line in such a way that the two shaded regions in Fig. 3.7 have the same area. The dashed curve in Fig. 3.7 is the locus of points ℓ and g when T is varied, and is called the **vapor-liquid coexisting curve** in the p versus V_m plane.

To investigate the shape of this curve in the neighborhood of the critical point, we assume that pressure p is an analytic function of molar volume V_m and temperature T in the one-phase and two-phase regions. We define δT and δV_m by

$$\delta T \equiv T - T_c, \qquad \delta V_m \equiv V_m - V_{mc} \tag{3.2.78}$$

and expand p in the vicinity of the critical point in powers of δT and δV_m

$$p - p_c = a(T) + b(T)(\delta V_m) + c(T)(\delta V_m)^2 + d(T)(\delta V_m)^3 + \ldots \tag{3.2.79}$$

where

$$a(T) = a_1(\delta T) + a_2(\delta T)^2 + \ldots \tag{3.2.80a}$$

$$b(T) = b_0 + b_1(\delta T) + b_2(\delta T)^2 + \ldots \tag{3.2.80b}$$

$$c(T) = c_0 + c_1(\delta T) + \ldots \tag{3.2.80c}$$

$$d(T) = d_0 + \ldots \tag{3.2.80d}$$

The first condition in Eq. (3.2.72) gives $b_0 = 0$, and the second, $c_0 = 0$. As T_c is approached from below, both δV_{mg} and $\delta V_{m\ell}$ vary strongly with δT. Therefore it is relevant for very small, negative δT to approximate Eq. (3.2.79) by [17]

$$p - p_c = a_1(\delta T) + b_1(\delta T)(\delta V_m) + d_0(\delta V_m)^3 \tag{3.2.81}$$

where a_1, b_1, and d_0 are represented by

$$a_1 = \left(\frac{\partial p}{\partial T}\right)_c, \quad b_1 = \left(\frac{\partial^2 p}{\partial V_m\, \partial T}\right)_c < 0, \quad d_0 = \frac{1}{6}\left(\frac{\partial^3 p}{\partial V_m{}^3}\right)_c < 0 \tag{3.2.82}$$

Though proof is omitted here, the negative signs of b_1 and d_0 are derived from the requirement that the system be in stable equilibrium at the critical point. Substituting Eq. (3.2.81) into Eq. (3.2.77) and noting that $V_{mg} - V_{m\ell} = \delta V_{mg} - \delta V_{m\ell}$, we obtain

$$\begin{aligned} p_0 &= p_c + a_1(\delta T) + \frac{1}{2} b_1(\delta T)(\delta V_{mg} + \delta V_{m\ell}) \\ &\quad + \frac{1}{4} d_0(\delta V_{mg} + \delta V_{m\ell})\,[(\delta V_{mg})^2 + (\delta V_{m\ell})^2] \end{aligned} \tag{3.2.83}$$

Therefore, the first approximation to the pressure p_0 for vapor-liquid phase equilibrium is given by

$$p_0 = p_c + a_1(\delta T) \tag{3.2.84}$$

The corresponding δV_m is obtained by inserting Eq. (3.2.84) into Eq. (3.2.81). We find three roots for δV_m:

$$\delta V_m = 0, \quad \delta V_m = \pm\,[-\,b_1(\delta T)/d_0]^{1/2} \tag{3.2.85}$$

The point defined by the first root and Eq. (3.2.84) for p_0 appears in the unstable region, where the system undergoes phase separation. On the other hand, the other two roots correspond to the desired V_{mg} and $V_{m\ell}$. Thus the first approximations to the molar volumes of the coexisting phases are

$$\begin{aligned} V_{mg} &= V_{mc} + (b_1/d_0)^{1/2}(T_c - T)^{1/2} \\ V_{m\ell} &= V_{mc} - (b_1/d_0)^{1/2}(T_c - T)^{1/2} \end{aligned} \tag{3.2.86}$$

Therefore, in the limit $T \to T_c^-$ (with the superscript indicating that T_c is approached from below), we have for the densities ρ_ℓ and ρ_g

$$\rho_\ell + \rho_g = M(V_{mg}{}^{-1} + V_{m\ell}{}^{-1}) = 2\rho_c \tag{3.2.87}$$

$$\frac{\rho_\ell - \rho_g}{2\rho_c} = \frac{1}{V_{mc}}\left(\frac{b_1}{d_0}\right)^{1/2}(T_c - T)^{1/2} \tag{3.2.88}$$

For the compressibility κ we obtain

$$\kappa^{-1} = -\,V_m(\partial p/\partial V_m)_T = (-\,b_1)\,V_m\,[\delta T + 3(d_0/b_1)(\delta V_m)^2] \tag{3.2.89}$$

Substitution of $\delta V_m = (-b_1 \delta T/d_0)^{1/2}$ yields, in the limit $T \to T_c^-$,

$$\kappa = \frac{B^-}{T_c - T}; \qquad B^- = -\frac{1}{2b_1 V_{mc}} \tag{3.2.90a}$$

Thus to this approximation we see $\kappa_g = \kappa_\ell$.

For $T > T_c$ no phase separation takes place, and δV_m varies only weakly with δT. Therefore, for small, positive δT we may approximate Eq. (3.2.79) by $p - p_c = a_1(\delta T) + b_1(\delta T)(\delta V_m)$. With this we can show that, in the limit $T \to T_c^+$ (i.e., approaching T_c from above), κ is represented by

$$\kappa = \frac{B^+}{T - T_c}; \qquad B^+ = -\frac{1}{b_1 V_{mc}} \tag{3.2.90b}$$

We note that $B^+ \neq B^-$. In a similar way, we find that the heat capacity at constant volume C_V in the vicinity of the critical point is expressed by

$$C_V(T) = C^\pm - D^\pm |T - T_c|; \quad C^- - C^+ > 0 \tag{3.2.91}$$

where C and D are constants, and the superscript signs are to be interpreted as above.

Equation (3.2.87) is consistent with the well-known rule of rectilinear diameters:

$$\rho_\ell + \rho_g = 2\rho_c + \text{constant}\,(T_c - T) \tag{3.2.92}$$

However, Eqs. (3.2.88), (3.2.90a), (3.2.90b), and (3.2.91) do not conform to experiment. For example, as is illustrated in Fig. 3.8, the exponent of $T_c - T$ in Eq. (3.2.88) should be 1/3 rather than 1/2 [18, 19].

To discuss the behavior of C_V, $\rho_\ell - \rho_g$, and κ in the vicinity of the critical point we introduce parameters α, α', β, γ, and γ' defined by

$$\alpha \equiv -\lim_{T \to T_c^+} \frac{\log C_V}{\log (T - T_c)}, \qquad \alpha' \equiv -\lim_{T \to T_c^-} \frac{\log C_V}{\log (T_c - T)} \tag{3.2.93a}$$

$$\beta \equiv \lim_{T \to T_c} \frac{\log (\rho_\ell - \rho_g)}{\log (T_c - T)} \tag{3.2.93b}$$

$$\gamma \equiv -\lim_{T \to T_c^+} \frac{\log \kappa}{\log (T - T_c)}, \qquad \gamma' \equiv -\lim_{T \to T_c^-} \frac{\log \kappa}{\log (T - T_c)} \tag{3.2.93c}$$

These are the **critical exponents**, and are associated fundamentally with analytic properties of the functions $A(T, V)$ and $p(T, V)$, the dimensionality of the system, etc. It can be shown that among α', β, and γ' there exists the relationship

$$\alpha' + 2\beta + \gamma' \geqslant 2 \qquad (3.2.94)$$

which is called the **Rushbrooke inequality** [20]. A proof, according to Fisher [19], is as follows.

The molar heat capacity of the system at fixed volume is denoted by C_{Vm}. When the system consists of coexisting vapor and liquid phases, three energy changes accompanying a temperature rise dT contribute to C_{Vm}: the increase in the respective phases and the increase due to evaporation of matter from the liquid phase into the vapor. If we denote the mole fractions of the vapor and liqud phases by x_g and x_ℓ, the molar volume of the system is $V_m = x_g V_{mg} + x_\ell V_{m\ell}$. Hence we have

$$x_g = \frac{V_m - V_{m\ell}}{V_{mg} - V_{m\ell}} \qquad (3.2.95)$$

If the change in x_g accompanying the temperature increase dT is denoted by dx_g, it follows from the conservation of mass and the condition of fixed volume that

$$x_\ell V_{m\ell} + x_g V_{mg} = (x_\ell - dx_g) V_{m\ell} \left[1 + \frac{1}{V_{m\ell}} \left(\frac{\partial V_{m\ell}}{\partial T} \right)_\sigma dT \right] + (x_g + dx_g) V_{mg} \left[1 + \frac{1}{V_{mg}} \left(\frac{\partial V_{mg}}{\partial T} \right)_\sigma dT \right] \qquad (3.2.96)$$

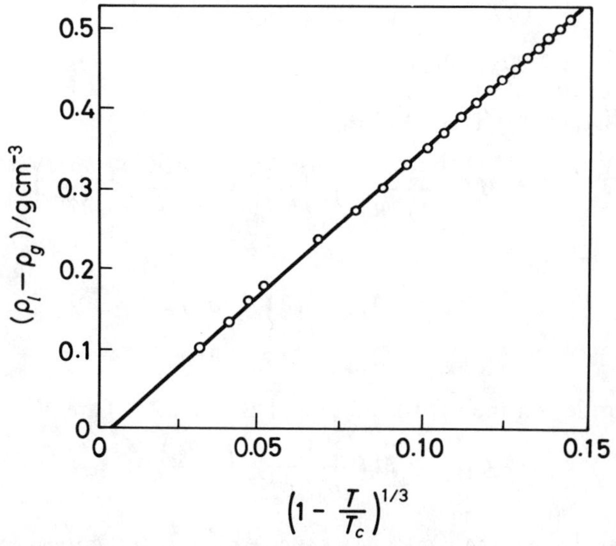

Fig. 3.8. Density difference between coexisting liquid and gas phases for xenon at temperatures near the critical point (Weinberger and Schneider [18]).

Here the subscript σ indicates that the derivative is to be evaluated along the vapor-liquid coexisting curve. If terms of second order are neglected, this equation gives

$$\left(\frac{\partial x_g}{\partial T}\right)_\sigma = -\frac{1}{V_{mg} - V_{m\ell}} \left[x_\ell \left(\frac{\partial V_{m\ell}}{\partial T}\right)_\sigma + x_g \left(\frac{\partial V_{mg}}{\partial T}\right)_\sigma\right] \tag{3.2.97}$$

The change dp_0 of equilibrium pressure with the temperature change dT is governed by the Clapeyron-Clausius equation

$$\frac{dp_0}{dT} = \left(\frac{\partial p}{\partial T}\right)_\sigma = \frac{H_{mg} - H_{m\ell}}{T(V_{mg} - V_{m\ell})} \tag{3.2.98}$$

where H_{mg} and $H_{m\ell}$ are the molar enthalpies of the vapor and liquid, respectively. Using these results we can express C_{Vm} as

$$C_{Vm} = x_\ell C_{\sigma\ell} + x_g C_{\sigma g} - T\left(\frac{\partial p}{\partial T}\right)_\sigma \left[x_\ell \left(\frac{\partial V_{m\ell}}{\partial T}\right)_\sigma + x_g \left(\frac{\partial V_{mg}}{\partial T}\right)_\sigma\right] \tag{3.2.99}$$

The first two terms are the heat capacities of the two phases, and the last term is derived from the heat of evaporation $(H_{mg} - H_{m\ell})(\partial x_g/\partial T)_\sigma$. Applying Eq. (1.2.31) to $C_{\sigma g}$ and $C_{\sigma\ell}$ and using the formula

$$\left(\frac{\partial V}{\partial T}\right)_p = \left(\frac{\partial V}{\partial T}\right)_\sigma - \left(\frac{\partial V}{\partial p}\right)_T \left(\frac{\partial p}{\partial T}\right)_\sigma \tag{3.2.100}$$

we can put Eq. (3.2.99) in the form

$$C_{Vm} = x_\ell C_{Vm\ell} + x_g C_{Vmg} + MT\left[\frac{x_\ell}{\rho_\ell^3 \kappa_\ell}\left(\frac{\partial \rho_\ell}{\partial T}\right)_\sigma^2 + \frac{x_g}{\rho_g^3 \kappa_g}\left(\frac{\partial \rho_g}{\partial T}\right)_\sigma^2\right] \tag{3.2.101}$$

where

$$\kappa_g = -V_{mg}^{-1}(\partial V_{mg}/\partial p)_T \tag{3.2.102a}$$

$$\kappa_\ell = -V_{m\ell}^{-1}(\partial V_{m\ell}/\partial p)_T \tag{3.2.102b}$$

All the quantities on the right-hand side of Eq. (3.2.101) are positive, so that

$$C_{Vm} \geqslant MT\frac{x_g}{\rho_g^3 \kappa_g}\left(\frac{\partial \rho_g}{\partial T}\right)_\sigma^2 \tag{3.2.103}$$

Since ρ_g may be written $[\rho_\ell + \rho_g - (\rho_\ell - \rho_g)]/2$, the critical exponent of $(\partial \rho_g/\partial T)_\sigma$ is $\beta - 1$. The critical exponents of C_{Vm} and κ_g^{-1} are $-\alpha'$ and γ', respectively. All other quantities in Eq. (3.2.103) converge to finite values as T approaches T_c. Thus the Rushbrooke inequality follows from Eq. (3.2.103).

Equations (3.2.88), (3.2.90a), and (3.2.90b) yield $\beta = 0.5$, $\gamma' = 1$, and $\alpha' = 0$, respectively. For these values the Rushbrooke relation becomes an equality. For the two-dimensional lattice gas, for which an exact statistical-mechanical solution is available, $C_V(T)$ diverges logarithmically as T approaches T_c

$$C_V(T) = -A^{\pm} \ln |1 - (T/T_c)| + B^{\pm} \tag{3.2.104}$$

so that $\alpha' = 0$. It is also known that $\beta = 1/8$ and $\gamma' = 7/4$ for the lattice gas. Hence the equality in Eq. (3.2.94) applies. These examples suggest that the Rushbrooke equality, rather than the inequality, might hold in general, and ideas including the scaling laws have been proposed to verify this conjecture. We will not go further into these matters here.

Finally, we touch upon Fisher's treatment of the critical exponents associated with light scattering. Since experimental observations in the neighborhood of the critical point deviate from the classical theory, the relation

$$g_2(r) = \langle \Delta\rho^{\alpha} \Delta\rho^{\beta} \rangle / \rho^2 = (kT/4\pi r) \exp [-r/(b\kappa)^{1/2}] \tag{3.2.105}$$

which is derived from Eqs. (3.1.102) and (3.1.75), is formally modified to

$$g_2(r) = (D/r^{1+\eta}) \exp(-qr) \tag{3.2.106}$$

with q assumed to depend on T according to

$$q \propto (T - T_c)^{\zeta} \tag{3.2.107}$$

as $T \to T^+$. Then the Ornstein-Zernike relation, which holds as a general equation in statistical mechanics, gives

$$\kappa kT \approx 4\pi D [\Gamma(2 - \eta)]/q^{2-\eta} \tag{3.2.108}$$

where Γ denotes the gamma function. Introduction of this into the first equation in Eq. (3.2.93c) yields

$$\zeta(2 - \eta) = \gamma \tag{3.2.109}$$

Substitution of Eq. (3.2.106) into Eq. (3.1.76) gives

$$L^2 \propto (T - T_c)^{-2\zeta} \tag{3.2.110}$$

Also, with Eq. (3.2.106), Eq. (3.2.68) gives

$$R(\theta) \propto G_n(\theta) \propto \frac{D'(T)}{(q^2 + s^2)^{1-(\eta/2)}} \approx \frac{R(0)}{(1 + q^{-2} s^2)^{1-(\eta/2)}} \tag{3.2.111}$$

where

$$R(0) = \frac{D'(T)}{q^{2-\eta}} \propto \frac{D'(T)}{(T - T_c)^{\gamma}} \propto (T - T_c)^{-\gamma} \qquad (3.2.112)$$

$D'(T)$ being a slowly varying function of T. These relations provide a means of investigating the pair correlation function $g_2(r)$ in the region near the critical point by light scattering measurements.

3.3. Light Scattering from Polymer Solutions

3.3.1. Solutions with a Single Solvent

We begin our discussion of light scattering from polymer solutions with a system which consists of one solvent (component 0) and r polymer solutes (components $1, 2, \ldots, r$). Two external conditions will be treated: constant temperature and pressure; constant temperature and μ_0.

a. System at Constant Pressure. In a system at constant T and p, a very small region contains a fixed amount n_0 of component 0. All quantities of species but n_0 in this region fluctuate about their mean values. It is evident that the set of state variables appropriate for the small system is T, p, and molalities m_i $(i = 1, 2, \ldots, r)$. We assume that the refractive index $\tilde{n}$ at fixed composition depends only on the solution density ρ. The fluctuation $\Delta\tilde{n}$ which occurs in the small system is then expressed, correct to the first order in ΔT, Δp, and Δm $(\Delta m_1, \Delta m_2, \ldots, \Delta m_r)$, by

$$\Delta\tilde{n} = \left(\frac{\partial\tilde{n}}{\partial\rho}\right)_m \left[\left(\frac{\partial\rho}{\partial T}\right)_{p,m} \Delta T + \left(\frac{\partial\rho}{\partial p}\right)_{T,m} \Delta p\right] + \sum_{i=1}^{r} \left(\frac{\partial\tilde{n}}{\partial m_i}\right)_{T,p,m_j} \Delta m_i \qquad (3.3.1)$$

By making use of the probability $w(\Delta T, \Delta p, \Delta m)$ given by Eq. (3.1.54) the mean square of $\Delta\tilde{n}$ can be evaluated. The result reads

$$\langle(\Delta\tilde{n})^2\rangle = \left(\rho\frac{\partial\tilde{n}}{\partial\rho}\right)_m^2 [\alpha^2\langle(\Delta T)^2\rangle - 2\alpha\kappa\langle\Delta T\,\Delta p\rangle + \kappa^2\langle(\Delta p)^2\rangle] + \sum_{i=1}^{r}\sum_{j=1}^{r} \Psi_i\Psi_j\langle\Delta m_i\,\Delta m_j\rangle \qquad (3.3.2)$$

where Ψ_i, defined by

$$\Psi_i \equiv (\partial\tilde{n}/\partial m_i)_{T,p,m_k}, \quad (i = 1, 2, \ldots, r) \qquad (3.3.3)$$

is the refractive index increment of component i on the molality scale. By substitution of Eqs. (3.1.55) and (3.1.56), Eq. (3.3.2) becomes

$$\langle(\Delta\tilde{n})^2\rangle = \frac{\kappa kT}{V}\left(\rho\,\frac{\partial\tilde{n}}{\partial\rho}\right)_m^2 + \frac{v_M kT}{V}\sum_{i=1}^{r}\sum_{j=1}^{r}\Psi_i\Psi_j\,\frac{\Delta_{ij}}{|\mu|} \tag{3.3.4}$$

where V is the mean volume of the small region considered. If Eq. (3.3.4) is introduced into Eq. (3.2.62) with σ set equal to V and the result is in turn inserted into Eq. (3.2.57), we obtain

$$R(0) = I(0)\,r^2/(2I_0) = R_\rho(0) + R_c(0) \tag{3.3.5}$$

where

$$R_\rho(0) = K\kappa RT\left(\rho\,\frac{\partial\tilde{n}}{\partial\rho}\right)_m^2 ; \quad K \equiv \frac{2\pi^2\tilde{n}^2}{N_A\lambda_0^{\,4}} \tag{3.3.6}$$

$$R_c(0) = K v_M RT\sum_{i=1}^{r}\sum_{j=1}^{r}\Psi_i\Psi_j\,\frac{\Delta_{ij}}{|\mu|} \tag{3.3.7}$$

As will be explained later, Eq. (3.2.62), which ignores interference effects of scattered light, is applicable only at zero scattering angle if the solutes are polymeric. For this reason we have presented here only the expression of $R(\theta)$ for $\theta = 0$. For simplicity, however, the notation (0) to indicate $\theta = 0$ will be omitted until the angular dependence of $R(\theta)$ is discussed in § 3.3.4. The quantity R_ρ is the contribution from density fluctuations (treated in § 3.2.3), while the quantity R_c is the contribution from fluctuations of solute concentrations, i.e., the **concentration scattering**. Under ordinary experimental conditions we may determine R_ρ to a satisfactory approximation from a scattering measurement on pure solvent.

According to Eq. (1.4.60), the concentration expansion of $\mu_{ii} = (\partial\mu_i/\partial m_i)_{T,p,m_j}$ starts with a term proportional to m_i^{-1}, while that of μ_{ij} $(i \neq j)$ has a constant leading term. With this in mind, we expand the determinant $|\mu|$ in powers of μ_{ii}^{-1}:

$$|\mu| = \left(\prod_{i=1}^{r}\mu_{ii}\right)\left(1 - \sum_{i<j}\sum\frac{\mu_{ij}\mu_{ji}}{\mu_{ii}\mu_{jj}} + \sum_{i<j<k}\sum\sum\frac{\mu_{ij}\mu_{jk}\mu_{ki} + \mu_{ik}\mu_{kj}\mu_{ji}}{\mu_{ii}\mu_{jj}\mu_{kk}} + \ldots\right) \tag{3.3.8}$$

The cofactor $\Delta_{ij} = (\partial |\mu| / \partial \mu_{ij})$ is then evaluated as

$$\Delta_{ii} = \frac{\Pi_k \mu_{kk}}{\mu_{ii}} \left(1 - \sum_{\substack{j<k \\ j,k \neq i}} \sum \frac{\mu_{jk}\mu_{kj}}{\mu_{jj}\mu_{kk}} + \ldots \right) \tag{3.3.9a}$$

$$\Delta_{ij} = - \frac{\Pi_k \mu_{kk}}{\mu_{ii}} \left(\frac{\mu_{ji}}{\mu_{jj}} - \frac{1}{\mu_{jj}} \sum_{k \neq i,j} \frac{\mu_{jk}\mu_{ki}}{\mu_{kk}} + \ldots \right) \tag{3.3.9b}$$

These expressions together with Eq. (1.4.60) yield

$$\Delta_{ii}/|\mu| = (m_i/RT) \left[1 - \beta_{ii} m_i - \sum_{j=1}^{r} (\beta_{iij} - \beta_{ij}\beta_{ji}) m_i m_j + \ldots \right] \tag{3.3.10a}$$

$$\Delta_{ij}/|\mu| = -(m_i m_j/RT) \left[\beta_{ji} + \sum_{k=1}^{r} (\beta_{jik} - \beta_{jk}\beta_{ki}) m_k + \ldots \right], \quad (i \neq j) \tag{3.3.10b}$$

Substitution of these relations into Eq. (3.3.7) yields

$$R_c/K\nu_M = \sum_{i=1}^{r} \sum_{j=1}^{r} \Psi_i \Psi_j \left[m_i \delta_{ij} - \beta_{ij} m_i m_j - \sum_{k=1}^{r} (\beta_{ijk} - \beta_{ik}\beta_{jk}) m_i m_j m_k + \ldots \right] \tag{3.3.11}$$

where δ_{ij} is the Kronecker delta. This expansion corresponds to Eq. (1.4.63) for osmotic pressure at constant T and p.

Since the molality scale is inconvenient for treating polymer solutions, we wish to convert to mass concentration. To do this, we define the **specific refractive index increment**

$$\psi_i \equiv \left(\frac{\partial \tilde{n}}{\partial C_i} \right)_{T,p,C_k}, \quad (i = 1, 2, \ldots, r) \tag{3.3.12}$$

and make the important assumption that ψ_i is independent of composition. It can be shown that ψ_i is related to Ψ_i by

$$\Psi_i = \psi_i \frac{M_i}{\nu_M} \left(1 - \sum_{j=1}^{r} \frac{\psi_j}{\psi_i} v_i C_j \right) \tag{3.3.13}$$

where v_i is the partial specific volume of component i. Inserting this into Eq. (3.3.11), converting from m_i to C_i, changing from the β coefficients to the B coefficients via Eqs. (1.4.78a) and (1.4.78b), and assuming that v_i is also independent of composition (thus approximating v_i by v_i^o), we arrive at

$$R_c/K = \sum_{i=1}^{r} \sum_{j=1}^{r} \psi_i \psi_j \Big[M_i C_i \delta_{ij} - M_i M_j B_{ij} C_i C_j - \sum_{k=1}^{r} M_i M_j (B_{ijk} - M_k B_{ik} B_{jk}) C_i C_j C_k + \ldots \Big] \tag{3.3.14}$$

It should be noted that the effect of the approximation $v_i = v_i^o$ appears only in the third and higher terms in the brackets.

b. System at Constant μ_0. We next consider the case with the solution under consideration brought to equilibrium at constant μ_0. If T, μ_0, and C_i ($i = 1, 2, \ldots, r$) are chosen as the intensive variables to describe the state of a very small region of fixed volume in the solution and to formulate the light scattering from this small region in a way similar to that described above, the concentration scattering R_c^* of the solution is given by

$$R_c^* = KRT \sum_{i=1}^{r} \sum_{j=1}^{r} \psi_i^* \psi_j^* \frac{(\Delta_C)_{ij}}{|\mu_C|} \tag{3.3.15}$$

where ψ_i^* is a specific refractive index increment defined by

$$\psi_i^* \equiv \left(\frac{\partial \tilde{n}}{\partial C_i} \right)_{T, \mu_0, C_k} \tag{3.3.16}$$

and $(\Delta_C)_{ij}$ and $|\mu_C|$ are defined as in § 3.1.3c. Equation (3.3.15) may be compared with Eq. (3.3.7) at constant T and p. If $(\Delta_C)_{ij}/|\mu_C|$ is expanded in powers of C_i, the result agrees with the C_i-expansion within brackets in Eq. (3.3.14). No approximation need be introduced in this derivation, while, as has been noted above, only the first and second terms in Eq. (3.3.14) are exact.

Under ordinary experimental conditions it is permissible to assume that the solution is incompressible. Then, with the basic assumption that $\tilde{n}$ at fixed composition depends only on the solution density, it can be shown that

$$\psi_i^* = \psi_i \tag{3.3.17}$$

The proof involves applying Eq. (A.13) and noting that the composition of an incompressible solution is determined by a set of variables $C_1, C_2, \ldots, C_r$.

Summarizing the results obtained, we find that the relation for concentration scattering of an incompressible polymer solution in a single solvent at constant pressure agrees with the relation for a solution at constant μ_0 up to the second order in the mass concentrations of the polymer components. The third-order term is correct if the approximation $v_i = v_i^o$ holds. It is interesting to recall that, in § 1.4.4, we arrived at the same conclusions in regard to the agreement between osmotic pressure at constant p and that at constant μ_0.

c. **Virial Expansion for Light Scattering.** For the case in which the solute components are a series of homologous polymers differing only in molecular weight, experimental data indicate that their ψ_i may be regarded as independent of molecular weight M_{ri} unless M_{ri} is very small. Denoting the specific refractive index increment common to all the polymer components by ψ_s and rewriting the series within brackets in Eq. (3.3.14) in terms of $C_i = \xi_i C_s$ (where C_s is the total mass concentration of polymer and ξ_i is the weight fraction of component i in the polymer), we obtain for R_c^{-1}

$$K(\psi_s)^2 C_s/R_c = M_w^{-1} + 2A_2^{LS} C_s + 3A_3^{LS} C_s^2 + \ldots \tag{3.3.18}$$

where M_w is the weight-average molar mass of the polymer and A_2^{LS} and A_3^{LS} are given by

$$A_2^{LS} = (2M_w^2)^{-1} \sum_{i=1}^{r} \sum_{j=1}^{r} \xi_i \xi_j M_i M_j B_{ij} \tag{3.3.19}$$

$$\begin{aligned} A_3^{LS} = {} & (3M_w^2)^{-1} \sum_{i=1}^{r} \sum_{j=1}^{r} \sum_{k=1}^{r} \xi_i \xi_j \xi_k M_i M_j B_{ijk} \\ & - (3M_w^3)^{-1} \sum_{i=1}^{r} \sum_{j=1}^{r} \sum_{k=1}^{r} \sum_{\ell=1}^{r} \xi_i \xi_j \xi_k \xi_\ell M_i M_j M_k M_\ell \\ & \times (B_{ik} B_{jk} - B_{ik} B_{j\ell}) \end{aligned} \tag{3.3.20}$$

Equation (3.3.18) is the virial expansion for light scattering for a polymer solution in a single solvent at constant T and p, and the quantities A_2^{LS} and A_3^{LS} are the **light-scattering second and third virial coefficients.** These coefficients reflect thermodynamic interactions of polymer components of different molecular weights in a given solvent at given T and p. Equation (3.3.19) for A_2^{LS} is exact, but Eq. (3.3.20) for A_3^{LS} is inexact in the sense that it lacks terms which arise from the concentration dependence of ν_i. The most important feature of Eq. (3.3.18) is that M_w of a polymer sample polydisperse in molecular weight can be determined from measurement of R_c as a function of C_s at constant T and p if an appropriate pure solvent is available. As will be shown in the next section, this is not always true when a mixed solvent must be used.

It should be noted that the light-scattering second virial coefficient A_2^{LS} differs from the corresponding osmotic quantity A_2^{OS}, expressed by Eq. (1.4.70), in the way in which binary interaction parameters B_{ij} are averaged over the polymer components. The two coefficients agree only when the polymer is monodisperse. The light-scattering third coefficient is also different from the osmotic third virial coefficient A_3^{OS} given by Eq. (1.4.71). Not only do these differ in the averaging of ternary interaction parameters B_{ijk}, but $A_3{}^{LS}$ also contains additional terms which involve products of B_{ij}. If we define $\langle B^k \rangle$ by

$$\langle B^k \rangle \equiv \frac{1}{M_w} \sum_{i=1}^{r} \xi_i M_i \langle B_i \rangle^k \tag{3.3.21}$$

with

$$\langle B_i \rangle = \frac{1}{M_w} \sum_{j=1}^{r} \xi_j M_j B_{ij} \tag{3.3.22}$$

these additional terms are expressed as $(M_w/3)\,(\langle B^2 \rangle - \langle B \rangle^2)$. This form suggests that they are associated with the dispersion of $M_i \langle B_i \rangle$ over the polymer solutes and, therefore, should become important for a polymer sample in which $M_i \langle B_i \rangle$ varies appreciably with M_i.

3.3.2. Solutions in Mixed Solvents

Light scattering from solutions in mixed solvents leads to very different results, depending on whether solute concentration is varied at constant p or constant μ_D; as before, μ_D denotes the set of chemical potentials of diffusible components, i.e., the components which form the mixed solvent. To explain this difference we consider a ternary solution composed of two solvents (a principal solvent 0 and a second solvent 1) and a monodisperse polymer solute (component 2).

a. System at Constant Pressure. The relevant expression for R_c is obtained from Eq. (3.3.7), because this equation is also valid if one of the solute components is taken as the second solvent. Thus we have for the system

$$\frac{R_c}{K\nu_M} = \frac{RT}{|\mu|} \left(\Psi_1{}^2 \mu_{22} - 2\Psi_1 \Psi_2 \mu_{12} + \Psi_2{}^2 \mu_{11} \right) \tag{3.3.23}$$

This may be written

$$\frac{R_c}{K\nu_M} = \frac{RT\mu_{11}}{|\mu|} \Psi_2{}^2 \Omega^2 + \frac{RT}{\mu_{11}} \Psi_1{}^2 \tag{3.3.24}$$

where

$$\Omega \equiv 1 - \frac{\Psi_1 \mu_{12}}{\Psi_2 \mu_{11}} = 1 + \frac{\Psi_1}{\Psi_2} \left(\frac{\partial m_1}{\partial m_2} \right)_{T, p, \mu_1} \tag{3.3.25}$$

In the limit $m_2 \to 0$, $|\mu|$ tends to $RT\mu_{11}{}^o/m_2$ [see Eq. (1.4.60)], so that we have

$$\frac{R_c{}^o}{K\nu_M{}^o} = (\Psi_1{}^o)^2 \frac{RT}{\mu_{11}{}^o} \tag{3.3.26}$$

where the superscript circle indicates the limit of vanishingly small m_2. If we define the excess concentration scattering ΔR_c by

$$\Delta R_c \equiv R_c - R_c{}^o \tag{3.3.27}$$

we obtain from Eqs. (3.3.24) and (3.3.26)

$$\lim_{m_2 \to 0} \frac{K\Psi_2{}^2\, m_2\, \nu_M}{\Delta R_c} = \frac{1}{(\Omega^o)^2} \tag{3.3.28}$$

We now change the state variables to the set T, p, m_1, C_2, where C_2 is the mass concentration of the polymer solute; and define a new specific refractive index increment $\psi_2{}'$ of component 2 by

$$\psi_2{}' \equiv \left(\frac{\partial \tilde{n}}{\partial C_2}\right)_{T,p,m_1} \tag{3.3.29}$$

This refractive index increment is what is actually determined in experiments with mixed-solvent systems, and is essentially independent of C_2 in most cases. Using Eqs. (1.1.40) and (1.1.71), we can show that Ψ_2 is related to $\psi_2{}'$ by

$$\Psi_2 = (\psi_2{}'M_2/\nu_M)(1 - v_2 C_2) \tag{3.3.30}$$

where v_2 is the partial specific volume of component 2. Substituting Eq. (3.3.30) into Eq. (3.3.28), we get

$$\lim_{C_2 \to 0} \frac{K(\psi_2{}')^2 C_2}{\Delta R_c} = \frac{1}{M_2(\Omega^o)^2} \tag{3.3.31}$$

This expression was derived independently by Brinkman and Hermans [21], Kirkwood and Goldberg [22], and Stockmayer [3]. It indicates that the light scattering experiment on a polymer solute in a mixed solvent at constant T and p allows only an apparent molar mass $M_{2app} \equiv M_2(\Omega^o)^2$ to be determined, unless $\Psi_1{}^o = 0$ or $(\partial m_1/\partial m_2)^o_{T,p,\mu_1} = 0$. The former condition obtains when the two solvents 0 and 1 happen to have the same refractive index, whereas the latter obtains when the second solvent has no thermodynamic interaction with the polymer solute at fixed T, p, and μ_1. However, when M_2 is known independently, we may use Eq. (3.3.31) to determine the interaction parameter $(\partial m_1 / \partial m_2)^o_{T,p,\mu_1}$.

The quantity Ω^o can be eliminated from Eq. (3.3.31) by introducing still another refractive index increment ψ_2^+:

$$\psi_2^+ \equiv \left(\frac{\partial \tilde{n}}{\partial C_2}\right)_{T,p,\mu_1} \tag{3.3.32}$$

Since $(\psi_2^+)^o$ is equal to $\psi_2'^o\, \Omega^o$, Eq. (3.3.31) reduces to

$$\lim_{C_2 \to 0} \frac{K(\psi_2^+)^2 C_2}{\Delta R_c} = \frac{1}{M_2} \tag{3.3.33}$$

This expression appears to afford a way of determining M_2 from the measurement of ΔR_c. However, the quantity ψ_2^+ is not accessible to experimental measurement, because the constraints involved in its definition do not correspond to the conditions for osmotic equilibrium.

The above analysis can be extended to a solution of s polymer components $(i = d+1, d+2, \ldots, d+s)$ in a mixed solvent consisting of $d+1$ low-molecular-weight components $(i = 0, 1, \ldots, d)$. Though we omit the actual analysis, we note that the specific refractive index increment of polymer component i pertinent to this system is defined by

$$\psi_i' \equiv \left(\frac{\partial \tilde{n}}{\partial C_i}\right)_{T,p,m,C_k} \tag{3.3.34}$$

Here the subscript m denotes the set of molalities $m_1, m_2, \ldots, m_d$ and the subscript C_k signifies that the mass concentrations of all polymer components other than component i are held fixed. The previously defined specific refractive index increment ψ_i for single solvent systems corresponds to a special case of ψ_i' in which $m = 0$.

Figure 3.9 illustrates the classic light scattering data of Ewart et al. [23] on the quasiternary solution composed of benzene (0), methanol (1), and polystyrene (2). It is seen that values of M_{2app}, the reciprocals of the intercepts of the plots, increase with the content of methanol in the mixed solvent. Since $\Psi_1{}^o$ for this system is negative, the data imply that $(\partial m_1 / \partial m_2)^o_{T,p,\mu_1}$ is negative and increases in absolute value with the methanol content. In other words, polystyrene rejects methanol in preference to benzene. A more detailed analysis of these data was carried out by Kirkwood and Goldberg [22].

As we have shown in § 3.1.3b, concentration fluctuations in a solution at constant T and p are interdependent. Therefore, in the ternary solution considered above, ΔR_c generally contains contributions from concentration

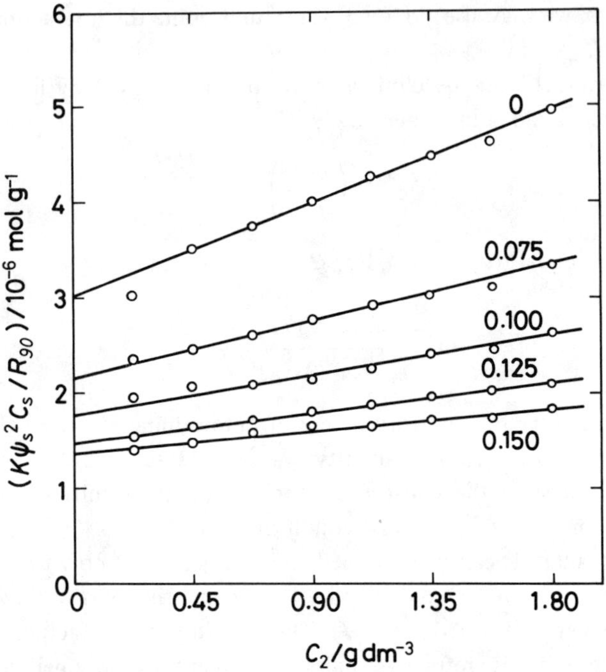

Fig. 3.9. Concentration dependence of light scattering for a polystyrene sample in mixtures of benzene and methanol at indicated volume fractions of methanol in the mixed solvent (Ewart et al. [23]). These data were derived from 90° scattering alone.

fluctuations of the polymer solute as well as concentration fluctuations of the second solvent. The latter effect determines the second term in Eq. (3.3.25) and is referred to as selective adsorption.

b. Systems at Constant μ_D. We let the ternary solution under consideration be at equilibrium with μ_D constant, where μ_D denotes the set of μ_0 and μ_1. We take a very small region of fixed volume V in the solution, and choose T, V, μ_0, μ_1, and C_2 as the state variables for the subsystem taken. The fluctuation of refractive index $\tilde{n}$ in the small region may be expressed by

$$\Delta \tilde{n} = \left(\frac{\partial \tilde{n}}{\partial T}\right)_{\mu_D, C_2} \Delta T + \left(\frac{\partial \tilde{n}}{\partial \mu_0}\right)_{T, \mu_1, C_2} \Delta \mu_0 + \left(\frac{\partial \tilde{n}}{\partial \mu_1}\right)_{T, \mu_0, C_2} \Delta \mu_1 + \left(\frac{\partial \tilde{n}}{\partial C_2}\right)_{T, \mu_D} \Delta C_2 \qquad (3.3.35)$$

It is reasonable to assume that $\tilde{n}$ depends only on the composition of the solution, i.e., $\tilde{n} = \tilde{n}(C_0, C_1, C_2)$. Since $C_0 + C_1 + C_2 = \rho$, where ρ is the density of the solution, this assumption is equivalent to considering $\tilde{n}$ as a function of ρ, C_1, and C_2. Equation (3.3.35) can now be transformed to

$$\Delta\tilde{n} = \left(\frac{\partial \tilde{n}}{\partial C_0}\right)_{C_1, C_2} \left[\left(\frac{\partial C_0}{\partial T}\right)_{\mu_D, C_2} \Delta T + \left(\frac{\partial C_0}{\partial \mu_0}\right)_{T, \mu_1, C_2} \Delta\mu_0\right.$$
$$\left. + \left(\frac{\partial C_0}{\partial \mu_1}\right)_{T, \mu_0, C_2} \Delta\mu_1\right] + \left(\frac{\partial \tilde{n}}{\partial C_1}\right)_{C_0, C_2} \left[\left(\frac{\partial C_1}{\partial T}\right)_{\mu_D, C_2} \Delta T\right.$$
$$\left. + \left(\frac{\partial C_1}{\partial \mu_0}\right)_{T, \mu_1, C_2} \Delta\mu_0 + \left(\frac{\partial C_1}{\partial \mu_1}\right)_{T, \mu_0, C_2} \Delta\mu_1\right]$$
$$+ \left(\frac{\partial \tilde{n}}{\partial C_2}\right)_{C_0, C_1} \Delta C_2 \tag{3.3.36}$$

The derivatives in the brackets may be expressed in terms of the characteristic function $\mathcal{A}$, defined by Eq. (2.3.49) for a system at constant μ_D; for example,

$$\left(\frac{\partial C_0}{\partial T}\right)_{\mu_D, C_2} = \frac{M_0}{V}\left(\frac{\partial n_0}{\partial T}\right)_{V, \mu_D, n_2} = -\frac{M_0}{V}\left(\frac{\partial^2 \mathcal{A}}{\partial \mu_0\, \partial T}\right)$$
$$\equiv -\frac{M_0}{V}\mathcal{A}_{0T} \tag{3.3.37}$$

$$\left(\frac{\partial C_0}{\partial \mu_1}\right)_{T, \mu_0, C_2} = \frac{M_0}{V}\left(\frac{\partial n_0}{\partial \mu_1}\right)_{T, V, \mu_0, n_2} = -\frac{M_0}{V}\left(\frac{\partial^2 \mathcal{A}}{\partial \mu_0\, \partial \mu_1}\right)$$
$$\equiv -\frac{M_0}{V}\mathcal{A}_{01} \tag{3.3.38}$$

Thus Eq. (3.3.36) can be written

$$\Delta\tilde{n} = -\frac{1}{V}\sum_{i=0}^{1} M_i \left(\frac{\partial \tilde{n}}{\partial C_i}\right)_{C_j, C_2} (\mathcal{A}_{iT}\,\Delta T + \mathcal{A}_{i0}\,\Delta\mu_0$$
$$+ \mathcal{A}_{i1}\,\Delta\mu_1) + \left(\frac{\partial \tilde{n}}{\partial C_2}\right)_{T, \mu_D} \Delta C_2 \tag{3.3.39}$$

The probability $w(\Delta T, \Delta\mu_0, \Delta\mu_1, \Delta C_2)$ that fluctuations ΔT, $\Delta\mu_0$, $\Delta\mu_1$, and ΔC_2 occur simultaneously in the small region may be obtained by the method described in § 3.1.3c. Using the resulting expression, we find for the mean-square fluctuation of $\tilde{n}$

$$\langle(\Delta\tilde{n})^2\rangle = \frac{1}{V^2}\left[\left(\frac{\partial\tilde{n}}{\partial C_0}\right)^2_{C_1,C_2} M_0^2\langle A_0^2\rangle + 2\left(\frac{\partial\tilde{n}}{\partial C_0}\right)_{C_1,C_2}\left(\frac{\partial\tilde{n}}{\partial C_1}\right)_{C_0,C_2} M_0M_1\langle A_0A_1\rangle + \left(\frac{\partial\tilde{n}}{\partial C_1}\right)^2_{C_0,C_2} M_1^2\langle A_1^2\rangle\right] + \left(\frac{\partial\tilde{n}}{\partial C_2}\right)^2_{T,\mu_D}\langle(\Delta C_2)^2\rangle \tag{3.3.40}$$

where

$$A_0 \equiv \mathcal{A}_{0T}\,\Delta T + \mathcal{A}_{00}\,\Delta\mu_0 + \mathcal{A}_{01}\,\Delta\mu_1 \tag{3.3.41a}$$

$$A_1 \equiv \mathcal{A}_{1T}\,\Delta T + \mathcal{A}_{10}\,\Delta\mu_0 + \mathcal{A}_{11}\,\Delta\mu_1 \tag{3.3.41b}$$

It can be shown [Eq. (3.1.64a) with $i = j = 2$ is applicable to the present case] that

$$\langle(\Delta C_2)^2\rangle = (kT/V)M_2(\partial\mu_2/\partial C_2)^{-1}_{T,\mu_D} \tag{3.3.42}$$

The following results are also derived:

$$\langle A_0^2\rangle = -(kT/|\mathcal{A}|)\,[\mathcal{A}_{0T}^2\,\Delta_{TT} + \mathcal{A}_{00}^2\,\Delta_{00} + \mathcal{A}_{01}^2\,\Delta_{11} + 2(\mathcal{A}_{0T}\mathcal{A}_{00}\,\Delta_{T0} + \mathcal{A}_{0T}\mathcal{A}_{01}\,\Delta_{T1} + \mathcal{A}_{00}\mathcal{A}_{01}\,\Delta_{01})] \tag{3.3.43a}$$

$$\langle A_0A_1\rangle = -(kT/|\mathcal{A}|)\,[\mathcal{A}_{0T}\mathcal{A}_{1T}\Delta_{TT} + \mathcal{A}_{00}\mathcal{A}_{01}\Delta_{00} + \mathcal{A}_{01}\mathcal{A}_{11}\Delta_{11} + (\mathcal{A}_{0T}\mathcal{A}_{10} + \mathcal{A}_{00}\mathcal{A}_{1T})\Delta_{T0} + (\mathcal{A}_{0T}\mathcal{A}_{11} + \mathcal{A}_{01}\mathcal{A}_{1T})\Delta_{T1} + (\mathcal{A}_{00}\mathcal{A}_{11} + \mathcal{A}_{01}\mathcal{A}_{10})\Delta_{01}] \tag{3.3.43b}$$

where

$$|\mathcal{A}| \equiv \begin{vmatrix} \mathcal{A}_{TT} & \mathcal{A}_{T0} & \mathcal{A}_{T1} \\ \mathcal{A}_{0T} & \mathcal{A}_{00} & \mathcal{A}_{01} \\ \mathcal{A}_{1T} & \mathcal{A}_{10} & \mathcal{A}_{11} \end{vmatrix} \tag{3.3.44}$$

and, for example, Δ_{T0} is the co-factor of element $\mathcal{A}_{T0}$ in the determinant $|\mathcal{A}|$. According to Eq. (A.76), the following relations hold:

$$\mathcal{A}_{0T}\Delta_{0T} + \mathcal{A}_{00}\Delta_{00} + \mathcal{A}_{01}\Delta_{01} = |\mathcal{A}|$$

$$\mathcal{A}_{0T}\Delta_{TT} + \mathcal{A}_{00}\Delta_{T0} + \mathcal{A}_{01}\Delta_{T1} = 0 \tag{3.3.45}$$

$$\mathcal{A}_{0T}\Delta_{1T} + \mathcal{A}_{00}\Delta_{10} + \mathcal{A}_{01}\Delta_{11} = 0$$

Using these relations, together with the symmetry of $|\mathcal{A}|$, we find

$$\langle A_0{}^2\rangle = -kT\mathcal{A}_{00} = (kTV/M_0)(\partial C_0/\partial\mu_0)_{T,\mu_1,C_2}$$

$$\langle A_0 A_1\rangle = -kT\mathcal{A}_{01} = (kTV/M_1)(\partial C_1/\partial\mu_0)_{T,\mu_1,C_2} \tag{3.3.46}$$

$$\langle A_1{}^2\rangle = -kT\mathcal{A}_{11} = (kTV/M_1)(\partial C_1/\partial\mu_1)_{T,\mu_0,C_2}$$

The expression for $\langle A_1{}^2\rangle$ is obtained from that for $\langle A_0{}^2\rangle$ with subscript 0 replaced by 1. Introduction of these relations into Eq. (3.3.40) yields

$$\begin{aligned}\langle(\Delta\tilde{n})^2\rangle = \frac{kT}{V}\Bigg[&\left(\frac{\partial\tilde{n}}{\partial C_0}\right)^2_{C_1,C_2} M_0\left(\frac{\partial C_0}{\partial\mu_0}\right)_{T,\mu_1,C_2}\\ &+2\left(\frac{\partial\tilde{n}}{\partial C_0}\right)_{C_1,C_2}\left(\frac{\partial\tilde{n}}{\partial C_1}\right)_{C_0,C_2} M_0\left(\frac{\partial C_1}{\partial\mu_0}\right)_{T,\mu_1,C_2}\\ &+\left(\frac{\partial\tilde{n}}{\partial C_1}\right)^2_{C_0,C_2} M_1\left(\frac{\partial C_1}{\partial\mu_1}\right)_{T,\mu_D,C_2}\Bigg]\\ &+\frac{kT}{V}\left(\frac{\partial\tilde{n}}{\partial C_2}\right)^2_{T,\mu_D} M_2\left(\frac{\partial\mu_2}{\partial C_2}\right)^{-1}_{T,\mu_D}\end{aligned} \tag{3.3.47}$$

If this is combined with Eqs. (3.2.57) and (3.2.62), the desired expression for $R(0)$ at constant T and μ_D is derived. Designating this $R(0)$ by R^*, we have

$$R^* = R_c{}^{*o} + \Delta R_c{}^* \tag{3.3.48}$$

where

$$\begin{aligned}R_c{}^{*o} = KRT\Bigg[&\left(\frac{\partial\tilde{n}}{\partial C_0}\right)^2_{C_1,C_2} M_0\left(\frac{\partial C_0}{\partial\mu_0}\right)_{T,\mu_1,C_2}\\ &+2\left(\frac{\partial\tilde{n}}{\partial C_0}\right)_{C_1,C_2}\left(\frac{\partial\tilde{n}}{\partial C_1}\right)_{C_0,C_2} M_0\left(\frac{\partial C_1}{\partial\mu_0}\right)_{T,\mu_1,C_2}\\ &+\left(\frac{\partial\tilde{n}}{\partial C_1}\right)^2_{C_0,C_2} M_1\left(\frac{\partial C_1}{\partial\mu_1}\right)_{T,\mu_0,C_2}\Bigg]\end{aligned} \tag{3.3.49}$$

and

$$\Delta R_c^* = KRT\,(\psi_2^*)^2\, M_2/(\partial\mu_2/\partial C_2)_{T,\mu_D} \tag{3.3.50}$$

with

$$\psi_2^* \equiv \left(\frac{\partial \tilde{n}}{\partial C_2}\right)_{T,\mu_D} \tag{3.3.51}$$

We note that the specific refractive index increment ψ_2^* is accessible to experimental measurement.

Since it follows from Eq. (2.3.65) that

$$(\partial\mu_2/\partial C_2)_{T,\mu_D} = RT\,[C_2^{-1} + M_2 B_2(T,\mu_D) + \ldots] \tag{3.3.52}$$

we see that, in the limit of vanishingly small C_2, ΔR_c^* diminishes to zero. The corresponding limiting value of R_c^{*o} includes contributions of density and concentration scattering of the mixed solvent. At nonzero C_2, interactions between the polymer solute and the two solvents contribute to R_c^{*o}, but we assume such coupling effects to be negligible. Thus we have

$$\Delta R_c^* = R^* - (R_c^{*o})_{C_2=0} = R^* - (R^*)^o \tag{3.3.53}$$

and we obtain from Eqs. (3.3.50) and (3.3.52)

$$\frac{K\,(\psi_2^*)^2\, C_2}{\Delta R_c^*} = \frac{1}{M_2} + B_2(T,\mu_D)\, C_2 + \ldots \tag{3.3.54}$$

This expression leads to a very important conclusion: even when a mixed solvent is used, the true molecular weight of a polymer solute can be determined by light scattering if the excess scattering intensity and the specific refractive index increment are measured at constant μ_D.

3.3.3. Light Scattering from Copolymer Solutions

In the case of a copolymer solution the distributions of both molecular weight and monomer composition over the solute molecules must be taken into consideration in formulating the theory of light scattering. Fractions having the same molecular weight may have different monomer compositions and thus different refractive index increments. For dilute copolymer solutions containing a single solvent, either at constant p or constant μ_0, we may, with this fact in mind, rearrange Eq. (3.3. 14) to the form

$$\frac{K\,(\psi_s)^2\, C_s}{R_c} = \frac{1}{M_{app}} + 2A_{2app}\, C_s + 3A_{3app}\, C_s^2 + \ldots \tag{3.3.55}$$

where

$$\psi_s \equiv \Sigma_i \xi_i \psi_i, \quad C_s \equiv \Sigma_i \xi_i C_i \tag{3.3.56}$$

$$M_{app} = \frac{1}{(\psi_s)^2} \Sigma_i \psi_i^2 \xi_i M_i \tag{3.3.57}$$

$$A_{2app} = \frac{1}{2\,(M_{app}\,\psi_s)^2} \Sigma_i \Sigma_j (\psi_i \xi_i M_i)(\psi_j \xi_j M_j) B_{ij} \tag{3.3.58}$$

$$\begin{aligned} A_{3app} = & \frac{1}{3\,(M_{app}\,\psi_s)^2} \Sigma_i \Sigma_j \Sigma_k (\psi_i \xi_i M_i)(\psi_j \xi_j M_j) \xi_k B_{ijk} \\ & - \frac{1}{3\,(\psi_s)^4\,(M_{app})^3} \Sigma_i \Sigma_j \Sigma_k \Sigma_\ell (\psi_i \xi_i M_i)(\psi_j \xi_j M_j) \\ & \times (M_k \xi_k)(M_\ell \xi_\ell)(\psi_\ell^2 B_{ik} B_{jk} - \psi_k \psi_\ell B_{ik} B_{j\ell}) \end{aligned} \tag{3.3.59}$$

As before, ξ_i denotes the weight fraction of the entire solute that consists of polymer component i having molecular weight M_{ri}. It must be noted that such a "component" is actually a mixture of species with a distribution of monomer composition. From Eq. (3.3.57) we see that the apparent molar mass M_{app} agrees with the true weight-average molar mass M_w of the solute only when all copolymer components have the same monomer composition so that $\psi_i = \psi_s$ for all $i \geqslant 1$. The problem of determining M_w from M_{app}, which is a measurable quantity, was solved by Stockmayer et al. [24] and also by Bushuk and Benoit [25].

To illustrate the argument we consider a binary copolymer which consists of monomers A and B and denote the weight fractions of A and B in the i-th component by f_{Ai} and f_{Bi}, respectively, so that $f_{Bi} = 1 - f_{Ai}$. The weight-average molar mass M_w and the overall weight fraction f_A of A monomers in the entire sample are expressed in terms of ξ_i as follows:

$$M_w = \Sigma_i M_i \xi_i, \quad f_A = \Sigma_i f_{Ai} \xi_i \tag{3.3.60}$$

The specific refractive index increments of homopolymers of A and B are denoted by ψ_A and ψ_B, respectively, and are assumed to be independent of chain length. It is further assumed that, regardless of the arrangement of A and B monomers in the individual copolymer molecules, the specific refractive index increment ψ_i of the i-th copolymer is expressed by

$$\psi_i = f_{Ai} \psi_A + f_{Bi} \psi_B \tag{3.3.61a}$$

Then, ψ_s is represented as

$$\psi_s = f_A \psi_A + f_B \psi_B \tag{3.3.61b}$$

We denote the deviation of the composition of copolymer i from the overall value by Δf_i, i.e.,

$$\Delta f_i \equiv f_{Ai} - f_A = f_B - f_{Bi} \tag{3.3.62}$$

and introduce two parameters σ and δ defined by

$$\sigma \equiv M_w^{-1} \Sigma_i \xi_i M_i \Delta f_i \tag{3.3.63a}$$

$$\delta^2 \equiv M_w^{-1} \Sigma_i \xi_i M_i (\Delta f_i)^2 \tag{3.3.63b}$$

where, by definition, $\Sigma_i \xi_i \Delta f_i = 0$. Therefore, σ is a measure of the correlation between composition distribution and mass distribution. This parameter is positive for a copolymer sample in which higher molecular weight fractions are richer in A monomer. The parameter δ is essentially a measure of the spread of the composition distribution. In terms of the various notations introduced above we can write

$$\Sigma_i \xi_i M_i \psi_i = [\psi_s + (\psi_A - \psi_B)\sigma]\, M_w \tag{3.3.64a}$$

$$\Sigma_i \xi_i M_i (\psi_i)^2 = [(\psi_s)^2 + 2\psi_s(\psi_A - \psi_B)\sigma + (\psi_A - \psi_B)^2 \delta^2]\, M_w \tag{3.3.64b}$$

Substitution of Eq. (3.3.64b) into Eq. (3.3.57) gives

$$M_{app} = M_w \left[1 + 2\frac{\Delta\psi}{\psi_s}\sigma + \left(\frac{\Delta\psi}{\psi_s}\right)^2 \delta^2\right] \tag{3.3.65}$$

with $\Delta\psi \equiv \psi_A - \psi_B$. Both $\Delta\psi$ and ψ_s, and hence the ratio $\Delta\psi/\psi_s$, may change as the solvent is varied. If experimental data for M_{app} are obtained as a function of $\Delta\psi/\psi_s$, it should be possible to evaluate M_w, σ, and δ by use of Eq. (3.3.65).

Similar expressions may be derived for A_{2app} and A_{3app} if appropriate expressions are assumed for B_{ij} and B_{ijk} as functions of f_{Ai} and M_i. We shall not go into the details of this problem, but simply note that A_{3app} can be negative even when A_{2app} is positive.

Figure 3.10 shows experimental values of M_{app} as a function of $\Delta\psi/\psi_s$ for two copolymers of styrene and methyl methacrylate in a number of solvents. The solid lines have been computed from Eq. (3.3.65) with $\sigma = 0$ for three different values of δ^2. Sample 1 (unfilled circles) has a broader distribution of composition than sample 2 (filled circles).

When a light scattering experiment is done with a solvent which is isorefractive to one monomer species, e.g., B in an A-B copolymer, we may put $\psi_B = 0$ and $\psi_s = f_A \psi_A$. In this case, if a plot of $K(\psi_s)^2 C_A / R_c$ against C_A (the mass concentration $f_A C_s$ of A monomers) is extrapolated to $C_A = 0$, the reciprocal of the ordinate intercept is

$$M_{wA} = \frac{1}{(\psi_A)^2 f_A} \Sigma_i (f_{Ai}\psi_A)^2 \xi_i M_i = \frac{1}{f_A} \Sigma_i (f_{Ai}\xi_i)(f_{Ai} M_i) \tag{3.3.66}$$

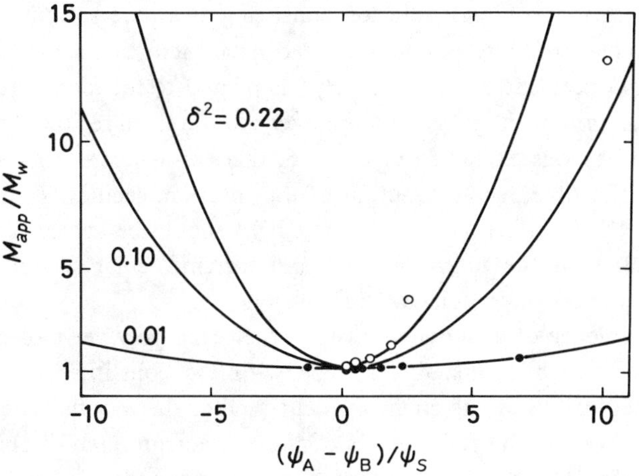

Fig. 3.10. Apparent molar masses M_{app} for two copolymers of styrene and methyl methacrylate as a function of the difference between refractive index increments ψ_A and ψ_B (Bushuk and Benoit [25]): ○, sample 1, ●, sample 2. The solid lines are calculated with $\sigma = 0$ and the indicated values of δ^2.

Here $f_{Ai}M_i$ is the mass of A monomers per mole of component i and $f_{Ai}\xi_i/f_A$ is the weight fraction of A monomers in the same component relative to the total weight of A. Therefore, M_{wA} represents the weight-average molar mass of A monomers in the entire copolymer sample. In a similar way, the weight-average molar mass M_{wB} of B monomers in the copolymer sample may be determined by using a solvent which has the same refractive index as the A monomer. The expressions for σ and δ^2 can be rewritten in terms of M_{wA} and M_{wB}:

$$2M_w\,\sigma = -f_A\,(M_w - M_{wA}) + f_B\,(M_w - M_{wB}) \tag{3.3.67a}$$

$$M_w\,\delta^2 = f_A\,f_B\,(M_{wA} + M_{wB} - M_w) \tag{3.3.67b}$$

These relations may be utilized to determine σ and δ separately.

3.3.4. Angular Dependence of Scattered Light – 1. General Theory

The intensity of light scattered from a polymer solution may be found to decrease with increasing scattering angle. This phenomenon occurs, because of destructive interference between light waves scattered from different portions of polymer molecules that are comparable in size with the wavelength of incident light. The present section is concerned with a general theoretical formulation.

We suppose that a binary solution consisting of a pure solvent and a monodisperse polymer solute is equilibrated at constant temperature and μ_0. For this solution the concentration scattering $R_c{}^*$ is related to the mean-square concentration fluctuation of the polymer component in the small region of volume V in the solution. As was explained in § 3.1.6, this mean-square value is expressed in terms of the distribution function of the solute molecules. Thus our aim in formulating the interference of scattered light should be achieved by knowing certain features of the intermolecular and intramolecular spatial distribution functions of the polymer molecules in solution.

For convenience of analysis we assume that each polymer molecule consists of P segments each comparable in size with the constituent monomer. No rigorous specification of the chain segments and of the way that they are linked is necessary, because the final results of the subsequent analysis are unaffected by this choice. The total volume of the solution and the total number of polymer molecules in it are denoted by V_t and N_t, respectively. These molecules are distinguished by subscript σ which runs from 1 to N_t. The segments in each polymer molecule are numbered according to a certain rule and expressed by a running index i from 1 to P. The configuration of polymer molecule σ in the solution is represented by the position vectors of its P segments, $\underset{\sim}{r}_{1_\sigma}, \underset{\sim}{r}_{2_\sigma}, \ldots, \underset{\sim}{r}_{P_\sigma}$, where the subscript i_σ indicates the i-th segment of polymer molecule σ. We introduce the notations (σ) and $(d\sigma)$ to represent the configuration of polymer molecule σ and a volume element of its $3P$-fold configuration space

$$(\sigma) \equiv \underset{\sim}{r}_{1_\sigma}, \underset{\sim}{r}_{2_\sigma}, \ldots, \underset{\sim}{r}_{P_\sigma}, \quad d(\sigma) \equiv d\underset{\sim}{r}_{1_\sigma}\, d\underset{\sim}{r}_{2_\sigma} \ldots d\underset{\sim}{r}_{P_\sigma} \tag{3.3.68}$$

which correspond to Eq. (3.1.78) for a monatomic molecule. The coordinates of one chain segment, for example, $\underset{\sim}{r}_{1_\sigma}$, can be regarded as the external coordinates which determine the location of the molecule σ in the solution, while the remaining (internal) coordinates determine the shape of the molecule σ.

a. Molecular Distribution Function and Segment-Pair Correlation Functions. By analogy with the case of monatomic molecules discussed in § 3.1.5, we define a one-body distribution function $F_1(\sigma)$ and a two-body distribution function $F_2(\sigma, \tau)$ for the polymer molecules under consideration. Thus the probability of finding polymer molecule σ in a configuration (position and shape) between (σ) and $(\sigma) + d(\sigma)$ is written $V_t^{-1} F_1(\sigma)\, d(\sigma)$. Also the joint probability that polymer molecule σ is in a configuration between (σ) and $(\sigma) + d(\sigma)$ and, at the same time, polymer τ is in a configuration between (τ) and $(\tau) + d(\tau)$ is written $V_t^{-2} F_2(\sigma, \tau) d(\sigma) d(\tau)$. The distribution functions $F_1(\sigma)$ and $F_2(\sigma, \tau)$ satisfy the normalization conditions

$$V_t^{-1} \int_{V_t} F_1(\sigma)\, d(\sigma) = 1 \tag{3.3.69a}$$

and

$$V_t^{-2} \int_{V_t} F_2(\sigma, \tau)\, d(\sigma)\, d(\tau) = 1 \tag{3.3.69b}$$

By definition we have the relation

$$F_1(\sigma) = V_t^{-1} \int_{V_t} F_2(\sigma, \tau)\, d(\tau) \tag{3.3.70}$$

which corresponds to Eq. (3.1. 81) for monatomic molecules. As in Eq. (3.1.82), we define the molecular pair correlation function $g_2(\sigma, \tau)$ by

$$g_2(\sigma, \tau) \equiv F_2(\sigma, \tau) - F_1(\sigma) F_1(\tau) \tag{3.3.71}$$

However, it should be noted that Eq. (3.1.83) does not hold for polymer molecules (or any molecules with internal degrees of freedom). Instead we have the relation for a homogeneous solution

$$f_1(i_\sigma) \equiv \int_{V_t} F_1(\sigma)\, d(\sigma)/d(i_\sigma) = 1 \tag{3.3.72}$$

where

$$(i_\sigma) \equiv \underset{\sim}{r}_{i_\sigma}, \quad d(i_\sigma) \equiv d\underset{\sim}{r}_{i_\sigma} \tag{3.3.73a}$$

and

$$d(\sigma)/d(i_\sigma) \equiv d\underset{\sim}{r}_{1_\sigma}\, d\underset{\sim}{r}_{2_\sigma} \ldots d\underset{\sim}{r}_{(i-1)_\sigma}\, d\underset{\sim}{r}_{(i+1)_\sigma} \ldots d\underset{\sim}{r}_{P_\sigma} \tag{3.3.73b}$$

If segment i_σ is fixed in a volume element $d(i_\sigma)$ around the position (i_σ) and $V_t^{-1} F_1(\sigma)$ is integrated over all positions of the other segments of the molecule σ, the result gives the probability that segment i_σ is in the volume element $d(i_\sigma)$. If this probability is written $V_t^{-1} f_1(i_\sigma) d(i_\sigma)$, the first equality of Eq. (3.3.72) is derived. If the solution is assumed to be homogeneous, $V_t^{-1} f_1(i_\sigma)$ $d(i_\sigma)$ is simply equal to $V_t^{-1} d(i_\sigma)$, so that $f_1(i_\sigma) = 1$. In this way, the second equality of Eq. (3.3.72) is obtained. The function $f_1(i_\sigma)$ could be called the one-body segment distribution function.

We integrate $F_1(\sigma)$ over all segments of polymer molecule σ, except segments i_σ and j_σ which are fixed in space. The result is denoted by $f_2(i_\sigma, j_\sigma)$:

$$f_2(i_\sigma, j_\sigma) \equiv \int_{V_t} F_1(\sigma)\, d(\sigma)/d(i_\sigma)\, d(j_\sigma) \tag{3.3.74a}$$

The function $f_2(i_\sigma, j_\sigma)$ is the probability density for finding segment i_σ at (i_σ) and segment j_σ simultaneously at (j_σ). Therefore, it may be termed the intra-molecular two-body segment distribution function. It is also possible to regard $f_2(i_\sigma, j_\sigma)$ as the product of the probability densities for finding segment i_σ at

(i_σ) and segment j_σ at a distance $\underset{\sim}{r} = \underset{\sim}{r}_j - \underset{\sim}{r}_i$ from the position (i_σ). If the latter probability density is designated by $P_{ij}(\underset{\sim}{r})$, we can write

$$f_2(i_\sigma, j_\sigma) = f_1(i_\sigma) P_{ij}(\underset{\sim}{r}) = P_{ij}(\underset{\sim}{r}) \tag{3.3.74b}$$

because $f_1(i_\sigma) = 1$. The quantity $P_{ij}(\underset{\sim}{r})$ is called the **intramolecular segment-pair radial distribution function**. Its form depends only on the segment numbers i and j. Since $f_1(i_\sigma) = 1$, $f_2(i_\sigma, j_\sigma)$ is often confused with $P_{ij}(\underset{\sim}{r})$, but the two functions have different physical meanings. It should be noted that $P_{ij}(\underset{\sim}{r})$ is subject to the normalization condition

$$\int P_{ij}(\underset{\sim}{r})\, d\underset{\sim}{r} = 1 \tag{3.3.75}$$

For segments belonging to different polymer molecules σ and τ, we may define the intermolecular two-body segment distribution function $f_2(i_\sigma, k_\tau)$ and the intermolecular segment-pair correlation function $Q_{ik}(i_\sigma, k_\tau)$ by

$$f_2(i_\sigma, k_\tau) \equiv \int_{V_t} F_2(\sigma, \tau)\, d(\sigma)\, d(\tau)/d(i_\sigma)\, d(k_\tau) \tag{3.3.76}$$

and

$$Q_{ik}(i_\sigma, k_\tau) \equiv f_2(i_\sigma, k_\tau) - f_1(i_\sigma) f_1(k_\tau) = f_2(i_\sigma, k_\tau) - 1 \tag{3.3.77a}$$

It can be shown that $Q_{ik}(i_\sigma, k_\tau)$ is expressed in terms of $g_2(\sigma, \tau)$ as

$$Q_{ik}(i_\sigma, k_\tau) = \int_{V_t} g_2(\sigma, \tau)\, d(\sigma)\, d(\tau)/d(i_\sigma)\, d(k_\tau) \tag{3.3.77b}$$

For a pair of polymer molecules sufficiently separated from each other, $g_2(\sigma, \tau)$ vanishes, so that $Q_{ik}(i_\sigma, k_\tau)$ vanishes for any i and k.

The total segment concentration η_s^α at position α labelled by a vector $\underset{\sim}{r}^\alpha$ is given, as in Eq. (3.1.98), by

$$\eta_s^\alpha = \sum_{\sigma=1}^{N_t} \sum_{i_\sigma=1}^{P} \delta(\underset{\sim}{r}_{i_\sigma} - \underset{\sim}{r}^\alpha) \tag{3.3.78}$$

where $\delta(\underset{\sim}{r})$ is a three-dimensional delta function. The average η_s^α is obtained by the following operation:

$$\begin{aligned} \langle \eta_s^\alpha \rangle &= \frac{1}{V_t} \sum_{\sigma=1}^{N_t} \sum_{i_\sigma=1}^{P} \int_{V_t} \delta(\underset{\sim}{r}_{i_\sigma} - \underset{\sim}{r}^\alpha) F_1(\sigma)\, d(\sigma) \\ &= \frac{1}{V_t} \sum_{\sigma=1}^{N_t} \sum_{i_\sigma=1}^{P} \left[\int_{V_t} F_1(\sigma) \frac{d(\sigma)}{d(i_\sigma)} \right]_{(i_\sigma) = \underset{\sim}{r}^\alpha} = \frac{N_t P}{V_t} \end{aligned} \tag{3.3.79}$$

As expected, $\langle \eta_s{}^\alpha \rangle$ does not depend on position. For simplicity, it is designated by η_s in the ensuing discussion.

The product of segment concentrations $\eta_s{}^\alpha$ and $\eta_s{}^\beta$ at different positions α and β, i.e., $\underset{\sim}{r}^\alpha$ and $\underset{\sim}{r}^\beta$, is

$$\eta_s{}^\alpha \eta_s{}^\beta = \sum_{\sigma=1}^{N_t} \sum_{\tau=1}^{N_t} \sum_{i_\sigma=1}^{P} \sum_{k_\tau=1}^{P} \delta(\underset{\sim}{r}_{i_\sigma} - \underset{\sim}{r}^\alpha)\,\delta(\underset{\sim}{r}_{k_\tau} - \underset{\sim}{r}^\beta)$$

$$= \sum_{\sigma=1}^{N_t} \sum_{i_\sigma=1}^{P} \sum_{j_\sigma=1}^{P} \delta(\underset{\sim}{r}_{i_\sigma} - \underset{\sim}{r}^\alpha)\,\delta(\underset{\sim}{r}_{j_\sigma} - \underset{\sim}{r}^\beta)$$

$$+ \sum\sum_{\sigma \neq \tau} \sum_{i_\sigma=1}^{P} \sum_{k_\tau=1}^{P} \delta(\underset{\sim}{r}_{i_\sigma} - \underset{\sim}{r}^\alpha)\,\delta(\underset{\sim}{r}_{k_\tau} - \underset{\sim}{r}^\beta) \qquad (3.3.80)$$

Considering that N_t polymer molecules ($N_t \gg 1$) are identical, we obtain for the average of $\eta_s{}^\alpha \eta_s{}^\beta$

$$\langle \eta_s{}^\alpha \eta_s{}^\beta \rangle = \frac{N_t}{V_t} \sum_{i_\sigma=1}^{P} \sum_{j_\sigma=1}^{P} \int_{V_t} \delta(\underset{\sim}{r}_{i_\sigma} - \underset{\sim}{r}^\alpha)\,\delta(\underset{\sim}{r}_{j_\sigma} - \underset{\sim}{r}^\beta)\, F_1(\sigma)\, d(\sigma)$$

$$+ \frac{N_t^2}{V_t^2} \sum_{i_\sigma=1}^{P} \sum_{\substack{k_\tau=1 \\ \sigma \neq \tau}}^{P} \int_{V_t} \delta(\underset{\sim}{r}_{i_\sigma} - \underset{\sim}{r}^\alpha)\,\delta(\underset{\sim}{r}_{k_\tau} - \underset{\sim}{r}^\beta)\, F_2(\sigma,\tau)\, d(\sigma)\, d(\tau)$$

$$= \frac{\eta_s}{P} \sum_{i=1}^{P} \sum_{j=1}^{P} P_{ij}(\underset{\sim}{r}^{\alpha\beta}) + \frac{\eta_s{}^2}{P^2} \sum_{i=1}^{P} \sum_{k=1}^{P} [1 + Q_{ik}(\underset{\sim}{r}^\alpha, \underset{\sim}{r}^\beta)] \qquad (3.3.81)$$

where $\underset{\sim}{r}^{\alpha\beta} \equiv \underset{\sim}{r}^\beta - \underset{\sim}{r}^\alpha$, and the property $P_{ii}(\underset{\sim}{r}) = \delta(\underset{\sim}{r})$ has been used.

The fluctuation of segment concentration at $\underset{\sim}{r}^\alpha$ is defined by

$$\Delta\eta_s{}^\alpha \equiv \eta_s{}^\alpha - \langle \eta_s{}^\alpha \rangle \qquad (3.3.82)$$

Hence we obtain

$$\langle \Delta\eta_s{}^\alpha \Delta\eta_s{}^\beta \rangle = \langle \eta_s{}^\alpha \eta_s{}^\beta \rangle - (\eta_s)^2$$

$$= \frac{\eta_s}{P} \sum_{i=1}^{P} \sum_{j=1}^{P} P_{ij}(\underset{\sim}{r}^{\alpha\beta}) + \frac{\eta_s{}^2}{P^2} \sum_{i=1}^{P} \sum_{k=1}^{P} Q_{ik}(\underset{\sim}{r}^\alpha, \underset{\sim}{r}^\beta) \qquad (3.3.83)$$

This shows that the correlation of fluctuations in segment concentrations at two points is expressed in terms of the intramolecular segment-pair radial distribution function and the intermolecular segment-pair correlation function.

For the binary solution under consideration, if the variation of refractive index $\tilde{n}$ with solvent concentration is ignored, the fluctuation of $\tilde{n}$ which occurs in a small region of volume V is expressed by

$$\Delta \tilde{n} = \left(\frac{\partial \tilde{n}}{\partial C_s}\right)_{T, \mu_0} \Delta C_s = \frac{M_s}{N_A} \left(\frac{\partial \tilde{n}}{\partial C_s}\right)_{T, \mu_0} \Delta \eta_s \tag{3.3.84}$$

where C_s is the mass concentration of the polymer and M_s is the molar mass of a segment. Here we have chosen T, μ_0, and C_s as the variables to describe properties in the small region. Substituting Eq. (3.3.84) into Eq. (3.2.58) and inserting the resulting $G_n(\theta)$ into Eq. (3.2.57), we obtain for the scattering intensity due to concentration fluctuations at specified μ_0

$$R_c^*(\theta) = \frac{2\pi^2 \tilde{n}^2}{N_A \lambda_0^4} \left(\frac{\partial \tilde{n}}{\partial C_s}\right)^2_{T, \mu_0} MC_s H(\theta) \tag{3.3.85a}$$

where

$$H(\theta) \equiv \frac{1}{NP^2} \int_V \langle \Delta \eta_s^{\alpha} \Delta \eta_s^{\beta} \rangle \exp(i\underset{\sim}{s} \cdot \underset{\sim}{r}^{\alpha\beta})\, d\underset{\sim}{r}^{\alpha}\, d\underset{\sim}{r}^{\beta} \tag{3.3.85b}$$

Here $M = PM_s$ is the molar mass of the polymer solute and N is the average number of polymer molecules in the small region, so that

$$N = \eta_s V/P = VN_t/V_t \tag{3.3.86}$$

Figure A.1 illustrates the geometric relationships among the scattering volume V, the point of detection of scattered light, and the scattering vector $\underset{\sim}{s}$.

Substitution of Eq. (3.3.83) into Eq. (3.3.85b) yields

$$H(\theta) = P_1(\theta) + (\eta_s/P)\, Q(\theta) \tag{3.3.87}$$

where

$$P_1(\theta) \equiv \frac{1}{VP^2} \sum_{i=1}^{P} \sum_{j=1}^{P} \int_V P_{ij}(\underset{\sim}{r}^{\alpha\beta}) \exp(i\underset{\sim}{s} \cdot \underset{\sim}{r}^{\alpha\beta})\, d\underset{\sim}{r}^{\alpha}\, d\underset{\sim}{r}^{\beta} \tag{3.3.88}$$

$$Q(\theta) \equiv \frac{1}{VP^2} \sum_{i=1}^{P} \sum_{k=1}^{P} \int_V Q_{ik}(\underset{\sim}{r}^{\alpha}, \underset{\sim}{r}^{\beta}) \exp(i\underset{\sim}{s} \cdot \underset{\sim}{r}^{\alpha\beta})\, d\underset{\sim}{r}^{\alpha}\, d\underset{\sim}{r}^{\beta} \tag{3.3.89}$$

The integration variable $\underset{\sim}{r}^{\beta}$ in Eq. (3.3.88) may be replaced by $\underset{\sim}{r}^{\alpha\beta}$. Since V is much larger than the size of a polymer molecule, it follows from Eq. (3.3.88) in conjunction with Eq. (3.3.75) that $P_1(\theta)$ tends to unity as θ approaches zero, which means that no interference occurs at $\theta = 0$. Deviations of $P_1(\theta)$ from unity thus represent interference effects due to the spatial distribution of segments within a single polymer molecule. We call $P_1(\theta)$ the **intramolecular interference factor**.

Use of Eq. (3.3.77b) for Q_{ik} in Eq. (3.3.89) leads to

$$\begin{aligned} Q(0) &= \frac{1}{VP^2} \sum_{i=1}^{P} \sum_{k=1}^{P} \int_V Q_{ik}(\underset{\sim}{r}^\alpha, \underset{\sim}{r}^\beta)\, d\underset{\sim}{r}^\alpha\, d\underset{\sim}{r}^\beta \\ &= \frac{1}{V_t} \int_{V_t} g_2(\sigma, \tau)\, d(\sigma)\, d(\tau) \end{aligned} \tag{3.3.90}$$

In deriving this result we first change the variables $\underset{\sim}{r}^\beta$ to $\underset{\sim}{r}^{\alpha\beta}$ and note the condition that $Q_{ik}(\underset{\sim}{r}^\alpha, \underset{\sim}{r}^{\alpha\beta})$ vanishes as $|\underset{\sim}{r}^{\alpha\beta}|$ increases indefinitely. Then the range of integration with respect to $\underset{\sim}{r}^{\alpha\beta}$ is expanded from V to V_t. Since the result of integration does not depend on $\underset{\sim}{r}^\alpha$, the range of integration with respect to $\underset{\sim}{r}^\alpha$ can also be extended from V to V_t if the integral is multiplied by the factor V/V_t. The sum over i and k simply cancels P^2 in the denominator. In this way, we obtain the second equality in Eq. (3.3.90).

It is customary to rewrite Eq. (3.3.87) in the form

$$H(\theta) = P_1(\theta) + \eta Q(0) P_2(\theta) \tag{3.3.91}$$

with

$$\eta = N_t/V_t \tag{3.3.92a}$$

$$P_2(\theta) \equiv Q(\theta)/Q(0) \tag{3.3.92b}$$

We call $P_2(\theta)$ the **intermolecular interference factor**. Some of the general properties of $P_1(\theta)$ and $P_2(\theta)$ are presented below.

b. The Intramolecular Interference Factor $P_1(\theta)$. The mean-square distance $\underset{\sim}{r}_{ij}$ between segments i and j in a polymer molecule is represented by

$$\langle r_{ij}^2 \rangle = \int r^2 P_{ij}(\underset{\sim}{r})\, d\underset{\sim}{r} \tag{3.3.93}$$

Equation (3.3.88) may be transformed [see Eq. (3.2.59)] to

$$P_1(\theta) = \frac{1}{P^2} \sum_{i=1}^{P} \sum_{j=1}^{P} \int_0^\infty P_{ij}(r) \frac{\sin(sr)}{sr} 4\pi r^2\, dr \tag{3.3.94}$$

where s is the magnitude of scattering vector $\underset{\sim}{s}$, i.e.,

$$s = (4\pi/\lambda) \sin(\theta/2) \tag{3.3.95}$$

λ being the wavelength of light in the solution. Substituting the expansion

$$\frac{\sin(sr)}{sr} = 1 - \frac{1}{3!}(sr)^2 + \frac{1}{5!}(sr)^4 - \ldots \tag{3.3.96}$$

and performing the integration, we obtain from Eq. (3.3.94)

$$P_1(\theta) = 1 - \frac{16\pi^2}{3\lambda^2} \langle S^2 \rangle \sin^2\left(\frac{\theta}{2}\right) + \ldots \tag{3.3.97}$$

where

$$\langle S^2 \rangle = \frac{1}{2P^2} \sum_{i=1}^{P} \sum_{j=1}^{P} \langle r_{ij}^2 \rangle \tag{3.3.98}$$

Equation (3.3.97) was first derived by Debye [26] in conjunction with X-ray scattering.

Denoting the position of the center of mass of a polymer molecule by $\underset{\sim}{r}_G$ and that of segment i in the same molecule by $\underset{\sim}{r}_i$, we have by definition

$$\underset{\sim}{r}_G = P^{-1} \sum_{i=1}^{P} \underset{\sim}{r}_i \tag{3.3.99}$$

and hence

$$\begin{aligned} \langle r_G^2 \rangle &= P^{-2} \sum_{i=1}^{P} \sum_{j=1}^{P} \langle \underset{\sim}{r}_i \cdot \underset{\sim}{r}_j \rangle \\ &= P^{-1} \Sigma_i \langle r_i^2 \rangle - (2P^2)^{-1} \Sigma_i \Sigma_j \langle r_{ij}^2 \rangle \end{aligned} \tag{3.3.100}$$

This can be used to derive the relation

$$\begin{aligned} P^{-1} \Sigma_i \langle \underset{\sim}{S}_i^2 \rangle &= P^{-1} \Sigma_i \langle (\underset{\sim}{r}_i - \underset{\sim}{r}_G)^2 \rangle = P^{-1} \Sigma_i \langle r_i^2 \rangle - \langle r_G^2 \rangle \\ &= (2P^2)^{-1} \Sigma_i \Sigma_j \langle r_{ij}^2 \rangle = \langle S^2 \rangle \end{aligned} \tag{3.3.101}$$

where $\underset{\sim}{S}_i \equiv \underset{\sim}{r}_i - \underset{\sim}{r}_G$. Thus $\langle S^2 \rangle$ is the **mean-square radius of gyration** of a polymer molecule. Equation (3.3.98) is valid regardless of solution concentration and molecular shape.

c. The Intermolecular Interference Factor $P_2(\theta)$. In order that the function $g_2(\sigma, \tau)$, which governs $Q_{ik}(i_\sigma, k_\tau)$, may have nonzero value the molecules σ and τ must be very close. At least one pair of segments of the two molecules must come within the range of intersegmental forces. We denote such paired segments by j_σ and ℓ_τ and assume that only when

$$\underset{\sim}{r}_{j_\sigma \ell_\tau} = \underset{\sim}{r}_{\ell_\tau} - \underset{\sim}{r}_{j_\sigma} = 0 \tag{3.3.102}$$

can these segments interact, since intersegmental forces act only at distances far shorter than the wavelength of incident light. Figure 3.11 illustrates this situation.

In general, the intramolecular segment-pair radial distribution function is affected by other molecules. Therefore, in the situation where there is a segment

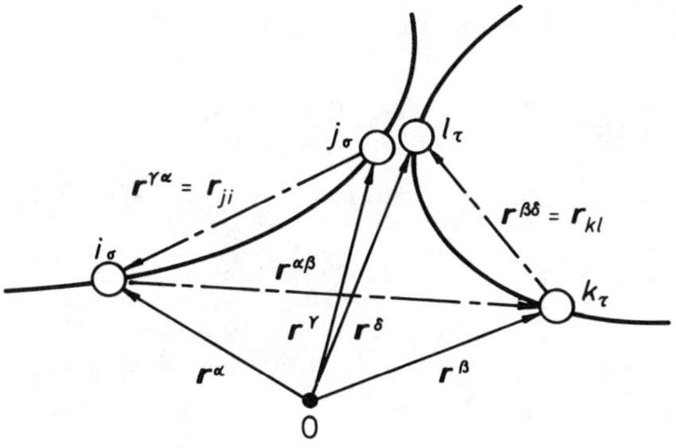

Fig. 3.11. Positions of segments i_σ, j_σ, ℓ_τ, and k_τ in the calculation of intermolecular interference factor $P_2(\theta)$.

ℓ_τ in the near vicinity of a segment j_σ, the distribution of $\underset{\sim}{r}_{ji}$ between segment j_σ and i_σ may no longer be given by P_{ji} defined by Eq. (3.3.74b). However, the P_{ji} function of a rigid polymer molecule is obviously intrinsic to the molecular structure, and that of a flexible chain molecule in a theta solvent is also intrinsic to the chain structure, because perturbations from other molecules vanish in such a solvent. As will be explained below, for a flexible polymer in a theta solvent, $P_2(\theta)$ is given simply by $[P_1(\theta)]^2$.

If the normalization condition for $P_{ij}(\underset{\sim}{r})$ is taken into account, $Q_{ik}(\underset{\sim}{r}^\alpha, \underset{\sim}{r}^\beta)$ can be written

$$Q_{ik}(\underset{\sim}{r}^\alpha, \underset{\sim}{r}^\beta) = P^{-2} \sum_{j_\sigma=1}^{P} \sum_{\ell_\tau=1}^{P} \int P_{ji}(\underset{\sim}{r}^{\gamma\alpha})\, Q_{ik}(\underset{\sim}{r}^\alpha, \underset{\sim}{r}^\beta)\, P_{k\ell}(\underset{\sim}{r}^{\beta\delta})\, d\underset{\sim}{r}^{\gamma\alpha}\, d\underset{\sim}{r}^{\beta\delta} \tag{3.3.103}$$

where $\underset{\sim}{r}^{\gamma\alpha}$ and $\underset{\sim}{r}^{\beta\delta}$ are the vectors illustrated in Fig. 3.11. Integration over $\underset{\sim}{r}^{\gamma\alpha}$ and $\underset{\sim}{r}^{\beta\delta}$ with $\underset{\sim}{r}^\alpha$ and $\underset{\sim}{r}^\beta$ fixed gives $Q_{ik}(\underset{\sim}{r}^\alpha, \underset{\sim}{r}^\beta)$ for the integral, which is independent of j_σ and ℓ_τ. Hence the double sum is simply equal to $P^2 Q_{ik}(\underset{\sim}{r}^\alpha, \underset{\sim}{r}^\beta)$.

Substitution of Eq. (3.3.103) into Eq. (3.3.89) yields

$$Q(\theta) = \frac{1}{VP^4} \Sigma_i \Sigma_j \Sigma_k \Sigma_\ell \int_V P_{ji}(\underset{\sim}{r}^{\gamma\alpha})\, Q_{ik}(\underset{\sim}{r}^\alpha, \underset{\sim}{r}^\beta)\, P_{k\ell}(\underset{\sim}{r}^{\beta\delta}) \times \exp(i\underset{\sim}{s} \cdot \underset{\sim}{r}^{\alpha\beta})\, d\underset{\sim}{r}^\alpha\, d\underset{\sim}{r}^\beta\, d\underset{\sim}{r}^{\gamma\alpha}\, d\underset{\sim}{r}^{\beta\delta} \tag{3.3.104}$$

Using the condition of Eq. (3.3.102) and referring to Fig. 3.11, we note that $\underset{\sim}{r}^{\alpha\beta}$ is represented by

$$\underset{\sim}{r}^{\alpha\beta} = \underset{\sim}{r}^{\alpha\gamma} + \underset{\sim}{r}^{\delta\beta} \tag{3.3.105}$$

With this, Eq. (3.3.104) may be transformed to

$$Q(\theta) = \left[\frac{1}{P^2} \Sigma_i \Sigma_j \int P_{ji}(\underset{\sim}{r}^{\gamma\alpha}) \exp (i\underset{\sim}{s} \cdot \underset{\sim}{r}^{\gamma\alpha}) d\underset{\sim}{r}^{\gamma\alpha} \right]$$
$$\times \left[\frac{1}{P^2} \Sigma_k \Sigma_\ell \int P_{k\ell}(\underset{\sim}{r}^{\beta\delta}) \exp (i\underset{\sim}{s} \cdot \underset{\sim}{r}^{\beta\delta}) d\underset{\sim}{r}^{\beta\delta} \right]$$
$$\times \frac{1}{V} \int Q_{ik}(\underset{\sim}{r}^{\alpha}, \underset{\sim}{r}^{\beta}) d\underset{\sim}{r}^{\alpha} d\underset{\sim}{r}^{\beta}$$

By referring to Eqs. (3.3.88) and (3.3.90) and noting that the last integral involving Q_{ik} does not depend on segment numbers i and k, we see that

$$Q(\theta) = [P_1(\theta)]^2 Q(0)$$

or

$$P_2(\theta) = [P_1(\theta)]^2 \qquad (3.3.106)$$

Zimm [27] was the first to derive this relation. Albrecht [28] has made a perturbation calculation of the influence exerted by the presence of other molecules on $P_{ij}(\underset{\sim}{r})$ of a flexible chain.

d. Zimm Plot. We find from comparison of Eqs. (3.1.130) and (3.3.90) that

$$Q^o(0) = -(2M^2/N_A)A_2 \qquad (3.3.107)$$

superscript zero indicating, as before, infinite dilution. Expanding the right-hand side of Eq. (3.3.91) in powers of $C_s = \eta_s M/N_A$, ignoring a term in C_s from $P_1(\theta)$, and substituting Eq. (3.3.107) for $Q^o(0)$, we obtain

$$H(\theta) = P_1{}^o(\theta) - 2MA_2 P_2{}^o(\theta) C_s + \ldots \qquad (3.3.108)$$

This is substituted into Eq. (3.3.85a), together with $P_2{}^o(\theta) = [P_1{}^o(\theta)]^2$ derived from Eq. (3.3.106), to obtain

$$K(\psi_s{}^*)^2 C_s/R_c{}^*(\theta) = [MP_1{}^o(\theta)]^{-1} + 2A_2 C_s + \ldots \qquad (3.3.109)$$

where $K = 2\pi^2 \tilde{n}^2/N_A \lambda_0{}^4$ and $\psi_s{}^* = (\partial \tilde{n}/\partial C_s)_{T,\mu_0}$. It should be noted that the coefficient of C_s in this expansion is inaccurate for two reasons: (1) neglect of the concentration dependence of $P_1(\theta)$ and (2) $P_2(\theta) \neq [P_1(\theta)]^2$. However, the neglected effects vanish at $\theta = 0$, so that the second virial coefficient A_2 can be correctly determined by analyzing experimental data in terms of Eq. (3.3.109) after extrapolation to zero scattering angle.

The $P_1(\theta)$ function for a polymer solute heterogeneous in molecular weight is shown to be represented by

$$P_1(\theta) = \sum_{k=1}^{r} \eta_k P_k{}^2 P_1{}^{(k)}(\theta) \Bigg/ \sum_{k=1}^{r} \eta_k P_k{}^2 \qquad (3.3.110)$$

where $P_1^{(k)}(\theta)$ is the intramolecular interference factor of polymer component k, P_k is the number of segments in its molecule, and η_k is its molecular concentration. Substitution of Eq. (3.3.97) for $P_1^{(k)}$, with $\langle S^2 \rangle$ replaced by $\langle S^2 \rangle_k$, yields

$$P_1(\theta) = 1 - \frac{16\pi^2}{3\lambda^2} \langle S^2 \rangle_z \sin^2 \frac{\theta}{2} + \ldots \tag{3.3.111}$$

where

$$\langle S^2 \rangle_z = \frac{\Sigma_k \eta_k P_k^2 \langle S^2 \rangle_k}{\Sigma_k \eta_k P_k^2} = \frac{\Sigma_k M_k \xi_k \langle S^2 \rangle_k}{M_w} \tag{3.3.112}$$

where M_k is the molar mass of component k, ξ_k is its weight fraction in the sample, and the subscript z indicates the z-average as was explained in § 2.2.1. As in Eq. (3.3.109), we find

$$K(\psi_s^*)^2 C_s / R_c^*(\theta) = [M_w P_1^o(\theta)]^{-1} + 2A_2^{LS} C_s + \ldots \tag{3.3.113}$$

Again, the coefficient of C_s is in error except at $\theta = 0$. At any rate it is concluded from these analyses that when C_s/R_c^* is expanded in powers of C_s, the coefficient for the linear term in C_s becomes almost independent of scattering angle and nearly equal to twice the second virial coefficient of the solution.

Substitution of Eq. (3.3.111) into Eq. (3:3.113) gives a double series in C_s and $\sin^2(\theta/2)$, i.e.,

$$\frac{K(\psi_s^*)^2 C_s}{R_c^*(\theta)} = \frac{1}{M_w} + \frac{16\pi^2}{3\lambda^2 M_w} \langle S^2 \rangle_z \sin^2 \frac{\theta}{2} + 2A_2^{LS} C_s + \ldots \tag{3.3.114}$$

It can be shown that up to the terms indicated here ψ_s^* and R_c^* may be replaced by the corresponding quantities ψ_s and R_c at constant T and p. Figure 3.12 illustrates some experimental values of $K(\psi_s)^2 C_s/R_c$ for acetone solutions of nitrocellulose at 25°C, with $\sin^2(\theta/2) + 2000(C_s/g\ \mathrm{cm}^{-3})$ as the abscissa [29]. The open and filled circles in the figure were determined respectively by extrapolating the data points at fixed C_s to $\theta = 0$ and those at fixed θ to $C_s = 0$. According to Eq. (3.3.114), the lines fitting the circles should meet at the same intercept which is equal to the reciprocal of M_w. Furthermore, the initial slopes of these lines afford values of $\langle S^2 \rangle_z$ and A_2^{LS}. The graphical representation of light scattering data as illustrated here for evaluation of M_w, $\langle S^2 \rangle_z$, and A_2^{LS} is called the **Zimm plot**. Figure 3.12 gives $M_{rw} = 1.27 \times 10^6$. On the other hand, osmotic pressure measurements gave $M_{rn} = 0.635 \times 10^6$ for this nitrocellulose fraction. Thus its M_{rw}/M_{rn} ratio is about 2. As will be

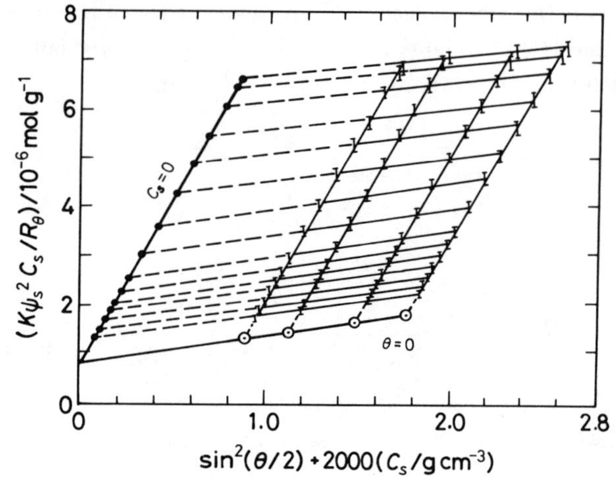

Fig. 3.12. Zimm plot for nitrocellulose in acetone at 25°C (Benoit et al. [29]).

explained in § 3.3.6, the line for $C_s = 0$ on the Zimm plot becomes linear for a polymer sample with a most probable distribution of molecular weight.

3.3.5. Angular Dependence of Scattered Light – 2. Relationships with Molecular Shape

a. Rigid Molecules. When the molecule is rigid, the relative positions of its segments are fixed. Hence we may write $P_{ij}(\underset{\sim}{r}^{\alpha\beta}) = \delta(\underset{\sim}{r}^{\alpha\beta} - \underset{\sim}{r}_{ij})$ for a pair of segments i and j in a rigid molecule which has a given orientation to the direction of scattering vector $\underset{\sim}{s}$. Substituting this P_{ij} into Eq. (3.3.88) and taking an average over all orientations of the molecule with equal probability, we obtain

$$P_1(\theta) = P^{-2} \sum_{i=1}^{P} \sum_{j=1}^{P} \langle \exp(i\underset{\sim}{s} \cdot \underset{\sim}{r}_{ij}) \rangle_{\mathrm{OR}} \tag{3.3.115}$$

where $\langle \ldots \rangle_{\mathrm{OR}}$ denotes the orientational average. Since $\langle \exp(i\underset{\sim}{s} \cdot \underset{\sim}{r}_{ij}) \rangle_{\mathrm{OR}}$ is equal to $\sin(sr_{ij})/(sr_{ij})$, Eq. (3.3.115) can be written

$$P_1(\theta) = P^{-2} \sum_{i=1}^{P} \sum_{j=1}^{P} \frac{\sin(sr_{ij})}{sr_{ij}} \tag{3.3.116}$$

If the positions of segments i and j relative to the center of mass of the molecule are denoted by $\underset{\sim}{S}_i$ and $\underset{\sim}{S}_j$, respectively, so that $\underset{\sim}{r}_{ij} = \underset{\sim}{S}_j - \underset{\sim}{S}_i$, Eq. (3.3.115) can also be expressed as

$$P_1(\theta) = \langle P^{-2} \sum_{j=1}^{P} \exp(i\underset{\sim}{s} \cdot \underset{\sim}{S}_j) \sum_{i=1}^{P} \exp(-i\underset{\sim}{s} \cdot \underset{\sim}{S}_i)\rangle_{\text{OR}} = \langle |U|^2 \rangle_{\text{OR}} \tag{3.3.117}$$

where $|U|$ is the absolute value of the complex quantity U defined by

$$U = P^{-1} \sum_{j=1}^{P} \exp(i\underset{\sim}{s} \cdot \underset{\sim}{S}_j) \tag{3.3.118a}$$

If the spatial distribution of segments may be considered continuous rather than discrete, U is given by

$$U = \frac{1}{P} \int_V \rho(\underset{\sim}{S}) \exp(i\underset{\sim}{s} \cdot \underset{\sim}{S})\, d\tau \tag{3.3.118b}$$

where $\rho(\underset{\sim}{S})$ is the segment density at the position specified by $\underset{\sim}{S}$, $d\tau$ is the volume element, and V is the region occupied by the molecule.

(1) Uniform sphere with radius a. In this case we have $\rho(\underset{\sim}{S}) = P/(4\pi a^3/3)$, and Eq. (3.3.118b) gives

$$U = 3(sa)^{-3} [\sin(sa) - sa\cos(sa)]$$

Because of the spherical symmetry, no orientational average is needed in applying Eq. (3.3.117), and it follows immediately that

$$P_1(\theta) = \left\{3(sa)^{-3} [\sin(sa) - sa\cos(sa)]\right\}^2 \tag{3.3.119}$$

(2) Uniform thin rod of length ℓ. In this case, we have $\rho(\underset{\sim}{S}) = P/\ell$, so that

$$U = \frac{2}{s\ell} \frac{\sin[(\ell s\cos\theta)/2]}{\cos\theta}$$

and, by Eq. (3.3.117),

$$P_1(\theta) = \frac{2}{s\ell} \int_0^{\ell s} \frac{\sin y}{y}\, dy - \left[\frac{\sin(\ell s/2)}{\ell s/2}\right]^2 \tag{3.3.120}$$

(3) Uniform thin disc with radius a. The final result of a somewhat more complicated calculation is

$$P_1(\theta) = \frac{2}{(sa)^2} \left[1 - \frac{J_1(2sa)}{sa}\right] \tag{3.3.121}$$

where J_1 is the Bessel function of the first kind of order one.

b. Gaussian Chains. A linear chain is called **Gaussian** when $P_{ij}(\underset{\sim}{r})$ is given by

$$P_{ij}(\underset{\sim}{r}) = \left(\frac{\beta_{ij}}{\pi^{1/2}}\right)^3 \exp(-\beta_{ij}^2 r^2), \quad \beta_{ij}^2 \equiv \frac{3}{2|j-i|b^2} \tag{3.3.122}$$

where b is the length of a segment. The Gaussian chain is important as a model of a flexible linear polymer molecule. It is outside the scope of this book to discuss its properties in detail.

Substitution of Eq. (3.3.122) into Eq. (3.3.93) yields

$$\langle r_{ij}^2 \rangle = |j - i| b^2, \quad \langle R^2 \rangle = Pb^2 \tag{3.3.123}$$

where $\langle R^2 \rangle$ is the mean-square end-to-end distance of the chain. With this $\langle r_{ij}^2 \rangle$, Eq. (3.3.98) becomes

$$\begin{aligned} \langle S^2 \rangle &= \frac{1}{P^2} \sum_{i<j}\sum (j - i) b^2 \approx Pb^2 \int_0^1 (1 - u) u\,du \\ &= Pb^2/6 = \langle R^2 \rangle / 6 \end{aligned} \tag{3.3.124}$$

This relation between $\langle S^2 \rangle$ and $\langle R^2 \rangle$ is one of the most fundamental properties of Gaussian chains.

Introducing Eq. (3.3.122) into Eq. (3.3.94), we get

$$P_1(\theta) = \frac{2}{P^2} \sum_{i<j}\sum \left[\frac{\beta_{ij}}{\pi^{1/2}} \int_{-\infty}^{\infty} \exp(-\beta_{ij}^2 x^2 + i s_x x)\, dx \right]^3 \tag{3.3.125}$$

x and s_x being the x-components of $\underset{\sim}{r}$ and $\underset{\sim}{s}$, respectively. Since

$$\int_{-\infty}^{\infty} \exp(-\beta_{ij}^2 x^2 + i s_x x)\, dx = \frac{\pi^{1/2}}{\beta_{ij}} \exp\left(-\frac{s_x^2}{4\beta_{ij}^2}\right) \tag{3.3.126}$$

Eq. (3.3.125) is written

$$P_1(\theta) = \frac{2}{P^2} \sum_{z=2}^{P} n_{z,P} [\exp(-s^2 b^2/6)]^{z-1} \tag{3.3.127}$$

where $n_{z,P}$ denotes the number of pairs of segments which are separated by a distance $z \equiv j - i + 1$ in a polymer chain of length P. For a linear chain we have

$$n_{z,P} = P - z + 1 = P - j + i, \quad (j > i) \tag{3.3.128}$$

Hence, with $u = (j - i)/P$, Eq. (3.3.127) yields for $P \gg 1$

$$P_1(\theta) = 2\xi^{-2}(e^{-\xi} - 1 + \xi) \tag{3.3.129}$$

where

$$\xi \equiv s^2 Pb^2/6 = s^2 \langle S^2 \rangle \tag{3.3.130}$$

This function of ξ is called the **Debye function**.

The behavior of $P_1(\theta)$ for small values of ξ is important in relation to light scattering determination of $\langle S^2 \rangle$. Equation (3.3.129) gives

$$[P_1(\theta)]^{-1} = 1 + (1/3)\,\xi + (1/36)\,\xi^2 - (1/540)\,\xi^3 + \ldots \quad (3.3.131)$$

$$[P_1(\theta)]^{-1/2} = 1 + (1/6)\,\xi - (1/1080)\,\xi^3 + \ldots \quad (3.3.132)$$

Therefore, when $\langle S^2 \rangle$ is relatively small so that terms of second and higher order in ξ in Eq. (3.3.131) may be neglected, the Zimm plot (at infinite dilution) against $\sin^2(\theta/2)$ gives a straight line, and $\langle S^2 \rangle$ may be evaluated accurately. As $\langle S^2 \rangle$ increases, the range of ξ for a linear Zimm plot becomes progressively narrower, and light scattering measurements down to lower angles become necessary for an accurate determination of $\langle S^2 \rangle$ (and of M_w as well). For example, if $\langle S^2 \rangle/\lambda^2 = 0.1$, ξ is already about unity at a scattering angle of 30°. In order to minimize this disadvantage of the Zimm plot, Berry [30] proposed to plot the square root of $K(\psi_s)^2 C_s/R_c(\theta)$ against $\sin^2(\theta/2)$, since, as Eq. (3.3.132) indicates, the expansion of $[P_1(\theta)]^{-1/2}$ does not contain a term in ξ^2. As the filled circles in Fig. 3.13 indicate, Berry's method is useful for the determination of $\langle S^2 \rangle$ of a fairly high molecular weight sample if accurate data points are available in the region of low scattering angles.

For large $\langle S^2 \rangle$ a method proposed by Fujita [32] is also useful. After the variables θ and ξ in Eq. (3.3.129) have been changed, respectively, to θ' and ξ', we multiply this equation by ξ'^2 and integrate with respect to ξ' from zero to ξ. The result is

$$\int_0^{\xi} P_1(\theta')\,\xi'^2\,d\xi' = 2\left(1 - e^{-\xi} - \xi + \frac{\xi^2}{2}\right) \quad (3.3.133)$$

This is combined with Eq. (3.3.129) to obtain

$$\frac{1}{P_1(\theta)} = 1 + \frac{1}{P_1(\theta)\,\xi^2}\int_0^{\xi} P_1(\theta')\,\xi'^2\,d\xi' \quad (3.3.134)$$

which may be transformed to

$$\frac{1}{P_1(\theta)} = 1 + \frac{32\pi^2}{3\lambda^2}\,\langle S^2 \rangle Z(\nu) \quad (3.3.135)$$

where

$$Z(\nu) = \frac{1}{P_1(\theta)\,\nu^{4/3}}\int_0^{\nu} P(\theta')\,\nu'\,d\nu' \quad (3.3.136a)$$

with

$$\nu \equiv \sin^3(\theta/2) \quad (3.3.136b)$$

The quantity $Z(\nu)$ can be evaluated by graphical integration from experimental values of $P_1(\theta)$. Equation (3.3.135) shows that plots of $1/P_1(\theta)$ against $Z(\nu)$

give a straight line over the entire range of Z if the polymer chain is Gaussian and that the slope of the line yields $\langle S^2 \rangle$. For non-Gaussian chains the initial slope of this plot may be used to determine $\langle S^2 \rangle$. Further, if the sample is polydisperse in molecular weight, the $\langle S^2 \rangle$ value so obtained is equal to $\langle S^2 \rangle_z$. The open circles in Fig. 3.13 show $1/P_1(\theta)$ versus Z for a narrow-distribution poly(α-methylstyrene) in benzene. In this system, the polymer chain must be non-Gaussian, because benzene is a good solvent; in fact, $\langle S^2 \rangle$ for this system has a molecular weight dependence that deviates markedly from Eq. (3.3.124). Nonetheless, the open circles follow a straight line over a wide range of the abscissa, suggesting that non-Gaussian character has little effect on the form of $P_1(\theta)$.

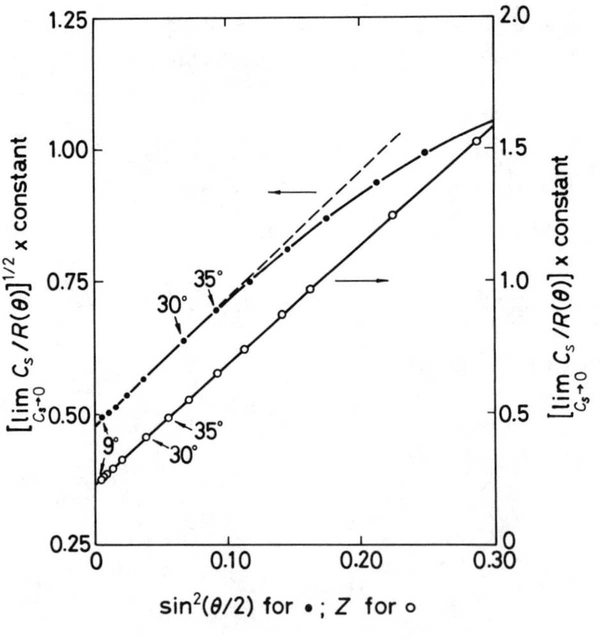

Fig. 3.13. Angular dependence of infinite-dilution values of scattered intensity for poly(α-methylstyrene) in benzene at 30°C (Utiyama et al. [31]): M_{rw} = $6.9_8 \times 10^6$, $\langle S^2 \rangle_z = 1.62 \times 10^{-10}$ cm^2. The upper plot is Berry's plot; the lower one, Fujita's plot.

3.3.6. Angular Dependence of Scattered Light – 3. Effects of Molecular Weight Distribution

Molecular weight distributions of polymer substances are closely related to the polymerization mechanisms by which they are produced. Although a discussion of this subject is outside the scope of the present treatise, some basic ideas are worth outlining.

We first consider polycondensation of a bifunctional monomer AB, as exemplified by the production of a polyester:

$$x\,(\mathrm{HO{-}R{-}COOH}) \rightleftarrows \mathrm{H(O{-}R{-}CO)}_x\,\mathrm{OH} + (x-1)\,\mathrm{H_2O}$$

We assume that the equilibrium constants for the condensation reactions

$$x\text{-mer} + y\text{-mer} \rightleftarrows (x+y)\text{-mer} + \mathrm{H_2O} \tag{3.3.137a}$$

$$(x-1)\text{-mer} + (y+1)\text{-mer} \rightleftarrows (x+y)\text{-mer} + \mathrm{H_2O} \tag{3.3.137b}$$

do not depend on the degree of polymerization of the species participating in the reactions. Then we have the relations

$$K = \frac{\eta_{x+y}\,\eta_{\mathrm{H_2O}}}{\eta_x\,\eta_y} = \frac{\eta_{x+y}\,\eta_{\mathrm{H_2O}}}{\eta_{x-1}\,\eta_{y+1}} \tag{3.3.138}$$

where K is a constant and η_x is the molecular concentration of x-mer. From the second relation we find that

$$\eta_x/\eta_{x-1} = \eta_{y+1}/\eta_y \tag{3.3.139}$$

which indicates that the ratio η_x/η_{x-1} does not depend on the subscript x. Thus if this ratio is denoted by α, we have

$$\eta_2 = \eta_1\alpha,\ \eta_3 = \eta_1\alpha^2, \ldots,\ \eta_x = \eta_1\alpha^{x-1} \tag{3.3.140}$$

and we recognize that α must be close to unity to obtain high molecular weight condensates. With Eq. (3.3.140), we get

$$\eta_s \equiv \sum_{x=1}^{\infty} \eta_x = \eta_1\,(1+\alpha+\alpha^2+\ldots) = \eta_1/(1-\alpha) \tag{3.3.141a}$$

$$g_1 \equiv \sum_{x=1}^{\infty} x\eta_x = \eta_1/(1-\alpha)^2 \tag{3.3.141b}$$

$$g_2 \equiv \sum_{x=1}^{\infty} x^2\eta_x = \eta_1\,(1+\alpha)/(1-\alpha)^3 \tag{3.3.141c}$$

$$g_3 \equiv \sum_{x=1}^{\infty} x^3\eta_x = \eta_1\,(1+4\alpha+\alpha^2)/(1-\alpha)^4 \tag{3.3.141d}$$

Therefore, when $\alpha \approx 1$, the number-average, weight-average, and z-average degrees of polymerization are given by

$$x_n = g_1/\eta_s = 1/(1-\alpha) \tag{3.3.142}$$

$$x_w = g_2/g_1 = (1+\alpha)/(1-\alpha) \approx 2x_n \tag{3.3.143}$$

$$x_z = g_3/g_2 = (1+4\alpha+\alpha^2)/(1-\alpha^2) \approx 3x_n \tag{3.3.144}$$

The number and weight distributions of x, denoted by $f(x)$ and $w(x)$, respectively, are represented by

$$f(x) = \eta_x/\eta_s = (1-\alpha)\alpha^{x-1} \tag{3.3.145}$$

$$w(x) = x\eta_x/g_1 = x(1-\alpha)^2\alpha^{x-1} \tag{3.3.146}$$

From its definition the quantity α is understood to be the probability that the B group of an AB monomer picked at random will have reacted with the A group of another AB monomer, i.e., the fraction of B groups reacted. Hence $f(x)$ is the probability that a polycondensate molecule taken at random from the reaction mixture has a degree of polymerization x, because the probability that it contains $(x-1)B$ units which have reacted is α^{x-1}, and the probability that it also has a terminal B group left unreacted is $1-\alpha$.

For $\alpha \approx 1$ and $x \gg 1$, we write

$$\alpha^{x-1} \approx \alpha^x = [1-(1-\alpha)]^x \approx \exp[-(1-\alpha)x] = \exp(-x/x_n) \tag{3.3.147}$$

so that Eq. (3.3.146) can be approximated as

$$w(x) \approx (x/x_n^{\,2}) \exp(-x/x_n), \quad (x_n \gg 1) \tag{3.3.148}$$

This agrees with Eq. (2.2.19) for $h = 1$, i.e., the most probable distribution of x.

The number of sequences $n_{z,x}$ of z monomers contained in an x-mer is 1 for a z-mer, 2 for a $(z+1)$-mer, and, in general,

$$n_{z,x} = x - z + 1 \tag{3.3.149}$$

The number q_z of such sequences contained in unit volume of the solution is given by

$$q_z = \sum_{x=z}^{\infty} n_{z,x}\eta_x = \eta_z/(1-\alpha)^2 = \eta_1\alpha^{z-1}/(1-\alpha)^2 \tag{3.3.150}$$

where Eq. (3.3.140) has been used for η_x.

The above analysis has been developed from the particular assumption leading to Eq. (3.3.138), but the results obtained are of more general applicability. For example, when free-radical polymerization terminates either by disproportionation alone or by chain transfer alone, the resulting product has a most probable distribution of molecular weight [33]. Random scission of a polymer gives, in the asymptotic limit where x_n is still much larger than unity, a sample with a most probable distribution, regardless of the molecular weight distribution of the original polymer [34]. The formulas following Eq. (3.3.140) apply to these reaction products, though the parameter α has a different physical meaning in each case.

Returning to the discussion of light scattering, we substitute Eq. (3.3.127) into Eq. (3.3.110) to obtain

$$P_1(\theta) = \frac{2}{\Sigma_k \eta_k P_k{}^2} \; \Sigma_k \eta_k \sum_{z=2}^{P_k} n_{z,P_k} \gamma^{z-1} \qquad (3.3.151)$$

where

$$\gamma \equiv \exp(-s^2 b^2/6) \qquad (3.3.152)$$

Equation (3.3.151) can be transformed to

$$P_1(\theta) = \frac{2}{\Sigma_k \eta_k P_k{}^2} \sum_{z=2}^{\infty} \left(\sum_{P_k=z}^{\infty} \eta_k n_{z,P_k} \right) \gamma^{z-1} \qquad (3.3.153)$$

The sum over P_k is just q_z in Eq. (3.3.150). Hence if η_k is given by Eq. (3.3.140), q_z equals $\eta_1 \alpha^{z-1}/(1-\alpha)^2$. In this case, the sum in the denominator of Eq. (3.3.153) is $(1+\alpha)\eta_1/(1-\alpha)^3$, according to Eq. (3.3.141c). Thus, for this η_k, Eq. (3.3.153) gives

$$\frac{1}{P_1(\theta)} = \frac{1+\alpha}{2(1-\alpha)} \left[\sum_{z=2}^{\infty} (\gamma\alpha)^{z-1} \right]^{-1} = \frac{(1+\alpha)(1-\gamma\alpha)}{2(1-\alpha)\gamma\alpha} \qquad (3.3.154)$$

On introduction of the condition $\alpha \approx 1$, the distribution of η_k assumes the most probable form; and it follows from Eq. (3.3.142) and Eq. (3.3.144) that

$$\alpha = 1 - (3/x_z) \qquad (3.3.155)$$

The length of a segment is much less than the wavelength of incident light, so that γ defined by Eq. (3.3.152) may be approximated by

$$\gamma = 1 - (s^2 b^2/6) \qquad (3.3.156)$$

With these expressions for α and γ, Eq. (3.3.154) may be written

$$1/P_1(\theta) \approx 1 + (1/3) s^2 \langle S^2 \rangle_z \qquad (3.3.157)$$

with $\langle S^2 \rangle_z = b^2 x_z/6$. The approximation involves neglect of quantities of order $b^2 s^2$ ($\propto b^2/\lambda^2$) and $b^2/\langle S^2 \rangle_z$ ($\propto 1/x_z$) in comparison with unity. Equation (3.3.157) shows that the Zimm plot of a polydisperse polymer sample with a most probable distribution of molecular weight becomes linear in $\sin^2(\theta/2)$

for $C_s = 0$, as is demonstrated in Fig. 3.12. This point was first clarified by Zimm [35] although the analysis presented here is due to Kajiwara et al. [36]. The method of the latter authors has the advantage of being capable of extension in a straightforward way to branched polymers. For such polymers Eq. (3.3.149) for n_{z, P_k} is no longer applicable, but, as in the case of linear polymers, the final result contains parameters α and γ combined in $s^2 \langle S^2 \rangle_z$.

Finally, we touch upon the asymptotic form of $P_1(\theta)$ for large scattering angles or, more precisely, for large ξ. It follows from Eq. (3.3.129) that for $\xi \gg 1$ we have

$$P_1(\theta) \approx 2\xi^{-1} - 2\xi^{-2} = 2(tP)^{-1} - 2(tP)^{-2} \tag{3.3.158}$$

where t denotes $s^2 b^2/6$. Introduction of this asymptotic expansion into Eq. (3.3.110) yields

$$P_1(\theta) = 2(\Sigma_k \eta_k P_k{}^2)^{-1} \Sigma_k P_k{}^2 \eta_k [(tP_k)^{-1} - (tP_k)^{-2}] \tag{3.3.159}$$

and therefore, asymptotically, [37]

$$\frac{1}{P_1(\theta)} \approx \frac{P_w}{2P_n}(1 + tP_n) = \frac{P_w}{2P_n}(1 + s^2 \langle S^2 \rangle_n) \tag{3.3.160}$$

where P_n and P_w are the weight-average and number-average relative chain length and $\langle S^2 \rangle_n = P_n b^2/6$ is the number-average mean-square radius of gyration. If the condition (3.3.158) holds for all components of a given polymer sample, two important quantities P_w/P_n and $\langle S^2 \rangle_n$ may be evaluated by applying Eq. (3.3.160) [29]. However, the condition is so restrictive that few systems permit this application.

3.3.7. Depolarization of Light Scattered from Polymer Solutions

In § 3.3.4 we calculated Rayleigh's ratio for dilute binary solutions of a monodisperse polymer solute for the case of isotropic local fluctuations of permittivity $\underset{\sim}{\epsilon}$, i.e., with $\Delta\underset{\sim}{\epsilon}$ as a scalar. The desired result was derived from Eq. (3.2.54) by substituting

$$\Delta\underset{\sim}{\epsilon}^{\alpha} = 2\epsilon_0 \tilde{n}\, \Delta\tilde{n}^{\alpha} = 2\epsilon_0 \left(\frac{\partial \tilde{n}}{\partial \eta_s}\right)_{T, \mu_0} \Delta\eta_s{}^{\alpha} \tag{3.3.161}$$

into Eq. (3.2.55). Here $\Delta\underset{\sim}{\epsilon}^{\alpha}$, $\Delta\tilde{n}^{\alpha}$, and $\Delta\eta_s{}^{\alpha}$ are the local fluctuations of permittivity, refractive index $\tilde{n}$, and segment density η_s at a point α, respectively, and ϵ_0 is the permittivity of vacuum. The subscripts attached to the derivative $\partial\tilde{n}/\partial\eta_s$ signify that the solution under consideration is at constant temperature and μ_0. The coefficient of $\Delta\eta_s{}^\alpha$ may be written $2\epsilon_0\tilde{n}(M_s/N_A)(\partial\tilde{n}/\partial C_s)_{T,\mu_0}$ with M_s and C_s the molar mass of a polymer segment and the mass concentration of segments, respectively. This coefficient is an effective (scalar) polarizability of a segment.

We now consider the case in which $\Delta\underset{\sim}{\epsilon}$ is anisotropic. The problem can be solved by again applying the basic equation (3.2.5), this time with $\Delta\underset{\sim}{\epsilon}^\alpha$ in Eq. (3.2.52) for $\underset{\sim}{p}^o$ treated as a tensor. It is worth remarking that Eq. (3.2.54) with Eq. (3.2.55) was obtained by treating $\Delta\underset{\sim}{\epsilon}$ as a scalar. As is shown in Fig. 3.6, we set an analyzer with its axis defined by the unit vector $\underset{\sim}{e}_a$ perpendicular to the vector $\underset{\sim}{r}$ which defines the point of observation P. From Eq. (3.2.5) we can derive, in the same way as for dilute gases, the relation

$$I_a = \frac{\pi^2 \epsilon}{2r^2 \epsilon_0{}^2 \lambda_0{}^4 V} \langle(\underset{\sim}{e}_a \cdot \underset{\sim}{p}^o)^2\rangle \qquad (3.3.162)$$

for the scattering light intensity I_a per unit scattering volume, observed at point P through the analyzer. In this expression for I_a, ϵ is the average permittivity of the medium, λ_0 is the wavelength of the incident beam in vacuum, and V is the scattering volume associated with the polarization $\underset{\sim}{p}^o$.

For the polymer solution under consideration, in which each chain segment acts as a discrete scatterer, Eq. (3.2.53) for $\underset{\sim}{p}^o$ should be written as a sum rather than an integral. Thus, by referring to Eq. (A.112), we have

$$\underset{\sim}{p}^o = \sum_{\sigma=1}^{N_t} \sum_{i=1}^{P} \underset{\sim}{p}_{i_\sigma}{}^o \exp(i\underset{\sim}{s} \cdot \underset{\sim}{r}_{i_\sigma})\, m(\underset{\sim}{r}_{i_\sigma}) \qquad (3.3.163)$$

Here, N_t is the number of polymer molecules, each consisting of P segments, in the entire solution of volume V_t; $\underset{\sim}{p}_i{}^o$ is the amplitude of the effective dipole moment of segment i in molecule σ; $\underset{\sim}{r}_{i_\sigma}$ is the instantaneous position of this segment; and $m(\underset{\sim}{r}_{i_\sigma})$ is the step function defined by Eq. (3.1.84). The symbol $\langle\ldots\rangle$ in Eq. (3.3.162) signifies an average over both segment position and orientation. At infinite dilution, the averaging over position can be performed, with the aid of the distribution function $F_1(\sigma)$ defined in § 3.3.4, to obtain

$$\frac{1}{V}\langle(\underset{\sim}{e}_a \cdot \underset{\sim}{p}^o)^2\rangle = \frac{1}{V}\Bigg[\sum_{i=1}^{P} \langle(\underset{\sim}{e}_a \cdot \underset{\sim}{p}_{i_\sigma}{}^o)^2\rangle \int_{V_t} m(\underset{\sim}{r}_{i_\sigma}) F_1(\sigma)\, d(\sigma)$$

$$+ \sum_{\substack{i_\sigma\, j_\sigma \\ i \neq j}}\sum \langle\underset{\sim}{e}_a \cdot \underset{\sim}{p}_{i_\sigma}{}^o\rangle\langle\underset{\sim}{e}_a \cdot \underset{\sim}{p}_{j_\sigma}{}^o\rangle \qquad (3.3.164)$$

$$\times \int_{V_t} \exp[i\underset{\sim}{s} \cdot (\underset{\sim}{r}_{i_\sigma} + \underset{\sim}{r}_{j_\sigma})]\, m(\underset{\sim}{r}_{i_\sigma})\, m(\underset{\sim}{r}_{j_\sigma}) F_1(\sigma)\, d(\sigma)\Bigg]$$

where the terms $\langle \underset{\sim}{e}_a \cdot \underset{\sim}{p}_{i_\sigma}{}^o\rangle\langle \underset{\sim}{e}_a \cdot \underset{\sim}{p}_{j_\tau}{}^o\rangle$ with $\sigma \neq \tau$ have been ignored, since they are proportional to the square of the average molecular concentration η:

$$\eta = N_t/V_t \tag{3.3.165}$$

The averages remaining on the right-hand side of Eq. (3.3.164) refer only to segment orientation, which is assumed to be independent of segment position. This assumption, however, will not be valid if the polymer molecule is stiff. With Eqs. (3.3.72), (3.3.74a), and (3.3.74b), Eq. (3.1.164) can be brought to the form

$$\frac{1}{V\eta}\langle(\underset{\sim}{e}_a \cdot \underset{\sim}{p}^o)^2\rangle = \sum_{i=1}^{P}\langle(\underset{\sim}{e}_a \cdot \underset{\sim}{p}_i{}^o)^2\rangle + \sum_{i=1}^{P}\sum_{\substack{j=1\\ i\neq j}}^{P}\langle \underset{\sim}{e}_a \cdot \underset{\sim}{p}_i{}^o\rangle\langle \underset{\sim}{e}_a \cdot \underset{\sim}{p}_j{}^o\rangle \int P_{ij}(\underset{\sim}{r})\,\frac{\sin(sr)}{sr}\,d\underset{\sim}{r} \tag{3.3.166}$$

When, as shown in Fig. 3.6, point P lies on the (y, z) plane and the analyzer is oriented in the x-direction, we have $\underset{\sim}{e}_a = \underset{\sim}{e}_V$, and Eq. (3.3.166) gives

$$\frac{1}{V\eta}\langle(\underset{\sim}{e}_V \cdot \underset{\sim}{p}^o)^2\rangle = [\alpha^2 P^2 P_1(\theta) + \tfrac{4}{5}\gamma^2 P]\,(E_{0x}{}^o)^2 + \tfrac{3}{5}\gamma^2 P(E_{0y}{}^o)^2 \tag{3.3.167}$$

Here, $P_1(\theta)$ is the intramolecular interference factor defined previously, and α and γ are the parameters defined according to Eqs. (3.2.21) and (3.2.22) with $\alpha_x{}'$, $\alpha_y{}'$, and $\alpha_z{}'$ replaced by the principal values of the effective polarizability tensor of each polymer segment.

When the analyzer is oriented in the direction $\underset{\sim}{e}_H$, we find from Eq. (3.3.166)

$$\frac{1}{V\eta}\langle(\underset{\sim}{e}_H \cdot \underset{\sim}{p}^o)^2\rangle = \tfrac{3}{5}P\gamma^2(E_{0x}{}^o)^2 + [P^2\alpha^2(\cos^2\theta)P_1(\theta) + \tfrac{1}{5}P\gamma^2(3+\cos^2\theta)]\,(E_{0y}{}^o)^2 \tag{3.3.168}$$

Substituting these expressions in Eq. (3.3.162), we obtain Rayleigh's ratios for four combinations of vertical and horizontal polarization of incident (v and h) and scattered (V and H) light [38]:

$$R_{Vv} = \frac{\pi^2}{\epsilon_0{}^2\lambda_0{}^4}\,[P^2\alpha^2 P_1(\theta) + \tfrac{4}{5}P\gamma^2]\,\eta \tag{3.3.169}$$

$$R_{Vh} = R_{Hv} = \frac{\pi^2}{\epsilon_0{}^2\lambda_0{}^4}\,(\tfrac{3}{5}P\gamma^2)\,\eta \tag{3.3.170}$$

$$R_{Hh} = \frac{\pi^2}{\epsilon_0{}^2 \lambda_0{}^4} [P^2 \alpha^2 (\cos^2 \theta) P_1(\theta) + \tfrac{1}{5} P \gamma^2 (3 + \cos^2 \theta)] \eta \qquad (3.3.171)$$

The degree of depolarization Δ_v defined by Eq. (3.2.45) is found to be

$$\Delta_v = \frac{(I_{Hv})_{\theta=\pi/2}}{(I_{Vv})_{\theta=0}} = \frac{3\gamma^2}{P(5\alpha^2 + P^{-1}\gamma^2)} \qquad (3.3.172)$$

Therefore, Δ_v vanishes for a sample of high molecular weight ($P \gg 1$). For example, Δ_v for an atactic polystyrene sample with $M_r = 5.1 \times 10^5$ in chlorobenzene is reported to be 2.6×10^{-3} [39], which indicates that, in this system, the anisotropy effect can be neglected. On the other hand, for an isotactic polystyrene sample with $M_r = 3.4 \times 10^5$ in the same solvent Δ_v is 0.191 [39]. For systems with such large Δ_v, adequate correction for the anisotropy becomes mandatory in order to obtain true molecular weights, molecular dimensions, and second virial coefficients from light scattering measurements.

Equations (3.3.169) through (3.3.171) may be rewritten

$$R_{Vv} = 2K(\psi_s)^2 MC_s \left[P_1(\theta) + \frac{4\Delta_v}{3 - 4\Delta_v} \right] \qquad (3.3.173)$$

$$R_{Hv} = R_{Vh} = 2K(\psi_s)^2 MC_s \left(\frac{3\Delta_v}{3 - 4\Delta_v} \right) \qquad (3.3.174)$$

$$R_{Hh} = 2K(\psi_s)^2 MC_s \left[(\cos^2 \theta) P_1(\theta) + \frac{\Delta_v}{3 - 4\Delta_v} (3 + \cos^2 \theta) \right] \qquad (3.3.175)$$

where $\psi_s = (\partial \tilde{n} / \partial C_s)_{T,p}$, and K is the constant defined by Eq. (3.2.18). When unpolarized incident light is used and no analyzer is inserted before the detector, Rayleigh's ratio R_{Uu} defined by Eq. (3.2.12) becomes

$$R_{Uu} = K(\psi_s)^2 MC_s \left[P_1(\theta) + \frac{\Delta_v}{3 - 4\Delta_v} \left(\frac{13 + \cos^2 \theta}{1 + \cos^2 \theta} \right) \right] \qquad (3.3.176)$$

Expansion in powers of $\sin^2 (\theta/2)$ yields

$$R_{Uu} = K(\psi_s)^2 C_s M_{app} \left[1 - \frac{16\pi^2}{3\lambda^2} \langle S^2 \rangle_{app} \sin^2 \left(\frac{\theta}{2} \right) + \ldots \right] \qquad (3.3.177)$$

where

$$M_{app} = M \left(\frac{3 + 3\Delta_v}{3 - 4\Delta_v} \right) \qquad (3.3.178)$$

$$\langle S^2 \rangle_{app} = \left(\frac{3 - 4\Delta_v}{3 + 3\Delta_v} \right) \langle S^2 \rangle - \left(\frac{\Delta_v}{3 + 3\Delta_v} \right) \left(\frac{9\lambda^2}{4\pi^2} \right) \qquad (3.3.179)$$

These expressions allow M and $\langle S^2 \rangle$ to be determined from experimental measurements of M_{app} and $\langle S^2 \rangle_{app}$ in conjunction with a separate determination of Δ_v. It is worth noting that the apparent mean-square radius $\langle S^2 \rangle_{app}$ can be negative if Δ_v is large. In fact, such an effect was observed in the system isotactic polystyrene + chlorobenzene mentioned above. A similar expression for A_{2app} is available, but we shall not discuss it here.

For semiflexible polymers it becomes necessary to consider correlations between orientations of segments, and intensity measurements in the (y, z) plane in Fig. 3.6 do not provide sufficient information for molecular characterization. Nagai [40] has made a theoretical calculation of the intensity of scattered light away from the (y, z) plane.

3.3.8. Light Scattering near the Critical Point

The determinant $|\mu|$ in the denominator of Eq. (3.3.7) vanishes at the critical point, as well as on the spinodal [see Eq. (1.3.65)]. For this reason critical opalescence due to concentration scattering should be observed in the vicinity of the critical point.

We have discussed density fluctuations in the vicinity of the gas-liquid critical point in § 3.1.4 and the consequent critical opalescence in § 3.2.3 and § 3.2.4. Critical opalescence in solution can be formulated simply by replacing density and pressure in the theory for a pure liquid by solute concentration and osmotic pressure, respectively. Accordingly, it might be deemed useless to introduce here the complex analysis of this subject worked out by Debye [41]. However, the Debye theory is very concrete in its formulation and may help the reader proceed to a study of a more general and sophisticated theory of critical phenomena in solutions. For this reason we offer a brief account of the Debye theory.

In the region near the critical point, concentration fluctuations become so large that spatial gradients of the local concentration can no longer be ignored. The problem here is to show that the probability of occurrence of a concentration fluctuation can be expressed in the same form as Eq. (3.1.67b) with $\Delta_p A$ given by Eq. (3.1.70). To demonstrate this we restrict our discussion to a binary solution and assume that the volume change on mixing is negligible. We denote the volume fraction of component i and its molecular concentration at a point α by ϕ_i^α and η_i^α, respectively. These quantities are related by

$$\phi_i^\alpha = \omega_i \eta_i^\alpha = (V_i/N_A)\eta_i^\alpha \tag{3.3.180}$$

where ω_i is the volume per molecule of component i and V_i is its molar volume. We express the molecular concentration of component j at a point β in the presence of a molecule of component i at point α by $\eta_j^\beta\,[1 + f_{ij}(\underset{\sim}{r}^{\alpha\beta})]$, where $\underset{\sim}{r}^{\alpha\beta} = \underset{\sim}{r}^\beta - \underset{\sim}{r}^\alpha$ and $f_{ij}(\underset{\sim}{r})$ is a molecular pair-correlation function, which approaches zero as $r = |\underset{\sim}{r}|$ increases indefinitely. Further, we denote the potential of the mean force acting between molecules i and j by $u_{ij}(r)$ and assume that it is short-range. In terms of these notations the potential of a molecule of component 0 or 1 at point α can be represented by

$$u_0^\alpha = \int [u_{00}\,\eta_0^\beta(1+f_{00}) + u_{01}\,\eta_1^\beta(1+f_{01})]\,d\underset{\sim}{r}^{\alpha\beta} \tag{3.3.181a}$$

$$u_1^\alpha = \int [u_{10}\,\eta_0^\beta(1+f_{10}) + u_{11}\,\eta_1^\beta(1+f_{11})]\,d\underset{\sim}{r}^{\alpha\beta} \tag{3.3.181b}$$

First, we consider the case in which concentration fluctuation can be neglected so that the solution may be regarded as homogeneous, not only macroscopically but also locally. Then the superscripts α and β are not needed, and we have

$$u_0 = \eta_0\,\bar{u}_{00} + \eta_1\,\bar{u}_{01}, \quad u_1 = \eta_0\,\bar{u}_{01} + \eta_1\,\bar{u}_{11} \tag{3.3.182}$$

where

$$\bar{u}_{ij} = \int u_{ij}(r)\,[1 + f_{ij}(\underset{\sim}{r})]\,d\underset{\sim}{r} = \bar{u}_{ji} \tag{3.3.183}$$

The intermolecular interaction energy U_ϕ per unit volume is, therefore, expressed by

$$\begin{aligned} U_\phi &= \tfrac{1}{2}\,(\eta_0\,u_0 + \eta_1\,u_1) \\ &= \tfrac{1}{2}\left(\frac{\bar{u}_{00}}{\omega_0^2}\,\phi_0^2 + \frac{2\bar{u}_{01}}{\omega_0\,\omega_1}\,\phi_0\,\phi_1 + \frac{\bar{u}_{11}}{\omega_1^2}\,\phi_1^2\right) \end{aligned} \tag{3.3.184}$$

The corresponding value of U_ϕ for the unmixed components is

$$U_\phi^o = \tfrac{1}{2}\left(\frac{\bar{u}_{00}}{\omega_0^2}\,\phi_0 + \frac{\bar{u}_{11}}{\omega_1^2}\,\phi_1\right) \tag{3.3.185}$$

if it is assumed that $f_{ij}(r)$ is not affected by the composition of the solution. This assumption corresponds to the Bragg-Williams approximation in the theory of lattice solutions. The energy of mixing per unit volume is thus given by

$$\Delta_m U_\phi = U_\phi - U_\phi^o = \Omega\,\phi_0\,\phi_1 \tag{3.3.186}$$

where

$$\Omega = \frac{\bar{u}_{01}}{\omega_0\,\omega_1} - \tfrac{1}{2}\left(\frac{\bar{u}_{00}}{\omega_0^2} + \frac{\bar{u}_{11}}{\omega_1^2}\right) \tag{3.3.187}$$

To obtain the corresponding expression for the entropy of mixing $\Delta_m S_\phi$ per unit volume, we assume that the Flory-Huggins lattice theory is applicable to continuous solutions, and thus write

$$\Delta_m S_\phi = -\frac{R}{V_0}\left(\phi_0 \ln \phi_0 + \frac{V_0}{V_1}\phi_1 \ln \phi_1\right) \tag{3.3.188}$$

Combination of Eqs. (3.3.186) and (3.3.188) gives the Helmholtz free energy of mixing $\Delta_m A_\phi$ per unit volume

$$\begin{aligned}\Delta_m A_\phi = \Delta_m U_\phi - T\Delta_m S_\phi = \frac{RT}{V_0}\Big[&(1-\phi_1)\ln(1-\phi_1) \\ &+ \frac{V_0}{V_1}\phi_1 \ln \phi_1 + \chi(1-\phi_1)\phi_1\Big]\end{aligned} \tag{3.3.189}$$

where

$$\chi \equiv V_0 \Omega/(RT) \tag{3.3.190}$$

If the solution is locally inhomogeneous due to concentration fluctuations we may proceed as follows. We denote the Cartesian coordinates of $\underset{\sim}{r}^{\alpha\beta}$ by x, y, and z, and expand η_j^β about η_j^α as (all derivatives referring to $x = y = z = 0$)

$$\begin{aligned}\eta_j^\beta &= \eta_j^\alpha + \left[x\left(\frac{\partial \eta_j^\alpha}{\partial x}\right) + y\left(\frac{\partial \eta_j^\alpha}{\partial y}\right) + z\left(\frac{\partial \eta_j^\alpha}{\partial z}\right)\right] \\ &\quad + \frac{1}{2}\left[x^2\left(\frac{\partial^2 \eta_j^\alpha}{\partial x^2}\right) + y^2\left(\frac{\partial^2 \eta_j^\alpha}{\partial y^2}\right) + z^2\left(\frac{\partial^2 \eta_j^\alpha}{\partial z^2}\right)\right. \\ &\quad \left. + 2xy\left(\frac{\partial^2 \eta_j^\alpha}{\partial x \partial y}\right) + 2yz\left(\frac{\partial^2 \eta_j^\alpha}{\partial y \partial z}\right) + 2zx\left(\frac{\partial^2 \eta_j^\alpha}{\partial z \partial x}\right)\right] + \cdots \\ &= \eta_j^\alpha + \underset{\sim}{r}^{\alpha\beta} \cdot \underset{\sim}{\nabla}\eta_j^\alpha + \frac{1}{2}\underset{\sim}{r}^{\alpha\beta}\underset{\sim}{r}^{\alpha\beta} : \underset{\sim}{\nabla}\underset{\sim}{\nabla}\eta_j^\alpha + \cdots\end{aligned} \tag{3.3.191}$$

where $\underset{\sim}{\nabla}$ is the gradient operator, $\underset{\sim}{\nabla}\underset{\sim}{\nabla}$ is the dyadic of $\underset{\sim}{\nabla}$, and $\underset{\sim}{A}:\underset{\sim}{B}$ is the double-dot product of tensors $\underset{\sim}{A}$ and $\underset{\sim}{B}$ (see § A.3 and A.4). Substituting Eq. (3.3.191) and recalling the assumption that $f_{ij}(\underset{\sim}{r})$ is independent of solution composition, and hence of concentration fluctuation, we can evaluate the integrals in Eqs. (3.3.181a) and (3.3.181b), e.g.,

$$\begin{aligned}\int u_{00}\,\eta_0^\beta(1+f_{00})\,d\underset{\sim}{r}^{\alpha\beta} &= \eta_0^\alpha \int u_{00}(1+f_{00})\,d\underset{\sim}{r} \\ &\quad + \tfrac{1}{6}(\nabla^2\eta_0^\alpha)\int r^2 u_{00}(1+f_{00})\,d\underset{\sim}{r} + \ldots\end{aligned} \tag{3.3.192}$$

where ∇^2 stands for the Laplacian operator $\partial^2/\partial x^2 + \partial^2/\partial y^2 + \partial^2/\partial z^2$. If we introduce a new parameter ℓ_{ij} defined by

$$\begin{aligned} \ell_{ij}^{\,2} &= \int r^2 u_{ij}(1+f_{ij})\, d\underset{\sim}{r} \Big/ \int u_{ij}(1+f_{ij})\, d\underset{\sim}{r} \\ &= (\bar{u}_{ij})^{-1} \int r^2 u_{ij}(1+f_{ij})\, d\underset{\sim}{r} = \ell_{ji}^{\,2} \end{aligned} \tag{3.3.193}$$

and neglect higher terms indicated by the dots in Eq. (3.3.192) and the corresponding equations, we obtain the following expressions for $u_0{}^\alpha$ and $u_1{}^\alpha$:

$$u_0{}^\alpha = \left[\eta_0{}^\alpha + \frac{\ell_{00}{}^2}{6}(\nabla^2\eta_0{}^\alpha)\right]\bar{u}_{00} + \left[\eta_1{}^\alpha + \frac{\ell_{01}{}^2}{6}(\nabla^2\eta_1{}^\alpha)\right]\bar{u}_{01} \tag{3.3.194a}$$

$$u_1{}^\alpha = \left[\eta_0{}^\alpha + \frac{\ell_{10}{}^2}{6}(\nabla^2\eta_0{}^\alpha)\right]\bar{u}_{10} + \left[\eta_1{}^\alpha + \frac{\ell_{11}{}^2}{6}(\nabla^2\eta_1{}^\alpha)\right]\bar{u}_{11} \tag{3.3.194b}$$

The local interaction energy per unit volume at point α is therefore

$$\begin{aligned} U_\phi{}^\alpha = \tfrac{1}{2}&\left[\frac{\bar{u}_{00}}{\omega_0{}^2}(\phi_0{}^\alpha)^2 + 2\frac{\bar{u}_{01}}{\omega_0\,\omega_1}\phi_0{}^\alpha\phi_1{}^\alpha + \frac{\bar{u}_{11}}{\omega_1{}^2}(\phi_1{}^\alpha)^2\right] \\ &+ \tfrac{1}{12}\left[\frac{\bar{u}_{00}}{\omega_0{}^2}\ell_{00}{}^2\phi_0{}^\alpha\nabla^2\phi_0{}^\alpha + \frac{\bar{u}_{01}}{\omega_0\,\omega_1}\ell_{01}{}^2(\phi_0{}^\alpha\nabla^2\phi_1{}^\alpha\right. \\ &\left. + \phi_1{}^\alpha\nabla^2\phi_0{}^\alpha) + \frac{\bar{u}_{11}}{\omega_{11}{}^2}\ell_{11}{}^2\phi_1{}^\alpha\nabla^2\phi_1{}^\alpha\right] \end{aligned} \tag{3.3.195}$$

where η_i has been replaced by ϕ_i according to Eq. (3.3.180). We define $\Delta\phi_1{}^\alpha$ by

$$\phi_1{}^\alpha = \langle\phi_1{}^\alpha\rangle + \Delta\phi_1{}^\alpha = \phi_1 + \Delta\phi_1{}^\alpha \tag{3.3.196}$$

and evaluate $\Delta_m U_\phi{}^\alpha = U_\phi{}^\alpha - U_\phi{}^o$ from Eqs. (3.3.195) and (3.3.185). The result reads

$$\begin{aligned} \Delta_m U_\phi{}^\alpha = \Omega\phi_0\phi_1 &+ \left[\left(\frac{\bar{u}_{11}}{\omega_1{}^2} - \frac{\bar{u}_{01}}{\omega_0\,\omega_1}\right)\phi_1 - \left(\frac{\bar{u}_{00}}{\omega_0{}^2} - \frac{\bar{u}_{01}}{\omega_0\,\omega_1}\right)\phi_0\right]\Delta\phi_1{}^\alpha \\ &- \Omega(\Delta\phi_1{}^\alpha)^2 + \tfrac{1}{12}\left\{\left[\left(\frac{\bar{u}_{11}}{\omega_1{}^2}\right)\ell_{11}{}^2 - \left(\frac{\bar{u}_{01}}{\omega_0\,\omega_1}\right)\ell_{01}{}^2\right]\phi_1\right. \\ &\left. - \left[\left(\frac{\bar{u}_{00}}{\omega_0{}^2}\right)\ell_{00}{}^2 - \left(\frac{\bar{u}_{01}}{\omega_0\,\omega_1}\right)\ell_{01}{}^2\right]\phi_0\right\}\nabla^2(\Delta\phi_1{}^\alpha) \\ &- H(\Delta\phi_1{}^\alpha)\nabla^2(\Delta\phi_1{}^\alpha) \end{aligned} \tag{3.3.197}$$

where H is defined by

$$H \equiv \tfrac{1}{6}\left[\frac{\bar{u}_{01}}{\omega_0\,\omega_1}\ell_{01}{}^2 - \tfrac{1}{2}\left(\frac{\bar{u}_{00}}{\omega_0{}^2}\ell_{00}{}^2 + \frac{\bar{u}_{11}}{\omega_1{}^2}\ell_{11}{}^2\right)\right] \tag{3.3.198}$$

Again we use the Flory-Huggins theory to find the local entropy of mixing per unit volume at point α. Neglecting terms of order higher than $(\Delta\phi_1{}^\alpha)^2$, we get

$$\begin{aligned}\Delta_m S_\phi{}^\alpha &= -\frac{R}{V_0}\left(\phi_0{}^\alpha \ln\phi_0{}^\alpha + \frac{V_0}{V_1}\phi_1{}^\alpha \ln\phi_1{}^\alpha\right)\\ &= -\frac{R}{V_0}\left(\phi_0 \ln\phi_0 + \frac{V_0}{V_1}\phi_1 \ln\phi_1\right)\\ &\quad -\left(\ln\phi_0 - \frac{V_0}{V_1}\ln\phi_1\right)\Delta\phi_1{}^\alpha\\ &\quad +\tfrac{1}{2}\left(\frac{1}{\phi_0} + \frac{V_0}{V_1\phi_1}\right)(\Delta\phi_1{}^\alpha)^2\end{aligned} \tag{3.3.199}$$

The increase in the Helmholtz free energy, $\Delta_\phi A$, which occurs in volume V as a result of concentration fluctuations can be calculated from

$$\Delta_\phi A = \int_V [(\Delta_m U_\phi{}^\alpha - T\Delta_m S_\phi{}^\alpha) - \Delta_m A_\phi]\, d\underset{\sim}{r}^\alpha \tag{3.3.200}$$

where $\Delta_m A_\phi$ is the Helmholtz free energy of mixing per unit volume of the corresponding fluctuation-free solution. Substituting $\Delta_m U_\phi{}^\alpha$, $\Delta_m S_\phi{}^\alpha$, and $\Delta_m A_\phi$ from Eqs. (3.3.197), (3.3.199), and (3.3.189), respectively, and imposing the boundary condition that $\underset{\sim}{\nabla}(\Delta\phi_1{}^\alpha)$ vanishes on the surface of the region V, we can transform Eq. (3.3.200) to

$$\Delta_\phi A = \tfrac{1}{2}\int_V \left\{A_{11}(\Delta\phi_1)^2 + 2H[\underset{\sim}{\nabla}(\Delta\phi_1)]^2\right\} d\underset{\sim}{r} \tag{3.3.201}$$

where

$$A_{11} = \left(\frac{\partial^2 A_\phi}{\partial\phi_1{}^2}\right)_T = \frac{RT}{V_0}\left(\frac{1}{1-\phi_1} + \frac{V_0}{V_1\phi_1} - 2\chi\right) \tag{3.3.202}$$

and A_ϕ is the equilibrium value of the Helmholtz free energy per unit volume. Here, use has been made of the formulas:

$$\int_V \nabla^2(\Delta\phi_1)\, d\underset{\sim}{r} = \int_S \underset{\sim}{\nabla}(\Delta\phi_1)\cdot d\underset{\sim}{S}$$

$$\int_V \Delta\phi_1 \nabla^2(\Delta\phi_1)\, d\underset{\sim}{r} = -\int_V [\underset{\sim}{\nabla}(\Delta\phi_1)]^2\, d\underset{\sim}{r} + \int_S \Delta\phi_1 \underset{\sim}{\nabla}(\Delta\phi_1)\cdot d\underset{\sim}{S}$$

Now it is seen that Eqs. (3.3.70) and (3.3.201) are of the same form inasmuch as replacement of $\Delta\rho/\rho$, κ, and b by $\Delta\phi_1$, $A_{11}{}^{-1}$, and $2H$ transforms the first equation to the second. It follows from Eq. (3.3.187) and Eq. (3.3.198) that

$$H = (\ell^2/6)\,\Omega \tag{3.3.203}$$

where

$$\ell^2 = \frac{(\bar{u}_{01}/\omega_0\,\omega_1)\ell_{01}{}^2 - (1/2)\,[(\bar{u}_{00}/\omega_0{}^2)\ell_{00}{}^2 + (\bar{u}_{11}/\omega_1{}^2)\ell_{11}{}^2]}{(\bar{u}_{01}/\omega_0\,\omega_1) - (1/2)\,[(\bar{u}_{00}/\omega_0{}^2) + (\bar{u}_{11}/\omega_1{}^2)]} \tag{3.3.204}$$

Referring to the definition of ℓ_{ij}, we see that ℓ represents an effective range of intermolecular forces. It is a more convenient parameter than H, because of its clearer physical meaning.

The probability of finding a concentration fluctuation $\Delta\phi_1$ is given by inserting Eq. (3.3.201) into the expression

$$w(\Delta\phi_1) = C \exp\left(-\frac{\Delta_\phi A}{kT}\right) \tag{3.3.205}$$

Thus the correlation of $\Delta\phi_1$ at two different points α and β can be calculated

$$\langle \Delta\phi_1{}^\alpha\,\Delta\phi_1{}^\beta \rangle = \frac{3\,kT}{4\pi\Omega\ell^2\,r^{\alpha\beta}} \exp\left[-\frac{r^{\alpha\beta}}{(\Omega\ell^2 A_{11}/3)^{1/2}}\right] \tag{3.3.206}$$

where $r^{\alpha\beta} = |\,\underset{\sim}{r}^{\alpha\beta}|$. Hence we obtain

$$L^2 \equiv \frac{\int (r^{\alpha\beta})^2 \langle \Delta\phi_1{}^\alpha\,\Delta\phi_1{}^\beta\rangle\, d\underset{\sim}{r}^{\alpha\beta}}{\int \langle \Delta\phi_1{}^\alpha\,\Delta\phi_1{}^\beta\rangle\, d\underset{\sim}{r}^{\alpha\beta}} = \frac{2\Omega}{A_{11}}\,\ell^2 \tag{3.3.207}$$

The length L is a measure of the correlation distance of concentration fluctuations. Equations (3.3.206) and (3.3.207) correspond, respectively, to Eqs. (3.1.75) and (3.1.77) for a pure liquid. An equation corresponding to Eq. (3.2.70) can also be derived, and yields

$$R(\theta) = K\left(\frac{\partial\tilde{n}}{\partial\phi_1}\right)^2 \frac{RT}{A_{11}\,[1 + (L^2 s^2/6)]} \tag{3.3.208}$$

The critical point of the solution under consideration can be calculated by use of $\Delta_m A_\phi$ (note that $\Delta_m G_\phi$ may be replaced by $\Delta_m A_\phi$ when no volume change occurs on mixing), and we find that

$$\phi_{1c} = V_0{}^{1/2}/(V_0{}^{1/2} + V_1{}^{1/2}) \tag{3.3.209}$$

$$2\chi_c = 2\,V_0\,\Omega/(RT_c) = (V_0{}^{1/2} + V_1{}^{1/2})^2/V_1 \tag{3.3.210}$$

Therefore, for a solution with the critical composition, Eq. (3.3.207) yields

$$L^2 = \frac{\ell^2\,T_c}{T - T_c} \tag{3.3.211}$$

and Eq. (3.3.208) gives

$$[R(\theta)]_c = K\left(\frac{\partial \tilde{n}}{\partial \phi_1}\right)^2 \frac{V_0 V_1}{(V_0^{1/2} + V_1^{1/2})^2} \frac{T/T_c}{(T/T_c) - 1 + (\ell^2 s^2/6)} \tag{3.3.212}$$

In solutions of small molecules, the effective range of intermolecular forces is of the order of 1 nm. In fact, a value of 1.47 nm is obtained for the parameter ℓ if Zimm's data [42] on the system carbon tetrachloride + perfluoromethylcyclohexane are analyzed by Eq. (3.3.212). When L/ℓ is about 50, so that the correlation length L is no longer negligible compared with the wavelength of light, $T - T_c$ is 0.1 K as estimated from Eq. (3.3.211) with T_c = 300 K.

In polymer solutions, $\langle S^2\rangle^{1/2}$ may be taken as a measure of the effective range of intermolecular forces, where, as before, $\langle S^2\rangle$ denotes the mean-square radius of gyration of a polymer molecule. Debye deduced that when the Flory-Krigbaum-Orofino equation is used for the polymer-polymer interaction potential, ℓ^2 could be equated to $\langle S^2\rangle$. However, later experiments [43] indicated $\langle S^2\rangle^{1/2}$ determined in this fashion with Debye's equation to be only about one-third of a presumably reasonable value. This aspect of Debye's theory is also unsatisfactory with respect to chain connectivity, but we will not discuss it further.

Recently, Lao et al. [44] performed accurate light scattering measurements on the system polystyrene + cyclohexane in the vicinity of its critical point. The open circles in Fig. 3.14 show their experimental values of $T/R(0)$ as a function of $(T/T_c) - 1$. The slope of the straight line fitting these data points gives

$$\gamma = 1.25 \tag{3.3.213}$$

for the critical exponent of A_{11}^{-1}, which corresponds to the isothermal compressibility in pure liquids. The large size of polymer molecules gives rise to an angular dependence of Rayleigh's ratio for their solutions, as was explained in § 3.3.4, but the term $L^2 s^2/6$ in Eq. (3.3.208) indicates that an additional angular dependence of scattered light arises from correlation of concentration fluctuations at different points. In the neighborhood of the critical point the contribution of the latter factor becomes overwhelming, because the correlation length L becomes indefinitely large. The filled circles in Fig. 3.14 show experimental values of L as a function of $(T/T_c) - 1$ for polystyrene in cyclohexane. The systematic downward deviation of the data points from the indicated straight line, when $(T/T_c) - 1$ is less than 10^{-3}, may be attributed to multiple scattering. The slope of the straight line gives 0.64 for the critical exponent ζ

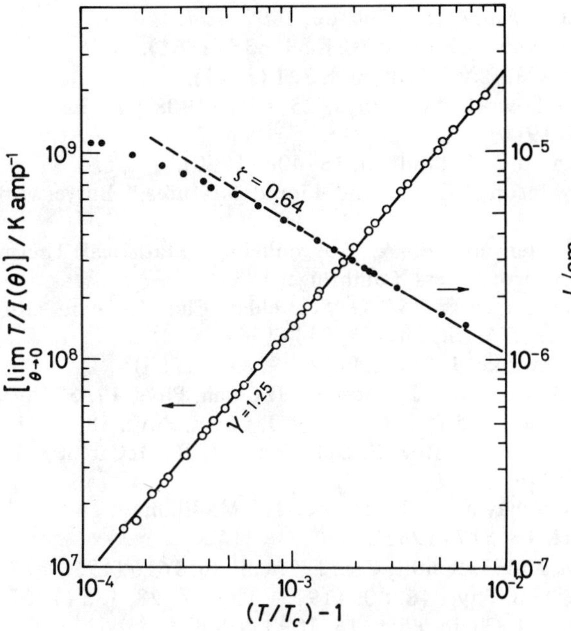

Fig. 3.14. Light scattering data for polystyrene in cyclohexane near the critical temperature (Lao et al. [44]): $M_{rn} = 1.96 \times 10^5$, $M_{rw}/M_{rn} < 1.02$, $T_c = 24.008°\text{C}$, $\phi_c = 0.0805$.

defined by $L \propto (T - T_c)^{-\zeta}$. Then, from Eq. (3.2.109), η is found to be 0.05. These critical exponents do not agree with the values $\gamma = 1.00$, $\zeta = 0.50$, and $\eta = 0$ that are predicted from Debye's theory described above.

References

1. S. Chandrasekhar, Rev. Mod. Phys. **15**, 1 (1943).
2. L. D. Landau and E. M. Lifshitz, "Statistical Physics," Chap. XII, Addison-Wesley, Reading, MA, 1958.
3. W. H. Stockmayer, J. Chem. Phys. **18**, 58 (1950).
4. H. Shogenji, Busseiron Kenkyu, **62** 1 (1953); T. Ooi, J. Polym. Sci. **28**, 459 (1958).
5. R. Kubo, "Thermodynamics and Statistical Physics," p. 446, Shokabo, Tokyo, 1961.
6. L. S. Ornstein and F. Zernike, Phys. Z. **19**, 134 (1918); **27**, 761 (1926).
7. J. G. Kirkwood and F. P. Buff, J. Chem. Phys. **19**, 774 (1951).
8. B. H. Zimm, J. Chem. Phys. **21**, 934 (1953).
9. B. H. Zimm and J. L. Lundberg, J. Phys. Chem. **60**, 425 (1956).
10. G. Gee and L. R. G. Treloar, Trans. Faraday Soc. **38**, 147 (1942).

11. H. B. Bull, J. Am. Chem. Soc. **66**, 1499 (1942).
12. M. Kurata, Ann. N.Y. Acad. Sci. **89**, 635 (1961).
13. J. Cabannes, J. Phys. Radium 8, 321 (1927).
14. M. Smoluchowski, Ann. Phys. **25**, 205 (1908); A. Einstein, Ann. Phys. **33**, 1275 (1910).
15. D. J. Coumou, J. Colloid Sci. **15**, 408 (1960).
16. J. S. Rowlinson, "Liquid and Liquid Mixtures," Butterworths, London, 1959.
17. R. H. Fowler and E. A. Guggenheim, "Statistical Thermodynamics," § 728, University Press, Cambridge, 1939.
18. M. A. Weinberger and W. G. Schneider, Can. J. Chem. **30**, 422 (1952).
19. M. E. Fisher, J. Math. Phys. **5**, 944 (1964).
20. G. S. Rushbrooke, J. Chem. Phys. **29**, 842 (1963).
21. H. C. Brinkman and J. J. Hermans, J. Chem. Phys. **17**, 574 (1949).
22. J. G. Kirkwood and R. J. Goldberg, J. Chem. Phys. **18**, 54 (1950).
23. R. H. Ewart, C. P. Roe, P. Debye, and J. R. McCartney, J. Chem. Phys. **14**, 687 (1946).
24. W. H. Stockmayer, L. D. Moore, Jr., M. Fixman, and B. N. Epstein, J. Polym. Sci. **16**, 517 (1955).
25. W. Bushuk and H. Benoit, Can. J. Chem. **36**, 1616 (1958).
26. P. Debye, Ann. Phys. **46**, 809 (1915); Phys. Z. **28**, 135 (1927).
27. B. H. Zimm, J. Chem. Phys. **16**, 1093 (1948).
28. A. C. Albrecht, J. Chem. Phys. **27**, 1014 (1957).
29. H. Benoit, A. M. Holtzer, and P. Doty, J. Phys. Chem. **58**, 635 (1954).
30. G. C. Berry, J. Chem. Phys. **44**, 4550 (1966).
31. H. Utiyama, Y. Tsunashima, and M. Kurata, J. Chem. Phys. **55**, 3133 (1971).
32. H. Fujita, Polym. J. **1**, 537 (1970).
33. See, for example, P. J. Flory, "Principles of Polymer Chemistry," Chap. VIII, Cornell Univ. Press, Ithaca, NY, 1953.
34. A. Charlesby, Proc. R. Soc. London Ser. A, **224**, 120 (1954).
35. B. H. Zimm, J. Chem. Phys. **16**, 1099 (1948).
36. K. Kajiwara, W. Burchard, and M. Gordon, Br. Polym. J. **2**, 110 (1970).
37. H. Benoit, J. Polym. Sci. **11**, 507 (1953).
38. H. Utiyama and M. Kurata, Bull. Inst. Chem. Res. Kyoto Univ. **42**, 128 (1964).
39. H. Utiyama, J. Phys. Chem. **69**, 4138 (1965).
40. K. Nagai, Polym. J. **3**, 67 (1972); **3**, 563 (1972).
41. P. Debye, J. Chem. Phys. **31**, 680 (1959).
42. B. H. Zimm, J. Phys. Colloid Chem. **54**, 1300 (1950).
43. P. Debye, H. Coll, and D. Woermann, J. Chem. Phys. **33**, 1746 (1960); P. Debye, B. Chu, and D. Woermann, J. Chem. Phys. **36**, 1803 (1962).
44. Q. H. Lao, B. Chu, and N. Kuwahara, J. Chem. Phys. **62**, 2039 (1975).

Sedimentation Equilibrium

CHAPTER 4

Sedimentation Equilibrium

4.1. Basic Equations

4.1.1. Equilibrium Conditions in a Centrifugal Force Field

Together with osmotic pressure and light scattering, sedimentation equilibrium in a centrifugal field is an important method for absolute determination of molecular weights of polymeric substances. When a solution rotates at constant angular velocity ω about a fixed axis, the centrifugal potential $\varphi(r)$ of unit mass of the solution at a radial distance r from the axis of rotation is given by

$$\varphi(r) = -(1/2)r^2\omega^2 \tag{4.1.1}$$

As before, we consider a solution containing $r + 1$ components and designate the molar mass of component i ($i = 0, 1, \ldots, r$) by M_i. Figure 4.1 depicts schematically the solution in a sector-shaped ultracentrifuge cell. The shaded area represents an air bubble introduced when the solution is injected into the cell for a sedimentation experiment. For simplicity, and to a good approximation, it can be assumed that no exchange of matter occurs between the solution and the air bubble. In other words, the solution is treated as a closed system. The cell is fixed in a rotor maintained at a given temperature T_0. Obviously, at sedimentation equilibrium, all portions of the solution as well as the air bubble are in thermal equilibrium with the rotor, and thus must be at temperature T_0. However, the pressure and composition in the solution vary with the radial distance r. It is from this composition distribution that the sedimentation equilibrium experiment affords information which leads to the molecular parameters of dissolved macromolecules.

Our first task is to find the conditions which govern sedimentation equilibrium. We start with Eq. (1.2.4), which is the most general criterion for thermodynamic equilibrium of a closed system. By virtue of the thermal equilibrium condition, the external temperature T_0 in this equation can be replaced by the temperature T of the solution, and the criterion reads

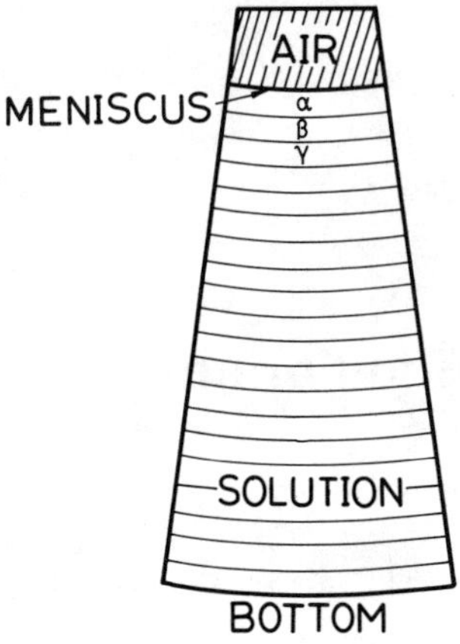

Fig. 4.1. Division of sector-shaped ultracentrifuge cell into a series of thin cylindrical shells of fixed volume. The shaded area represents an air bubble remaining when the cell is filled with solution.

$$\delta U - T\delta S + p_0\,\delta V \geqslant 0 \tag{4.1.2}$$

where the external pressure p_0 can be taken as the pressure exerted by the air bubble confined in the cell.

Now, as in Fig. 4.1, we divide the cell into a number of very thin cylindrical shells and take the centrifugal potential and all thermodynamic properties to be constant within each shell. We denote the internal energy, entropy, pressure, and volume of the α-th shell by U^α, S^α, p^α, and V^α, with α a running index to indicate the position of the shell. Equation (4.1.2) is then written

$$\Sigma_\alpha\,\delta U^\alpha - T\,\Sigma_\alpha\,\delta S^\alpha + p_0\,\Sigma_\alpha\,\delta V^\alpha \geqslant 0 \tag{4.1.3}$$

Since the solution in each shell is an open system, the variation δU^α must be expressed by Eq. (1.2.33) with the centrifugal potential $M_i\varphi^\alpha$ per unit amount of component i added to the chemical potential μ_i^α. Thus Eq. (4.1.3) becomes

$$\Sigma_\alpha \sum_{i=0}^{r} (\mu_i^\alpha + M_i\varphi^\alpha)\,\delta n_i^\alpha - \Sigma_\alpha\,(p^\alpha - p_0)\,\delta V^\alpha \geqslant 0 \tag{4.1.4}$$

Since the cylindrical shells are fixed relative to the cell, δV^α is zero for all shells but the one that faces the air bubble and thus includes the meniscus of the solution. Since the volume of the solution in this particular shell may undergo a change when components are virtually displaced to or from the adjacent shell, the volume variation for this shell cannot simply be equated to zero. However, at the meniscus, p^α equals p_0, because the pressure must be continuous.

Finally, the last term on the left-side of Eq. (4.1.4) vanishes, and we are left with

$$\Sigma_\alpha \sum_{i=0}^{r} (\mu_i^\alpha + M_i \varphi^\alpha)\, \delta n_i^\alpha \geqslant 0 \tag{4.1.5}$$

Since δn_i^α can be either positive or negative, it follows from this equation that the necessary and sufficient condition for sedimentation equilibrium is

$$\begin{aligned} \mu_i^\alpha + M_i \varphi^\alpha &= \text{constant for all shells,} \\ (i &= 0, 1, \ldots, r) \end{aligned} \tag{4.1.6}$$

Strictly speaking, this conclusion is valid only in the limit of vanishingly thin shells. In this limit, μ_i^α and φ^α are better represented as continuous variables $\mu_i(r)$ and $\varphi(r)$, and we write the desired condition in the form

$$\frac{d\mu_i(r)}{dr} + M_i \frac{d\varphi(r)}{dr} = 0, \quad (i = 0, 1, \ldots, r) \tag{4.1.7}$$

4.1.2. Differential Equations for Concentration Distributions

The next step is to transform Eqs. (4.1.7) to differential equations which govern concentration distributions at sedimentation equilibrium. For this purpose we must first choose $r + 2$ intensive state variables for μ_i. Among the variety of choices, we shall be concerned here with two cases: (a) T, p, and m_i ($i = 1, 2, \ldots, r$); (b) $T, \mu_0, \mu_1, \ldots, \mu_d, c_{d+1}, c_{d+2}, \ldots, c_{d+s}$ ($d + s = r$). In the latter choice we implicitly regard components 0 through d as small molecules constituting a mixed solvent (i.e., diffusible components) and components $d + 1$ through $d + s$ as macromolecules (i.e., nondiffusible components).

We derive the corresponding sedimentation equilibrium equations taking as an example a solution of r polymer components in a single solvent component 0.

With $T, \mu_0, c_1, c_2, \ldots, c_r$ as state variables, $d\mu_i$ ($i = 1, 2, \ldots, r$) at constant T is expressed by

$$d\mu_i = \left(\frac{\partial \mu_i}{\partial \mu_0}\right)_{T,c'} d\mu_0 + \sum_{j=1}^{r} \left(\frac{\partial \mu_i}{\partial c_j}\right)_{T,\mu_0,c_k} dc_j \tag{4.1.8}$$

where c' denotes the set of molar concentrations $c_1, c_2, \ldots, c_r$. At equilibrium both μ_0 and c' are functions of r only, so that from Eq. (4.1.8)

$$\frac{d\mu_i}{dr} = \left(\frac{\partial \mu_i}{\partial \mu_0}\right)_{T,c'} \frac{d\mu_0}{dr} + \sum_{j=1}^{r} \left(\frac{\partial \mu_i}{\partial c_j}\right)_{T,\mu_0,c_k} \frac{dc_j}{dr} \tag{4.1.9}$$

This is substituted into Eq. (4.1.7) to give

$$\left[M_i \frac{d\varphi}{dr} + \left(\frac{\partial \mu_i}{\partial \mu_0}\right)_{T,c'} \frac{d\mu_0}{dr}\right] = \sum_{j=1}^{r} \left(\frac{\partial \mu_i}{\partial c_j}\right)_{T,\mu_0,c_k} \frac{dc_j}{dr} \tag{4.1.10}$$

Since, again by Eq. (4.1.7), $d\mu_0/dr = -M_0\, d\varphi/dr$, the preceding equation can be written

$$\left[-M_i + M_0 \left(\frac{\partial \mu_i}{\partial \mu_0}\right)_{T,c'}\right] \frac{d\varphi}{dr} = \sum_{j=1}^{r} \left(\frac{\partial \mu_i}{\partial c_j}\right)_{T,\mu_0,c_k} \frac{dc_j}{dr} \tag{4.1.11}$$

The solution density ρ is expressed by

$$\rho = V^{-1} \sum_{j=0}^{r} n_j M_j = \sum_{j=0}^{r} c_j M_j \tag{4.1.12}$$

It follows that

$$\left(\frac{\partial \rho}{\partial c_i}\right)_{T,\mu_0,c_k} = M_i + M_0 \left(\frac{\partial c_0}{\partial c_i}\right)_{T,\mu_0,c_k} \tag{4.1.13}$$

but

$$\left(\frac{\partial c_0}{\partial c_i}\right)_{T,\mu_0,c_k} = \left(\frac{\partial n_0}{\partial n_i}\right)_{T,V,\mu_0,n_k} \tag{4.1.14}$$

The characteristic function appropriate for the set of state variables T, V, μ_0, $n_1, n_2, \ldots, n_r$ is $\mathcal{A} = A - \mu_0 n_0$ [see Eq. (1.4.79)]. It follows from Eq. (1.4.80) that

$$n_0 = -(\partial \mathcal{A}/\partial \mu_0)_{T,V,n_1,\ldots,n_r} \tag{4.1.15}$$

which may be combined with Eq. (1.4.81) to obtain

$$\left(\frac{\partial n_0}{\partial n_i}\right)_{T,V,\mu_0,n_k} = -\left(\frac{\partial^2 \mathcal{A}}{\partial n_i \partial \mu_0}\right)_{T,V,n_k} = -\left(\frac{\partial \mu_i}{\partial \mu_0}\right)_{T,c'} \tag{4.1.16}$$

With Eqs. (4.1.14) and (4.1.16), Eq. (4.1.13) can be rewritten

$$\left(\frac{\partial \rho}{\partial c_i}\right)_{T,\mu_0,c_k} = M_i - M_0 \left(\frac{\partial \mu_i}{\partial \mu_0}\right)_{T,c'} \tag{4.1.17}$$

which allows Eq. (4.1.11) to be expressed as

$$-\left(\frac{\partial \rho}{\partial c_i}\right)_{T,\mu_0,c_k} \frac{d\varphi}{dr} = \sum_{j=1}^{r} \left(\frac{\partial \mu_i}{\partial c_j}\right)_{T,\mu_0,c_k} \frac{dc_j}{dr} \tag{4.1.18}$$

where $i = 1, 2, \ldots, r$. The derivative $(\partial\rho/\partial c_i)_{T,\mu_0,c_k}$ is called the **density increment of component** i **at constant** μ_0 (on the molar concentration scale).

Next, when T, p, and m (denoting the set of molalities $m_1, m_2, \ldots, m_r$) are chosen as the state variables, we have at constant T

$$d\mu_i = \left(\frac{\partial\mu_i}{\partial p}\right)_{T,m} dp + \sum_{j=1}^{r} \left(\frac{\partial\mu_i}{\partial m_j}\right)_{T,p,m_k} dm_j \tag{4.1.19}$$

where $i = 0, 1, \ldots, r$, so that

$$\frac{d\mu_i}{dr} = V_i \frac{dp}{dr} + \sum_{j=1}^{r} \left(\frac{\partial\mu_i}{\partial m_j}\right)_{T,p,m_k} \frac{dm_j}{dr} \tag{4.1.20}$$

The Gibbs-Duhem relation at constant T gives

$$V dp = \sum_{j=0}^{r} n_j d\mu_j$$

which yields

$$\frac{dp}{dr} = \sum_{j=0}^{r} c_j \frac{d\mu_j}{dr} \tag{4.1.21}$$

Using this relation with Eqs. (4.1.7) and (4.1.12), we get

$$\frac{dp}{dr} = -\sum_{j=0}^{r} M_j c_j \frac{d\varphi}{dr} = -\rho \frac{d\varphi}{dr} \tag{4.1.22}$$

Substitution for φ from Eq. (4.1.1) then gives

$$\frac{dp}{dr} = \rho\omega^2 r \tag{4.1.23}$$

which is sometimes called the pressure equation. If Eqs. (4.1.20) and (4.1.22) are substituted into Eq. (4.1.7), then we get

$$-M_i(1 - v_i\rho)\frac{d\varphi}{dr} = \sum_{j=1}^{r} \left(\frac{\partial\mu_i}{\partial m_j}\right)_{T,p,m_k} \frac{dm_j}{dr} \tag{4.1.24}$$

where $i = 0, 1, \ldots, r$. The quantity $1 - v_i\rho$, where v_i is the partial specific volume of component i, is the **buoyancy factor** of component i.

Equation (4.1.24) can be expressed in a form similar to Eq. (4.1.18). We express ρ as

$$\rho = v_M^{-1}\left(1 + \sum_{j=1}^{r} M_j m_j\right) \tag{4.1.25}$$

where v_M is the volume of the solution per unit mass of component 0. Hence, using Eq. (1.1.71), we obtain

$$\left(\frac{\partial \rho}{\partial m_i}\right)_{T,p,m_k} = \frac{M_i}{v_M} - \frac{\rho}{v_M}\left(\frac{\partial v_M}{\partial m_i}\right)_{T,p,m_k} = \frac{M_i}{v_M}(1 - v_i \rho) \quad (4.1.26)$$

where $i = 1, 2, \ldots, r$. This relation allows Eq. (4.1.24) to be written

$$-v_M\left(\frac{\partial \rho}{\partial m_i}\right)_{T,p,m_k} \frac{d\varphi}{dr} = \sum_{j=1}^{r}\left(\frac{\partial \mu_i}{\partial m_j}\right)_{T,p,m_k} \frac{dm_j}{dr} \quad (4.1.27)$$

where $i = 0, 1, \ldots, r$. The quantity $(\partial \rho/\partial m_i)_{T,p,m_k}$ may be termed the **density increment of component i on the molality scale** [$(\partial \rho/\partial C_i)_{T,p,C_k}$ is usually called the **specific density increment of component i**].

In developing sedimentation equilibrium theory it has been customary to choose T, p, and c' as state variables. For this set of variables we have at constant T

$$d\mu_i = \left(\frac{\partial \mu_i}{\partial p}\right)_{T,c'} dp + \sum_{j=1}^{r}\left(\frac{\partial \mu_i}{\partial c_j}\right)_{T,p,c_k} dc_j \quad (4.1.28)$$

for $i = 0, 1, \ldots, r$. Hence if it is permissible to put

$$(\partial \mu_i/\partial p)_{T,c'} = V_i \quad (4.1.29)$$

the following set of sedimentation equilibrium equations can be derived from Eqs. (4.1.7) and (4.1.28):

$$-M_i(1 - v_i \rho)\frac{d\varphi}{dr} = \sum_{j=1}^{r}\left(\frac{\partial \mu_i}{\partial c_j}\right)_{T,p,c_k} \frac{dc_j}{dr} \quad (4.1.30)$$

where $i = 0, 1, \ldots, r$. Equation (1.1.38) holds among the molar concentrations $c_0, c_1, \ldots, c_r$ of all components. Hence even if c' is held constant, c_0 changes with p since V_i depends on p. This means that $(\partial \mu_i/\partial p)_{T,c'}$ does not refer to a fixed composition, so that Eq. (4.1.29) is invalid in general. The only exception is an incompressible solution for which V_i is independent of p for all components. Thus, although Eq. (4.1.28) is thermodynamically exact, Eq. (4.1.30) is valid only for an incompressible solution. Usually, sedimentation equilibrium experiments are carried out at relatively low rotor speed ($<$ 10,000 revolutions per min.), so that no large pressure is generated in the solution. In such a case, the solution may be regarded, in a very good approximation, as incompressible. For this reason the conventional sedimentation equilibrium equations given by Eq. (4.1.30) are useful if applied with caution.

Readers accustomed to Eqs. (4.1.24) and (4.1.30), which involve the buoyancy factor, find unfamiliar the form of Eqs. (4.1.18) and (4.1.27) in which the density increment appears. However, as was mentioned in § 1.1.3, and is evident from Eq. (4.1.26), the density increment must be measured to determine ν_i in the buoyancy factor. In other words, the density increment is the directly measured quantity, while the partial specific volume is a derived quantity.

4.2. Applications to Typical Systems

4.2.1. Polymer Homologs in a Single Solvent

The ultracentrifuge is an instrument in which a solution rotates about a fixed axis at a precisely controlled angular velocity, and concentration or concentration gradient distributions set up in the solution are recorded by an optical system [1]. The cell accommodating the test solution has the form of a circular sector of constant thickness. Figure 4.2 illustrates the geometrical arrangement of the cell and solution relative to the axis of rotation. The distances from the axis of rotation to the meniscus and bottom of the solution column are designated by r_a and r_b, respectively. In conventional sedimentation equilibrium experiments, $(r_b - r_a)$ is adjusted to about 2 mm, and r_a is about 6 cm.

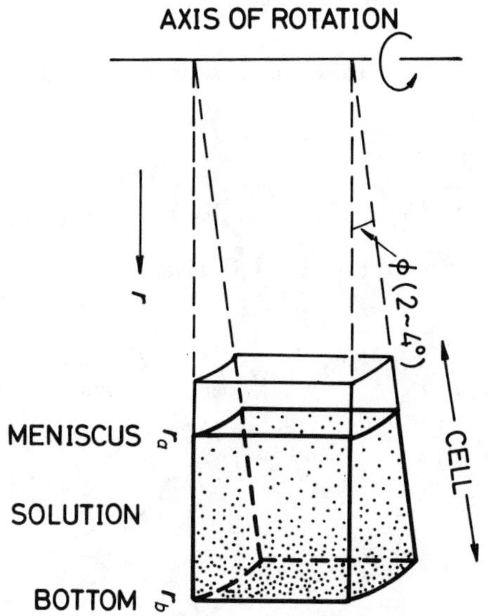

Fig. 4.2. The ultracentrifuge cell.

We now consider an incompressible system of a single solvent and r polymer solutes. We assume that the solution is so dilute that the ν_i for all polymer solutes may be replaced by their infinite-dilution values ν_i^o. Then we obtain from Eqs. (1.1.41a) and (1.1.41b)

$$\rho = \rho_0 + \sum_{j=1}^{r} (1 - \nu_j^o \rho_0) C_j \tag{4.2.1}$$

where C_j is the mass concentration of component j and ρ_0 is the density of the solvent (component 0). When compressibility is neglected, we find that $(\partial \rho / \partial C_j)_{T, \mu_0, C_k} = (\partial \rho / \partial C_j)_{T, p, C_k}$ and, therefore, that

$$(\partial \rho / \partial C_j)_{T, \mu_0, C_k} = 1 - \nu_j^o \rho_0 \tag{4.2.2}$$

which is independent of r.

Under the assumption $\nu_i = \nu_i^o$, we conclude that no volume change occurs in the solution when the components are mixed. It follows that r_a remains unchanged by centrifugation of the solution. The solution is a closed system and, thus by application of the law of conservation of mass, we find for all i

$$\int_{r_a}^{r_b} C_i(r) r dr = \int_{r_a}^{r_b} C_i^o r dr = \frac{C_i^o}{2} (r_b^2 - r_a^2) \tag{4.2.3}$$

where C_i^o is the mass concentration of component i in the solution before centrifugation.

On conversion from c_i to C_i and substitution for φ from Eq. (4.1.1), Eq. (4.1.18) becomes

$$M_i \left(\frac{\partial \rho}{\partial C_i} \right)_{T, \mu_0, C_k} \omega^2 r = \sum_{j=1}^{r} \left(\frac{\partial \mu_i}{\partial C_j} \right)_{T, \mu_0, C_k} \frac{dC_j}{dr} \tag{4.2.4}$$

which, upon introduction of Eq. (1.4.65), gives

$$\begin{aligned} \frac{\omega^2 r}{RT} \left(\frac{\partial \rho}{\partial C_i} \right)_{T, \mu_0, C_k} M_i C_i &= \frac{dC_i}{dr} + M_i C_i \sum_{j=1}^{r} B_{ij} \frac{dC_j}{dr} \\ &\quad + M_i C_i \sum_{j=1}^{r} \sum_{k=1}^{r} B_{ijk} C_k \frac{dC_j}{dr} + \ldots \end{aligned} \tag{4.2.5}$$

Now by defining

$$x \equiv \frac{r^2 - r_a^2}{r_b^2 - r_a^2} \tag{4.2.6}$$

$$\lambda_i \equiv \frac{\omega^2 (r_b{}^2 - r_a{}^2)}{2RT} \left(\frac{\partial \rho}{\partial C_i} \right)_{T, \mu_0, C_k} \tag{4.2.7}$$

we can write Eqs. (4.2.5) and (4.2.3) as

$$\begin{aligned} \lambda_i M_i C_i = \frac{dC_i}{dx} &+ M_i C_i \sum_{j=1}^{r} B_{ij} \frac{dC_j}{dx} \\ &+ M_i C_i \sum_{j=1}^{r} \sum_{k=1}^{r} B_{ijk} C_k \frac{dC_j}{dx} + \ldots \end{aligned} \tag{4.2.8}$$

and

$$\int_0^1 C_i(x)\, dx = C_i^o \tag{4.2.9}$$

Our task is to integrate Eqs. (4.2.8) with the condition of Eq. (4.2.9). In this case, we must note that λ_i is independent of r.

First, we consider a two-component system. Equations (4.2.8) reduce to a single equation

$$\lambda_1 M_1 C_1 = (1 + M_1 B_{11} C_1 + M_1 B_{111} C_1{}^2 + \ldots) \frac{dC_1}{dx} \tag{4.2.10}$$

Integrating this from $x = 0$ to $x = 1$ and using Eq. (4.2.9) for $i = 1$, we obtain

$$\begin{aligned} \lambda_1 M_1 C_1{}^o = (C_1{}^b - C_1{}^a) &+ \tfrac{1}{2} M_1 B_{11} [(C_1{}^b)^2 - (C_1{}^a)^2] \\ &+ \tfrac{1}{3} M_1 B_{111} [(C_1{}^b)^3 - (C_1{}^a)^3] + \ldots \end{aligned} \tag{4.2.11}$$

where $C_1{}^a$ and $C_1{}^b$ are the equilibrium values of C_1 at the meniscus and the bottom of the solution column, respectively. If the apparent molar mass M_{app} is defined by

$$M_{app} \equiv \Delta C_1 / (\lambda_1 C_1{}^o), \quad \Delta C_1 \equiv C_1{}^b - C_1{}^a \tag{4.2.12}$$

Eq. (4.2.11) yields

$$\begin{aligned} M_{app}{}^{-1} = M_1{}^{-1} &+ (1/2) B_{11} (C_1{}^a + C_1{}^b) \\ &+ (1/3) B_{111} [(C_1{}^a)^2 + C_1{}^a C_1{}^b + (C_1{}^b)^2] + \ldots \end{aligned} \tag{4.2.13}$$

This relation indicates that the molar mass M_1 of the solute and the binary interaction coefficient $B_{11}(T, \mu_0)$ may be determined by plotting the experimental quantity $M_{app}{}^{-1}$ against the mean concentration $\bar{C}_1 = (C_1{}^a + C_1{}^b)/2$ and extrapolating to $\bar{C}_1 = 0$ [2]. In the following, the relation between $\bar{C}_1$ and the initial solute concentration $C_1{}^o$ will be treated generally for a polymer solute heterogeneous in molecular weight.

What can be measured when a polydisperse sample dissolved in a single solvent is centrifuged is the distribution of the total mass concentration (or its gradient) of the solute. We denote the mass distribution by

$$C_s(x) = \sum_{i=1}^{r} C_i(x) \tag{4.2.14}$$

and the weight fraction of polymer component i in the sample by

$$\xi_i = C_i^o / C_s^o \tag{4.2.15}$$

where the superscript indicates the initial solution. For convenience we introduce a reduced concentration $\theta_i(x)$ defined by

$$\theta_i(x) = C_i(x)/C_i^o = C_i(x)/(\xi_i C_s{}^o) \tag{4.2.16}$$

Under the assumption that the system is incompressible and that $v_i = v_i{}^o$ ($i = 1, 2, \ldots, r$), Eq. (4.2.2) holds. If $v_i{}^o$ does not depend on M_i, as is known to be true to a very good approximation for polymer homologs, the parameter λ_i also becomes independent of M_i. Thus, putting $\lambda_1 = \lambda_2 = \ldots = \lambda_r = \lambda$ and rewriting Eq. (4.2.8) in terms of θ_i and ξ_i, we have

$$\lambda M_i \theta_i = \frac{d\theta_i}{dx} + C_s{}^o M_i \theta_i \sum_{j=1}^{r} B_{ij} \xi_j \frac{d\theta_j}{dx} + \ldots \tag{4.2.17}$$

where $i = 1, 2, \ldots, r$. For dilute solutions we can expand θ_i in powers of $C_s{}^o$ as

$$\theta_i = \theta_{i0} + \theta_{i1} C_s^o + \theta_{i2} (C_s{}^o)^2 + \ldots \tag{4.2.18}$$

Introduction of this into Eq. (4.2.17), followed by comparison of terms of the same order in $C_s{}^o$, yields a set of differential equations for $\theta_{i0}, \theta_{i1}, \ldots$. The first two of these equations are

$$d\theta_{i0}/dx - \lambda M_i \theta_{i0} = 0 \tag{4.2.19}$$

$$d\theta_{i1}/dx - \lambda M_i \theta_{i1} = -\lambda M_i \theta_{i0} \sum_{j=1}^{r} \xi_j M_j B_{ij} \theta_{j0} \tag{4.2.20}$$

By a similar operation we can derive from Eq. (4.2.9)

$$\int_0^1 \theta_{i0}\, dx = 1 \tag{4.2.21}$$

$$\int_0^1 \theta_{i1}\, dx = \int_0^1 \theta_{i2}\, dx = \ldots = 0 \tag{4.2.22}$$

Solution of Eqs. (4.2.19) and (4.2.20) subject to these boundary conditions yields

$$\theta_{i0}(x) = \frac{\lambda M_i \exp(\lambda M_i x)}{\exp(\lambda M_i) - 1} \tag{4.2.23}$$

$$\theta_{i1}(x) = M_i \theta_{i0}(x) \sum_{j=1}^{r} \xi_j B_{ij} \left[F_{ij}(\lambda, M_i, M_j) - \theta_{j0}(x)\right] \tag{4.2.24}$$

where

$$\begin{aligned} F_{ij}(\lambda, M_i, M_j) &= \frac{\lambda M_i M_j \left\{\exp\left[\lambda(M_i + M_j)\right] - 1\right\}}{(M_i + M_j)\left[\exp(\lambda M_i) - 1\right]\left[\exp(\lambda M_j) - 1\right]} \\ &= 1 + (1/12)\lambda^2 M_i M_j + 0(\lambda^4) \end{aligned} \tag{4.2.25}$$

Thus, we have

$$\theta_{i0}(1) - \theta_{i0}(0) = \lambda M_i \tag{4.2.26}$$

$$\theta_{i1}(1) - \theta_{i1}(0) = -\lambda M_i \sum_{j=1}^{r} \xi_j M_j B_{ij} F_{ij} \tag{4.2.27}$$

Introducing these expressions into $[\theta_i(1) - \theta_i(0)]$ as derived from Eq. (4.2.18) and converting from the reduced concentration θ_i to the actual mass concentration C_i by Eq. (4.2.16), we get

$$C_i^b - C_i^a = \lambda \xi_i M_i C_s^o - \lambda \xi_i M_i (C_s^o)^2 \sum_{j=1}^{r} \xi_j M_j B_{ij} F_{ij} + \ldots \tag{4.2.28}$$

Summation over all polymer components gives

$$C_s^b - C_s^a = \lambda M_w C_s^o - \lambda (C_s^o)^2 \sum_{i=1}^{r} \sum_{j=1}^{r} \xi_i \xi_j M_i M_j B_{ij} F_{ij} + \ldots \tag{4.2.29}$$

The generalization of Eq. (4.2.12) for M_{app} to the system under consideration is

$$M_{app} \equiv \Delta C_s / (\lambda C_s^o), \quad \Delta C_s \equiv C_s^b - C_s^a \tag{4.2.30}$$

Introduction of Eq. (4.2.29) leads to [3]

$$M_{app}^{-1} = M_w^{-1} + (C_s^o / M_w) \sum_{i=1}^{r} \sum_{j=1}^{r} \xi_i \xi_j M_i M_j B_{ij} F_{ij} + \ldots \tag{4.2.31}$$

which may be compared with the osmostic-pressure virial expansion and the light-scattering virial expansion for a polymer polydisperse in molecular weight dissolved in a single solvent. The point to note is that the coefficient of C_s^o, and the higher coefficients as well, depend on the parameter λ through F_{ij}.

The mean concentration $\bar{C}_s$ defined by $\bar{C}_s = (C_s{}^a + C_s{}^b)/2$ is given by

$$\begin{aligned} \bar{C}_s &= \tfrac{1}{2}\,\lambda C_s{}^o \sum_{i=1}^{r} \frac{\xi_i M_i\,[\exp(\lambda M_i) + 1]}{\exp(\lambda M_i) - 1} + O\,[(C_s{}^o)^2] \\ &= C_s{}^o\,[1 + (1/12)\,\lambda^2 M_z M_w + O\,(\lambda^4)] + O\,[(C_s{}^o)^2] \end{aligned} \quad (4.2.32)$$

where M_z is the z-average molar mass of the sample. In terms of $\bar{C}_s$ we can put Eq. (4.2.31) in the form [4]

$$M_{app}{}^{-1} = M_w{}^{-1} + 2A_2{}^{SE}\,\bar{C}_s + O\,(\bar{C}_s{}^2) \quad (4.2.33)$$

where $A_2{}^{SE}$ is defined by

$$A_2{}^{SE} = (2M_w{}^2)^{-1} \sum_{i=1}^{r} \sum_{j=1}^{r} \xi_i \xi_j M_i M_j B_{ij} f_{ij}\,(\lambda, M_i, M_j) \quad (4.2.34)$$

with

$$f_{ij}\,(\lambda, M_i, M_j) = 1 + (1/12)\,\lambda^2\,(M_i M_j - M_w M_z) + O\,(\lambda^4) \quad (4.2.35)$$

Ignoring terms of higher order than λ^2, we find that f_{ij} is much less dependent on λ than is F_{ij}. For small λ, $A_2{}^{SE}$ becomes approximately equal to $A_2{}^{LS}$, the light-scattering second virial coefficient.

The expansion coefficients in Eq. (4.2.31) or (4.2.33) increase with λ. Therefore, when experiments are done at large λ, the contribution from the higher expansion coefficients make it difficult to determine M_w by extrapolation of $M_{app}{}^{-1}$ to $\bar{C}_s = 0$. The **short-column method** of Van Holde and Baldwin [5], who originally proposed it for rapid attainment of sedimentation equilibrium, is also effective for reducing the magnitude of λ. In this method, the length of the solution column, $r_b - r_a$, is less than 2 mm. For an accurate determination of M_w and $A_2{}^{LS}$ by sedimentation equilibrium it is advisable to use the following double-extrapolation method. First, at a given $C_s{}^o$, M_{app} is determined with different rotor speeds and the resulting plot of $M_{app}{}^{-1}$ versus λ^2 is extrapolated to $\lambda^2 = 0$. Similar determinations of $(M_{app})_{\lambda^2=0}$ are made at different values of $C_s{}^o$. The desired M_w and $A_2{}^{LS}$ are then evaluated by extrapolating $1/(M_{app})_{\lambda^2=0}$ versus $C_s{}^o$ to $C_s{}^o = 0$; the intercept is $M_w{}^{-1}$ and the initial slope can be set equal to $2A_2{}^{LS}$. Actual applications of this method are found in reports by Utiyama et al. [6] and by Albright and Williams [7].

The variable λ method fails to be useful for polymer samples with very broad distributions of molecular weight. If we suppose that the molecular weight dependence of B_{ij} can be ignored so that $B_{ij} = B_2$ for $i, j = 1, 2, \ldots, r$, Eq. (4.2.34) becomes

$$A_2{}^{SE} = \frac{B_2}{2}\left[1 + \tfrac{1}{12}\,\lambda^2 M_w M_z \left(\frac{M_z}{M_w} - 1\right) + O\,(\lambda^4)\right] \quad (4.2.36)$$

If M_z/M_w is very large, the coefficient of λ^2 is large and higher-order terms cannot be neglected. Then it is not possible to choose a value of λ such that the series can be truncated without introducing an error greater than the experimental uncertainty in M_{app}.

For very large λ, Eq. (4.2.25) asymptotically approaches

$$F_{ij}(\lambda, M_i, M_j) = \frac{\lambda M_i M_j}{M_i + M_j} \tag{4.2.37}$$

Then the second term on the right-hand side of Eq. (4.2.31) is proportional to λ. Gordon et al. [8] thus proposed to evaluate M_w by extrapolating a plot of $M_{app}{}^{-1}$ versus λ to $\lambda = 0$ and applied this idea to branched polycondensates which had been prepared by advancing the reaction almost to the gel point. Their method may be compared to the high-angle extrapolation method [see Eq. (3.3.160)] in the theory of light scattering. It will be of some help in characterizing samples with very broad distributions, since at present few methods are available for this purpose, but in view of the approximation involved its use for samples with narrower distributions is problematical.

From Eqs. (4.2.19) and (4.2.23) we can derive

$$\left(\frac{d\theta_{i0}}{dx}\right)_{x=1} - \left(\frac{d\theta_{i0}}{dx}\right)_{x=0} = \lambda^2 M_i^2 \tag{4.2.38}$$

and also from Eqs. (4.2.20), (4.2.23), and (4.2.24)

$$\left(\frac{d\theta_{i1}}{dx}\right)_{x=1} - \left(\frac{d\theta_{i1}}{dx}\right)_{x=0} = -\lambda^2 \sum_{j=1}^{r} \xi_j M_i M_j (2M_i + M_j) B_{ij} F_{ij} \tag{4.2.39}$$

If we introduce these relations into $(d\theta_i/dx)_{x=1} - (d\theta_i/dx)_{x=0}$ and, after replacing θ_i by C_i, sum the result over all polymer components, we obtain

$$\Delta\left(\frac{dC_s}{dx}\right) \equiv \left(\frac{dC_s}{dx}\right)^b - \left(\frac{dC_s}{dx}\right)^a = \lambda^2 M_w M_z C_s{}^o - 3\lambda^2 (C_s{}^o)^2 \sum_{i=1}^{r} \sum_{j=1}^{r} \xi_i \xi_j M_i^2 M_j B_{ij} F_{ij} + O[(C_s{}^o)^3] \tag{4.2.40}$$

Here, in obtaining the coefficient of $(C_s{}^o)^2$, we have used the fact that subscripts i and j can be exchanged in the double summation. This series may be combined with Eq. (4.2.29) to give

$$\frac{\lambda \Delta C_s}{\Delta (dC_s/dx)} = \frac{1}{M_z} + \frac{C_s{}^o}{M_z M_w} \sum_{i=1}^{r} \sum_{j=1}^{r} \xi_i \xi_j M_i M_j B_{ij} F_{ij} \left(\frac{3M_i}{M_z} - 1 \right) + \ldots$$

$$= \frac{1}{M_z} + \frac{\bar{C}_s}{M_z M_w} \sum_{i=1}^{r} \sum_{j=1}^{r} \xi_i \xi_j M_i M_j B_{ij} f_{ij} \left(\frac{3M_i}{M_z} - 1 \right) + \ldots \tag{4.2.41}$$

It follows that

$$\lim_{\bar{C}_s \to 0} \frac{\lambda \Delta C_s}{\Delta (dC_s/dx)} = \frac{1}{M_z} \tag{4.2.42}$$

Finally, we touch upon the treatment of the set of conventional sedimentation equilibrium equations (4.1.30), in which T, p, and c' are used as the state variables. On conversion from c_i to C_i, these equations become

$$M_i (1 - v_i \rho) \omega^2 r = \sum_{j=1}^{r} \left(\frac{\partial \mu_i}{\partial C_j} \right)_{T, p, C_k} \frac{dC_j}{dr} \tag{4.2.43}$$

where $i = 0, 1, \ldots, r$. Using Eqs. (4.2.1), (1.4.56a), and (1.4.72), together with the approximation $v_i = v_i{}^o$, we get

$$(\omega^2 r/RT) \left[1 - v_i{}^o \rho_0 - v_i{}^o \sum_{j=1}^{r} (1 - v_j{}^o \rho_0) C_j \right] M_i C_i$$

$$= \frac{dC_i}{dr} + M_i C_i \sum_{j=1}^{r} \mathscr{B}_{ij} \frac{dC_j}{dr} + \ldots \tag{4.2.44}$$

Since for a polymer sample polydisperse in molecular weight we may set $v_1{}^o = v_2{}^o = \ldots = v_r{}^o \equiv v_s{}^o$, the terms in the brackets on the left-hand side are written $(1 - v_s{}^o \rho_0) \left(1 - v_s{}^o \sum_{j=1}^{r} C_j \right)$. Transfer of this second factor to the right-hand side and expansion in powers of C_j give

$$\frac{\omega^2 r (1 - v_s{}^o \rho_0)}{RT} M_i C_i = \frac{dC_i}{dr} + v_s{}^o \left(\sum_{j=1}^{r} C_j \right) \frac{dC_i}{dr}$$

$$+ M_i C_i \sum_{j=1}^{r} \mathscr{B}_{ij} \frac{dC_j}{dr} + \ldots \tag{4.2.45}$$

where $i = 1, 2, \ldots, r$. Equations (4.2.45) can be transformed to dimensionless form by Eqs. (4.2.6), (4.2.15), and (4.2.16), together with

$$\lambda \equiv \frac{\omega^2 (r_b{}^2 - r_a{}^2)(1 - v_s{}^o \rho_0)}{2RT} \tag{4.2.46}$$

Then, with Eq. (4.2.18), the differential equations for $\theta_{i0}, \theta_{i1}, \ldots$ are derived. The equation for θ_{i0} is identical with Eq. (4.2.19), while that for θ_{i1} is

$$\frac{d\theta_{i1}}{dx} - \lambda M_i \theta_{i1} = - v_s{}^o \frac{d\theta_{i0}}{dx} \sum_{j=1}^{r} \xi_j \theta_{j0} - M_i \theta_{i0} \sum_{j=1}^{r} \mathscr{B}_{ij} \xi_j \frac{d\theta_{j0}}{dx}$$

$$= - \lambda M_i \theta_{i0} \sum_{j=1}^{r} \xi_j M_j \left(\mathscr{B}_{ij} + \frac{v_s{}^o}{M_j} \right) \theta_{j0} \qquad (4.2.47)$$

If Eq. (1.4.77a) is used to convert from $\mathscr{B}_{ij}$ to B_{ij}, Eq. (4.2.47) is seen to conform to Eq. (4.2.20). As can be easily shown, λ defined by Eq. (4.2.46) is identical with λ in Eq. (4.2.20). This agreement has its root in the fact that both Eqs. (4.2.20) and (4.2.47) are based on the assumptions that the solution is incompressible and that $v_i = v_i{}^o$ for all polymer components.

4.2.2. Macromolecular Solutes in a Mixed Solvent – 1. Low-Speed Centrifugation

In formulating sedimentation equilibrium with T, p, and m as the state variables, we treat all components on an equal footing. Therefore, although Eq. (4.1.24) or (4.1.27) was derived on the assumption that all components other than component 0 are macromolecular, these relations are also applicable, as they stand, to solutions for macromolecular solutes in mixed solvents. Thus, for a mixed-solvent solution some of the concentration gradients appearing in each of these sedimentation-equilibrium equations would refer to the low-molecular-weight components which, together with a principal solvent (component 0), make up the mixed solvent. Though we do not demonstrate this here, because of this feature Eq. (4.1.24) or (4.1.27) seems to lead to the disturbing conclusion that when a mixed solvent is used, sedimentation equilibrium measurements on a polymer solution do not allow the true molecular weight of the polymer solute to be determined. This difficulty was first recognized in relation to molecular weight determinations of proteins, which, as was mentioned at the beginning of this book, are usually studied in aqueous solutions containing a supporting electrolyte and/or a hydrogen-ion buffer. Ways to circumvent the difficulty were sought by a number of investigators, and, in the late 1950's, there came a solution to the problem. It was to formulate sedimentation equilibrium by choosing as the state variables the temperature, a set of chemical potentials μ_D of the components making up the mixed solvent (diffusible components), and a set of molar concentration c' of the coexisting macromolecular (nondiffusible) components. This new formulation has already been described in § 4.1.2 for a special case in which the solvent is a single component.

However, its features are more clearly revealed by considering the general mixed solvent.

We suppose that a solution consists of $d+1$ diffusible components ($i=0, 1, \ldots, d$) and s nondiffusible polymer components ($i = d+1, d+2, \ldots, d+s$) and choose the quantities T, $\mu_D = (\mu_0, \mu_1, \ldots, \mu_d)$, and $c' = (c_{d+1}, c_{d+2}, \ldots, c_{d+s})$ as the state variables. At constant T, the total differential of the chemical potential of a nondiffusible component i is expressed by

$$d\mu_i = \sum_{j=0}^{d} \left(\frac{\partial \mu_i}{\partial \mu_j}\right)_{T,\, \mu_{Dj},\, c'} d\mu_j + \sum_{j=d+1}^{d+s} \left(\frac{\partial \mu_i}{\partial c_j}\right)_{T, \mu_D, c_k} dc_j \tag{4.2.48}$$

where $i = d+1, d+2, \ldots, d+s$ and the subscript μ_{Dj} denotes constancy of the chemical potentials μ_D with the exception of μ_j. It is evident that

$$\frac{d\mu_i}{dr} = \sum_{j=0}^{d} \left(\frac{\partial \mu_i}{\partial \mu_j}\right)_{T, \mu_{Dj}, c'} \frac{d\mu_j}{dr} + \sum_{j=d+1}^{d+s} \left(\frac{\partial \mu_i}{\partial c_j}\right)_{T, \mu_D, c_k} \frac{dc_j}{dr} \tag{4.2.49}$$

Substitution of Eq. (4.1.7) gives

$$\left[-M_i + \sum_{j=0}^{d} M_j \left(\frac{\partial \mu_i}{\partial \mu_j}\right)_{T, \mu_{Dj}, c'}\right] \frac{d\varphi}{dr} = \sum_{j=d+1}^{d+s} \left(\frac{\partial \mu_i}{\partial c_j}\right)_{T, \mu_D, c_k} \frac{dc_j}{dr} \tag{4.2.50}$$

From Eq. (4.1.12) it follows that

$$\left(\frac{\partial \rho}{\partial c_i}\right)_{T, \mu_D, c_k} = \left[\frac{\partial\left(\sum_{j=0}^{d+s} n_j M_j\right)}{\partial n_i}\right]_{T, V, \mu_D, n_k}$$

$$= \sum_{j=0}^{d} M_j \left(\frac{\partial n_j}{\partial n_i}\right)_{T, V, \mu_D, n_k} + M_i = M_i - \sum_{j=0}^{d} M_j \left(\frac{\partial \mu_i}{\partial \mu_j}\right)_{T, \mu_{Dj}, c'} \tag{4.2.51}$$

To obtain the last equality we can apply relevant properties of the characteristic function $\mathcal{A}\,(T,\ V,\ \mu_D,\ n')$. With Eq. (4.2.51), Eq. (4.2.50) can be written

$$-\left(\frac{\partial \rho}{\partial c_i}\right)_{T, \mu_D, c_k} \frac{d\varphi}{dr} = \sum_{j=d+1}^{d+s} \left(\frac{\partial \mu_i}{\partial c_j}\right)_{T, \mu_D, c_k} \frac{dc_j}{dr}\,,$$

$$(i = d+1, d+2, \ldots, d+s) \tag{4.2.52}$$

The only difference from Eq. (4.1.18) for a single-solvent system is that the subscript μ_0 in Eq. (4.1.18) is here replaced by μ_D. Like Eq. (4.1.18), Eq. (4.2.52) contains only concentration gradients of the nondiffusible components, in contrast to Eqs. (4.1.24) and (4.1.30) which, in the mixed-solvent case, contain concentration gradients of both diffusible and nondiffusible components. These features suggest that if T, μ_D, and m or c' are chosen as the state variables, the difficulty associated with Eqs. (4.1.24) and (4.1.30) will disappear and an equation which leads to molecular weight determination of nondiffusible components in a mixed solvent can be derived by an appropriate transcription of the theory of sedimentation equilibrium of single-solvent solutions.

For actual work with polymer solutes, for which molecular weights are generally unknown in advance, it is desirable to reformulate Eq. (4.2.52) in terms of C_i. With Eq. (4.1.1) for φ, we then obtain as the appropriate set of sedimentation-equilibrium equations for mixed-solvent solutions

$$M_i \left(\frac{\partial \rho}{\partial C_i}\right)_{T, \mu_D, C_k} \omega^2 r = \sum_{j=d+1}^{d+s} \left(\frac{\partial \mu_i}{\partial C_j}\right)_{T, \mu_D, C_k} \frac{dC_j}{dr} \tag{4.2.53}$$

where $i = d+1, d+2, \ldots, d+s$.

In § 1.4.4, we showed that $(\partial \mu_i / \partial C_j)_{T, \mu_D, C_k}$ can be expanded in powers of C_i as

$$\left(\frac{\partial \mu_i}{\partial C_j}\right)_{T, \mu_D, C_k} = RT \left[\frac{\delta_{ij}}{C_i} + M_i \left(B_{ij} + \sum_{k=d+1}^{d+s} B_{ijk} C_k + \ldots\right)\right] \tag{4.2.54}$$

To discuss sedimentation equilibrium of dilute polymer solutions, this expansion can be substituted into Eq. (4.2.53). In this case, the point to note is that the expansion coefficients B_{ij}, B_{ijk}, etc., refer to infinite dilution of nondiffusible polymer components. Therefore, these coefficients can be treated as functions of T and μ_D. To a good approximation, μ_D may be assumed to be independent of pressure. Thus, when we wish to integrate Eqs. (4.2.53), we may treat B_{ij}, B_{ijk}, etc., as independent of r. The μ_D values which control these expansion coefficients can be adjusted by the operation of **dialysis**, i.e., by osmotically equilibrating a solution against a large volume of mixed solvent held at specified composition and temperature.

As we have noted in the preceding section, when we are primarily concerned with measuring M_w by sedimentation equilibrium, it is desired to use the lowest rotor speed compatible with the required accuracy in measurement of $\Delta C_s =$

$C_s^b - C_s^a$. No detectable concentration gradient will be produced when a solution containing only diffusible low-molecular-weight components is centrifuged at such speeds. Therefore, the concentration gradients appearing when a polymer solution is centrifuged are those of the polymer solutes and gradients of diffusible components induced as a result of their adsorption onto the polymer molecules. This implies that the solution density varies with r only through the concentrations of the nondiffusible polymer components. Then if the polymer concentrations are sufficiently low throughout the solution so that terms of order higher than C_i $(i = d + 1, d + 2, \ldots, d + s)$ may be ignored, the derivative $(\partial \rho / \partial C_i)_{T, \mu_D, C_k}$ can be treated as independent of C_i and hence of r, and becomes a function of T and μ_D. Its value is then determinable from independent measurements of ρ as a function of C_i at fixed T and μ_D. We may define a parameter λ_i^* by

$$\lambda_i^* \equiv \frac{\omega^2 (r_b^2 - r_a^2)}{2RT} \left(\frac{\partial \rho}{\partial C_i}\right)_{T, \mu_D, C_k} \tag{4.2.55}$$

which, for the above reason, may be treated as independent of r in conventional low-speed centrifuge experiments.

With Eq. (4.2.55), Eq. (4.2.53) becomes formally identical with Eq. (4.2.4) for single-solvent solutions. Hence, all the equations derived from Eq. (4.2.4) can be applied, as they stand, to mixed-solvent solutions, with the understanding that λ_i^* and the B-coefficients (B_{ij}, B_{ijk}, etc.) here refer to fixed μ_D, instead of fixed μ_0. The most important conclusion from this result is that true molecular weights of polymer solutes can be derived from low-speed sedimentation equilibrium measurements on mixed-solvent solutions if the density increment necessary to compute λ is measured on dialyzed solutions. It must be noted that although, under the reasonable assumptions that the density increment in single-solvent solutions, i.e., $(\partial \rho / \partial C_i)_{T, \mu_0, C_k}$, can be equated to the buoyancy factor $1 - v_i^o \rho_0$ [see Eq. (4.2.2)], this is not the case with the density increment $(\partial \rho / \partial C_i)_{T, \mu_D, C_k}$ in mixed-solvent dialyzed solutions.

Casassa and Eisenberg [9] arrived at the same conclusions, developing their sedimentation equilibrium theory on the basis of Eqs. (4.1.27); but the analysis given above is more straightforward than theirs. At an intermediate step of the Casassa-Eisenberg formulation there appears a density increment $(\partial \rho / \partial C_i)_{T, p, \mu_1, \ldots, \mu_d, C_k}$. For incompressible solutions the following relations hold:

$$\left(\frac{\partial \rho}{\partial C_i}\right)_{T, \mu_D, C_k} = \left(\frac{\partial \rho}{\partial C_i}\right)_{T, p, \mu_1, \ldots, \mu_d, C_k} \neq \left(\frac{\partial \rho}{\partial C_i}\right)_{T, p, C_1, \ldots, C_{i-1}, C_{i+1}, \ldots, C_{d+s}} \tag{4.2.56}$$

It should be noted that in the second derivative the chemical potential of one diffusible component, the principal solvent, is not held constant. Furthermore, this density increment is not accessible to experimental measurement, as the constraints do not correspond to those for osmotic equilibrium.

Thermodynamically, a characteristic feature of polymer or macromolecular solutions is that their components can be divided into two groups in terms of their permeability through a semipermeable membrane. We have termed the components diffusible and nondiffusible. The former make up the solvent or mixed solvent and include low-molecular-weight liquids or solids, while the latter are macromolecular substances, synthetic and naturally-occurring, or biopolymers. Division of the components into two groups makes it possible to study nondiffusible components in a liquid medium of diffusible components whose chemical potentials are fixed externally by means of osmotic equilibration. It is a relatively recent finding that the difficulties associated with molecular weight determinations by light scattering and sedimentation equilibrium on mixed-solvent solutions can be circumvented by use of refractive index increment and density increment data obtained at constant μ_D.

4.2.3. Macromolecular Solutes in a Mixed Solvent – 2. Density Gradient Method

One notable application of sedimentation equilibrium in mixed-solvent systems is the **density gradient method** developed by Meselson et al. [10]. When a concentrated aqueous solution of a heavy electrolyte, such as cesium chloride, is subjected to a very strong centrifugal field, there is set up in the solution a concentration gradient of the electrolyte and hence a density gradient. At equilibrium, a polymer species introduced into the system accumulates in a band about the level at which the effective density of the polymer matches the solution density. The width of the band depends on the diffusivity of the polymer and hence on its molecular weight. The higher the molecular weight, the narrower is the band. Thus it should be possible to evalute both the partial specific volume and molecular weight of a polymer from the position and width of the band, respectively. To formulate this density gradient method in thermodynamic terms, we consider a three-component system in which component 0 is water, component 1 is a heavy electrolyte, and component 2 is a monodisperse polymer solute. We use Eqs. (4.2.53) and (4.2.54) to obtain, at very low concentration of the polymer component,

$$\frac{M_2 \omega^2 r}{RT} \left(\frac{\partial \rho}{\partial C_2} \right)^{o}_{T, \mu_D} = \frac{d \ln C_2}{dr} \tag{4.2.57}$$

where the superscript zero refers to the limit $C_2 = 0$. It is assumed that there is a radial position r_c in the solution at which the density increment $(\partial\rho/\partial C_2)^o_{T,\mu_D}$ $[\mu_D = (\mu_0, \mu_1)]$ vanishes. We expand this density increment in powers of $\Delta r = r - r_c$ as

$$(\partial\rho/\partial C_2)^o_{T,\mu_D} = -\chi\Delta r + O\,[(\Delta r)^2] \tag{4.2.58}$$

where

$$\chi \equiv -\left[\frac{d}{dr}\left(\frac{\partial\rho}{\partial C_2}\right)^o_{T,\mu_D}\right]_{r=r_c} \tag{4.2.59}$$

Substituting Eq. (4.2.58) into Eq. (4.2.57) and integrating, we obtain

$$\ell n\,\frac{C_2(r)}{C_2(r_c)} = -\frac{M_2\omega^2 r_c\chi}{2RT}(\Delta r)^2 + O\,[(\Delta r)^3] \tag{4.2.60}$$

Therefore, in the approximation that terms of order higher than $(\Delta r)^2$ may be neglected, we get

$$C_2(r) = C_2(r_c)\exp\left[-\frac{(r-r_c)^2}{2\sigma^2}\right] \tag{4.2.61}$$

where

$$\sigma^2 = RT/(M_2\omega^2 r_c\chi) \tag{4.2.62}$$

Thus if $\chi > 0$, and hence $\sigma^2 > 0$, the concentration distribution of the polymer component at the limit $C_2 = 0$ becomes Gaussian with its centroid at $r = r_c$. From the band width σ, M_2 may be determined.

It will be left as an instructive exercise for the reader to show that if the solution is incompressible, we have

$$\left(\frac{\partial\rho}{\partial C_2}\right)^o_{T,\mu_D} = 1 - v_2{}^o\rho^o + \left(\frac{M_1}{M_2}\right)\left(\frac{\partial m_1}{\partial m_2}\right)^o_{T,\mu_D}(1 - v_1{}^o\rho^o) \tag{4.2.63}$$

The left-hand side vanishes at $r = r_c$. Hence, if $(\partial m_1/\partial m_2)^o_{T,\mu_D} = 0$, i.e., if there is no selective binding of the electrolyte by the polymer on the molality basis at fixed μ_D, we have $\rho_c{}^o = 1/v_2{}^o$, where $\rho_c{}^o$ is the value of ρ^o at $r = r_c$. The electrolytes used in density gradient sedimentation equilibrium experiments are heavy ones such as cesium chloride and cesium sulfate; hence $1 - v_1{}^o\rho_c{}^o$ is positive. Thus, if $(\partial m_1/\partial m_2)^o_{T,\mu_D} > 0$, we may have $1 - v_2{}^o\rho_c{}^o < 0$, so that polymer molecules form a band about a point at which the solution density is higher than the "density" of the polymer itself.

Attempts have been made to extend the theory of density gradient sedimentation to solutions of a macromolecular solute which has a continuous distribution of molecular weight and/or partial specific volume. However, contrary to expectations, not a great deal has been achieved in quantitative characterization of such heterogeneous macromolecular substances by density gradient sedimentation equilibrium. Needless to say, this method is very effective for separating and analyzing a mixture of polymers with discrete differences of density. The classic example is a study by Meselson and Stahl [11], who used the method to verify the Watson-Crick hypothesis on the mechanism of replication of DNA.

4.2.4. Sedimentation Equilibrium of Concentrated Polymer Solutions

Finally, we give a brief account of Scholte's treatment [12] of concentrated polymer solutions at sedimentation equilibrium. We assume that the solution consists of a single solvent and r homologous polymer components and that it is incompressible. We start with the sedimentation equilibrium equations (4.1.30) and rewrite them in terms of the weight fractions w_i $(i = 1, 2, \ldots, r)$ of the polymer components:

$$M_i (1 - v_i \rho) \omega^2 r = \sum_{j=1}^{r} \left(\frac{\partial \mu_i}{\partial w_j} \right)_{T, p, w_k} \frac{dw_j}{dr} \tag{4.2.64}$$

From Eqs. (2.2.49), (2.2.50), and (2.2.54) we have

$$\Delta_m \mu_0 \equiv \mu_0 - \mu_0{}^o = RT \left\{ \ln (1 - w_s) + \left(1 - \frac{M_0}{M_n} \right) w_s + \left[g_w - (1 - w_s) \frac{\partial g_w}{\partial w_s} \right] w_s{}^2 \right\} \tag{4.2.65}$$

and

$$\Delta_m \mu_i = RT \left\{ \ln w_i - \left(\frac{M_i}{M_0} - 1 \right) + \frac{M_i}{M_0} \left(1 - \frac{M_0}{M_n} \right) w_s + \left(\frac{M_i}{M_0} \right) \left(g_w + w_s \frac{\partial g_w}{\partial w_s} \right) (1 - w_s)^2 \right\}, \quad (i = 1, 2, \ldots, r) \tag{4.2.66}$$

At sedimentation equilibrium, w_i, w_s (the weight fraction of the polymer sample), and M_n are all functions of r.

Substitution of Eqs. (4.2.66) into Eqs. (4.2.64) gives

$$\frac{M_i(1-v_s\rho)\omega^2 r}{RT} = \frac{1}{w_i}\frac{dw_i}{dr} + M_i\left[-\frac{1}{M_n} + \frac{1}{M_0}F(g_w, w_s)\right]\frac{dw_s}{dr} + \frac{M_i}{M_n^2}w_s\frac{dM_n}{dr} \tag{4.2.67}$$

where

$$F(g_w, w_s) \equiv 1 + (1-w_s)\frac{\partial^2[w_s(1-w_s)g_w]}{\partial w_s^2} \tag{4.2.68}$$

and v_s has the same meaning as in previous sections, i.e., $v_1 = v_2 = \ldots = v_r \equiv v_s$. Multiplying Eq. (4.2.67) by $w_i/w_s M_i$ and summing over all polymer components, we obtain

$$\frac{(1-v_s\rho)\omega^2 r}{RT} = \left[\frac{1}{w_s M_n} - \frac{1}{M_n} + \frac{1}{M_0}F(g_w, w_s)\right]\frac{dw_s}{dr} + \left(-\frac{1}{M_n^2} + \frac{w_s}{M_n^2}\right)\frac{dM_n}{dr} \tag{4.2.69}$$

Similarly, multiplying Eq. (4.2.67) by $w_i/w_s M_w = w_i/\Sigma_j w_j M_j$ and summing over all polymer components, we obtain

$$\frac{(1-v_s\rho)\omega^2 r}{RT} = \left[\frac{1}{w_s M_w} - \frac{1}{M_n} + \frac{1}{M_0}F(g_w, w_s)\right]\frac{dw_s}{dr} + \frac{w_s}{M_n^2}\frac{dM_n}{dr} \tag{4.2.70}$$

Equating the last two relations, we get

$$\frac{dM_n}{dr} = \frac{M_n(M_w - M_n)}{M_w w_s}\frac{dw_s}{dr} \tag{4.2.71}$$

which we use to eliminate dM_n/dr from Eq. (4.2.67) or (4.2.69):

$$\frac{(1-v_s\rho)\omega^2 r}{RT} = \left[\left(\frac{1-w_s}{w_s}\right)\frac{1}{M_w} + \frac{1}{M_0}F(g_w, w_s)\right]\frac{dw_s}{dr} \tag{4.2.72}$$

On the other hand, it follows from Eq. (4.2.65) that

$$\frac{1}{RT}\left(\frac{\partial \Delta_m \mu_0}{\partial w_s}\right)_{M_n} = -\frac{1}{1-w_s} + \left(1 - \frac{M_0}{M_n}\right) + \frac{w_s}{1-w_s}[1 - F(g_w, w_s)]$$

$$= -\frac{M_0}{M_n} - \frac{w_s}{1-w_s} F(g_w, w_s) \tag{4.2.73}$$

If we assume both ν_0 and ν_s to be independent of pressure and composition, we have $1/\rho = (1 - w_s)(1/\rho_0) + \nu_s w_s$, and hence

$$1 - \nu_s \rho = \frac{1-w_s}{\rho}\left(\frac{\partial \rho}{\partial w_s}\right)_{T,p} \tag{4.2.74}$$

Equations (4.2.73) and (4.2.74) allow us to eliminate $1 - \nu_s \rho$ and $F(g_w, w_s)$ from Eq. (4.2.72), and thus to arrive at Scholte's formula:

$$-M_0 \omega^2 r \frac{w_s}{\rho}\left(\frac{\partial \rho}{\partial w_s}\right)_{T,p} = \left[\left(\frac{\partial \Delta_m \mu_0}{\partial w_s}\right)_{M_n} + RT\left(\frac{M_0}{M_n} - \frac{M_0}{M_w}\right)\right] \frac{dw_s}{dr} \tag{4.2.75}$$

The second term in the brackets may be ignored for concentrated solutions. Thus the derivative $(\partial \Delta_m \mu_0 / \partial w_s)_{M_n}$ can be evaluated from measurements of w_s and dw_s/dr at various points in the solution at sedimentation equilibrium. From such measurements on solutions of different initial concentrations it is possible to obtain this derivative as a function of w_s. Integration of this function at a fixed M_n from $w_s = 0$ yields $\Delta_m \mu_0 (w_s)$. For dilute solutions, M_n of the polymer solute at the point where $w_s(r)$ coincides with the initial concentration $w_s{}^o$ may be identified, to a good approximation, with the number-average molar mass $M_n{}^o$ of the sample. At higher concentrations, $|w_s(r) - w_s{}^o|$ is much smaller than $w_s{}^o$ for all r and the approximation $M_n \approx M_n{}^o$ holds at the mid-point between the meniscus and bottom of the solution. We may take advantage of these facts to determine $\Delta_m \mu_0 (w_s)$, which corresponds to $M_n{}^o$.

The χ_Θ data for cyclohexane solutions of polystyrene shown in Fig. 2.15 and the χ data for toluene solutions of polystyrene in Fig. 2.18 were obtained by Scholte using the method described above.

References

1. T. Svedberg and K. O. Pedersen, "The Ultracentrifuge," Oxford Univ. Press, London and New York (Johnson Reprint Corp., New York), 1940.

2. J. W. Williams, K. E. Van Holde, R. L. Baldwin, and H. Fujita, Chem. Rev. **58**, 715 (1958).
3. H. Fujita, "Foundations of Ultracentrifugal Analysis," Chap. 5, John-Wiley, New York, London, Sydney, and Toronto, 1975.
4. R. C. Deonier and J. W. Williams, Proc. Natl. Acad. Sci. U.S.A. **64**, 828 (1969).
5. K. E. Van Holde and R. L. Baldwin, J. Phys. Chem. **62**, 734 (1958).
6. H. Utiyama, N. Tagata, and M. Kurata, J. Phys. Chem. **73**, 1448 (1969).
7. D. A. Albright and J. W. Williams, J. Phys. Chem. **71**, 2780 (1967); J. W. Williams, "Ultracentrifugation of Macromolecules," Chap. I, Academic Press, New York and London, 1972.
8. M. Gordon, C. G. Leonis, and H. Suzuki, Proc. R. Soc. London, Ser. A, **345**, 207 (1975).
9. E. F. Casassa and H. Eisenberg, Adv. Protein Chem. **19**, 287 (1964).
10. M. Meselson, F. W. Stahl, and J. Vinograd, Proc. Natl. Acad. Sci. U.S.A. **43**, 581 (1958).
11. M. Meselson and F. W. Stahl, Proc. Natl. Acad. Sci. U.S.A. **44**, 671 (1958).
12. Th. G. Scholte, J. Polym. Sci. A-2, **8**, 841 (1970).

Appendix

Appendix

A.1. Partial Derivatives

Let z be a single-valued function of q independent real variables $x_1, x_2, \ldots, x_q$:

$$z = z(x_1, x_2, \ldots, x_q) \tag{A.1}$$

For simplicity we denote by x_j a set of $q - 1$ variables other than x_i, and express the partial derivative of z with respect to x_i by $(\partial z/\partial x_i)_{x_j}$. Thus we have

$$\left(\frac{\partial z}{\partial x_i}\right)_{x_j} = \lim_{\Delta x_i \to 0} \frac{z(x_1, \ldots, x_{i-1}, x_i + \Delta x_i, x_{i+1}, \ldots, x_q) - z(x_1, x_2, \ldots, x_q)}{\Delta x_i} \tag{A.2}$$

We assume that derivatives of z of all order exist. Then

$$\left[\frac{\partial}{\partial x_j}\left(\frac{\partial z}{\partial x_i}\right)_{x_j}\right]_{x_k} = \left[\frac{\partial}{\partial x_i}\left(\frac{\partial z}{\partial x_j}\right)_{x_k}\right]_{x_j} \tag{A.3}$$

The total differential of z is given by

$$dz = \sum_{i=1}^{q} \left(\frac{\partial z}{\partial x_i}\right)_{x_j} dx_i \tag{A.4}$$

For example, when z is a function of two variables x and y, Eq. (A.4) is

$$dz = \left(\frac{\partial z}{\partial x}\right)_y dx + \left(\frac{\partial z}{\partial y}\right)_x dy \tag{A.5}$$

For those changes in x and y which keep z constant, this equation gives

$$0 = \left(\frac{\partial z}{\partial x}\right)_y + \left(\frac{\partial z}{\partial y}\right)_x \left(\frac{\partial y}{\partial x}\right)_z \tag{A.6}$$

or

$$\left(\frac{\partial y}{\partial x}\right)_z = -\frac{(\partial z/\partial x)_y}{(\partial z/\partial y)_x} \tag{A.7}$$

For the general case in which there are more than two independent variables the corresponding equation is

$$\left(\frac{\partial x_j}{\partial x_i}\right)_{z, x_k} = - \frac{(\partial z/\partial x_i)_{x_j, x_k}}{(\partial z/\partial x_j)_{x_i, x_k}} \tag{A.8}$$

where the subscript x_k indicates $q - 2$ variables other than x_i and x_j. Unless special attention is to be called to it, this additional subscript will be omitted.

Returning to z as a function of x and y, we consider the case in which z varies with y at fixed x. Then Eq. (A.5) gives

$$\left(\frac{\partial z}{\partial y}\right)_x \left(\frac{\partial y}{\partial z}\right)_x = 1 \quad \text{or} \quad \left(\frac{\partial z}{\partial y}\right)_x = \frac{1}{(\partial y/\partial z)_z} \tag{A.9}$$

If we substitute this into Eq. (A.7), we get

$$\left(\frac{\partial z}{\partial x}\right)_y \left(\frac{\partial x}{\partial y}\right)_z \left(\frac{\partial y}{\partial z}\right)_x = -1 \tag{A.10}$$

Let $z(x, y)$ and $u(x, y)$ be two functions of x and y. The partial derivative of z with respect to x at fixed u is obtained from Eq. (A.5) as

$$\left(\frac{\partial z}{\partial x}\right)_u = \left(\frac{\partial z}{\partial x}\right)_y + \left(\frac{\partial z}{\partial y}\right)_x \left(\frac{\partial y}{\partial x}\right)_u \tag{A.11}$$

If we replace z in Eq. (A.7) by u, we have

$$\left(\frac{\partial y}{\partial x}\right)_u = - \frac{(\partial u/\partial x)_y}{(\partial u/\partial y)_x} \tag{A.12}$$

Combining Eqs. (A.11) and (A.12), we obtain

$$\left(\frac{\partial z}{\partial x}\right)_u = \left(\frac{\partial z}{\partial x}\right)_y - \left(\frac{\partial z}{\partial y}\right)_x \frac{(\partial u/\partial x)_y}{(\partial u/\partial y)_x} \tag{A.13}$$

Equations (A.11) and (A.13) are often applied in this book.

A.2. Homogeneous Functions

If, for any given constant λ, there holds a relation

$$z(\lambda x_1, \lambda x_2, \ldots, \lambda x_q) = \lambda^m z(x_1, x_2, \ldots, x_q) \tag{A.14}$$

z is called a **homogeneous function** of order m of the variables $x_1, x_2, \ldots, x_q$. Differentiating Eq. (A.14) with respect to λ, with all x_i fixed, we obtain

$$\sum_{i=1}^{q} \left[\frac{\partial z(\lambda x_1, \lambda x_2, \ldots, \lambda x_q)}{\partial(\lambda x_i)}\right]_{\lambda x_j} x_i = m\lambda^{m-1} z(x_1, x_2, \ldots, x_q) \tag{A.15}$$

Since this must be valid for any λ, we may put $\lambda = 1$, so that

$$mz(x_1, x_2, \ldots, x_q) = \sum_{i=1}^{q} x_i \left(\frac{\partial z}{\partial x_i}\right)_{x_j} \tag{A.16}$$

This relation is **Euler's theorem** for homogeneous functions.

Every extensive state variable in thermodynamics is a first-order homogeneous function of the amounts n_i of the component substances at fixed temperature and pressure. Hence if a state variable for an $(r+1)$-component system is denoted by $Y(T, p, n_0, n_1, \ldots, n_r)$, Eq. (A.16) with $m = 1$ gives

$$Y(T, p, n_0, n_1, \ldots, n_r) = \sum_{i=0}^{r} n_i Y_i \tag{A.17}$$

where

$$Y_i \equiv \left(\frac{\partial Y}{\partial n_i}\right)_{T, p, n_j} \tag{A.18}$$

The quantity Y_i defined by Eq. (A.18) is called the partial molar Y of component i.

Differentiation of Eq. (A.17) with respect to n_k gives

$$\left(\frac{\partial Y}{\partial n_k}\right)_{T, p, n_\ell} = Y_k + \sum_{i=0}^{r} n_i \left(\frac{\partial Y_i}{\partial n_k}\right)_{T, p, n_\ell} \tag{A.19}$$

The left-hand side is Y_k, so that we have

$$\sum_{i=0}^{r} n_i \left(\frac{\partial Y_i}{\partial n_k}\right)_{T, p, n_\ell} = 0 \tag{A.20}$$

Since $(\partial Y_i / \partial n_k)_{T, p, n_\ell}$ is the second derivative of Y with respect to n_i and n_k, it is equal to $(\partial Y_k / \partial n_i)_{T, p, n_\ell}$. Hence Eq. (A.20) is equivalent to

$$\sum_{i=0}^{r} n_i \left(\frac{\partial Y_k}{\partial n_i}\right)_{T, p, n_\ell} = 0 \tag{A.21}$$

which, by reference to Eq. (A.16), indicates that Y_k at constant T and p is a zero-order homogeneous function of $n_0, n_1, \ldots, n_r$; in other words, any partial molar quantity is an intensive quantity, i.e., one that is not changed by a uniform change in the amounts of the components in the system at fixed T and p.

A. 3. Vectors and Vector Operator

A vector is defined as a quantity which has both magnitude and direction. In this book it is designated by a letter with a wavy underline. A vector $\underset{\sim}{A}$ can be specified by its three components (A_x, A_y, A_z) which are the projections of $\underset{\sim}{A}$ on the three axes of a Cartesian coordinate system (x, y, z).

The scalar product $\underset{\sim}{A} \cdot \underset{\sim}{B}$ of two vectors $\underset{\sim}{A}$ and $\underset{\sim}{B}$ is defined by

$$\underset{\sim}{A} \cdot \underset{\sim}{B} = AB \cos\theta = A_x B_x + A_y B_y + A_z B_z \tag{A.22}$$

where A and B represent the magnitudes of $\underset{\sim}{A}$ and $\underset{\sim}{B}$, respectively, and θ is the angle between the two vectors. Let $\underset{\sim}{i}, \underset{\sim}{j}$, and $\underset{\sim}{k}$ be unit vectors which point in the directions of the x, y, and z axes, respectively. These satisfy the relations

$$\underset{\sim}{i} \cdot \underset{\sim}{i} = \underset{\sim}{j} \cdot \underset{\sim}{j} = \underset{\sim}{k} \cdot \underset{\sim}{k} = 1, \quad \underset{\sim}{i} \cdot \underset{\sim}{j} = \underset{\sim}{j} \cdot \underset{\sim}{k} = \underset{\sim}{k} \cdot \underset{\sim}{i} = 0 \tag{A.23}$$

and $\underset{\sim}{A}$ can be expressed as

$$\underset{\sim}{A} = A_x \underset{\sim}{i} + A_y \underset{\sim}{j} + A_z \underset{\sim}{k} \tag{A.24}$$

The vector product $\underset{\sim}{A} \times \underset{\sim}{B}$ of $\underset{\sim}{A}$ and $\underset{\sim}{B}$ is defined by

$$\underset{\sim}{A} \times \underset{\sim}{B} = (A_y B_z - A_z B_y)\underset{\sim}{i} + (A_z B_x - A_x B_z)\underset{\sim}{j} + (A_x B_y - A_y B_x)\underset{\sim}{k} \tag{A.25}$$

From this definition it follows that

$$\begin{gathered} \underset{\sim}{i} \times \underset{\sim}{j} = \underset{\sim}{k}, \quad \underset{\sim}{j} \times \underset{\sim}{k} = \underset{\sim}{i}, \quad \underset{\sim}{k} \times \underset{\sim}{i} = \underset{\sim}{j}, \\ \underset{\sim}{i} \times \underset{\sim}{i} = \underset{\sim}{j} \times \underset{\sim}{j} = \underset{\sim}{k} \times \underset{\sim}{k} = 0 \end{gathered} \tag{A.26}$$

Among many others, the following formula is frequently used in this book:

$$\underset{\sim}{A} \times (\underset{\sim}{B} \times \underset{\sim}{C}) = \underset{\sim}{B}(\underset{\sim}{A} \cdot \underset{\sim}{C}) - \underset{\sim}{C}(\underset{\sim}{A} \cdot \underset{\sim}{B}) \tag{A.27}$$

The components of a vector change when the coordinate system to which it refers is changed. Considering two Cartesian coordinate systems (x, y, z) and (x', y', z'), we have

$$\underset{\sim}{i}' = \ell_{xx'} \underset{\sim}{i} + \ell_{yx'} \underset{\sim}{j} + \ell_{zx'} \underset{\sim}{k} \tag{A.28}$$

and

$$\underset{\sim}{i} = \ell_{xx'} \underset{\sim}{i}' + \ell_{xy'} \underset{\sim}{j}' + \ell_{xz'} \underset{\sim}{k}' \tag{A.29}$$

where $\underset{\sim}{i}'$, $\underset{\sim}{j}'$, and $\underset{\sim}{k}'$ are the unit vectors along the x', y', and z' axes, respectively, and $\ell_{xx'}$, $\ell_{yx'}$, etc., are the direction cosines (cosines of the angles between x and x', y and x', etc.).

Substituting Eq. (A.29) into Eq. (A.24), we obtain

$$A_x = A_{x'} \underset{\sim}{i}' + A_{y'} \underset{\sim}{j}' + A_{z'} \underset{\sim}{k}', \text{etc.} \tag{A.30}$$

where

$$A_{x'} = \ell_{xx'} A_x + \ell_{yx'} A_y + \ell_{zx'} A_z, \text{etc.} \tag{A.31}$$

Conversely we may get

$$A_x = \ell_{xx'} A_{x'} + \ell_{xy'} A_{y'} + \ell_{xz'} A_{z'}, \text{etc.} \tag{A.32}$$

The quantities $A_{x'}$, $A_{y'}$, and $A_{z'}$ defined by Eq. (A.31) represent the components of vector $\underset{\sim}{A}$ in the coordinate system (x', y', z'). A vector may be defined as a set of three components (or three numbers) which transform according to Eq. (A.31) when the coordinate system is rotated.

As is well known from calculus, the differentiation operator $\partial/\partial x$ for a scalar point function $\phi(x, y, z)$ is transformed to $\partial/\partial x'$ by the formula:

$$\frac{\partial}{\partial x'} = \left(\frac{\partial x}{\partial x'}\right) \frac{\partial}{\partial x} + \left(\frac{\partial y}{\partial x'}\right) \frac{\partial}{\partial y} + \left(\frac{\partial z}{\partial x'}\right) \frac{\partial}{\partial z} \tag{A.33}$$

Since $\partial x/\partial x' = \ell_{xx'}$, $\partial y/\partial x' = \ell_{yx'}$, and $\partial z/\partial x' = \ell_{zx'}$, this transformation is formally identical to Eq. (A.31), meaning that a set of differential operators $(\partial/\partial x, \partial/\partial y, \partial/\partial z)$ may be regarded as a kind of vector. This vector operator, called the gradient vector, is designated by $\underset{\sim}{\nabla}$:

$$\underset{\sim}{\nabla} = \underset{\sim}{i} \frac{\partial}{\partial x} + \underset{\sim}{j} \frac{\partial}{\partial y} + \underset{\sim}{k} \frac{\partial}{\partial z} \tag{A.34}$$

For a scalar point function $\phi(x, y, z)$ we can derive

$$\underset{\sim}{\nabla}\phi = \left(\frac{\partial \phi}{\partial x}\right)\underset{\sim}{i} + \left(\frac{\partial \phi}{\partial y}\right)\underset{\sim}{j} + \left(\frac{\partial \phi}{\partial z}\right)\underset{\sim}{k} \equiv \text{grad}\,\phi \tag{A.35}$$

$$\underset{\sim}{\nabla} \cdot \underset{\sim}{\nabla}\phi \equiv \nabla^2 \phi = \frac{\partial^2 \phi}{\partial x^2} + \frac{\partial^2 \phi}{\partial y^2} + \frac{\partial^2 \phi}{\partial z^2} \tag{A.36}$$

and for a vector point function $\underset{\sim}{A}(x, y, z)$

$$\underset{\sim}{\nabla} \cdot \underset{\sim}{A} = \frac{\partial A_x}{\partial x} + \frac{\partial A_y}{\partial y} + \frac{\partial A_z}{\partial z} \tag{A.37}$$

$$\underset{\sim}{\nabla} \times \underset{\sim}{A} = \left(\frac{\partial A_z}{\partial y} - \frac{\partial A_y}{\partial z}\right)\underset{\sim}{i} + \left(\frac{\partial A_x}{\partial z} - \frac{\partial A_z}{\partial x}\right)\underset{\sim}{j} + \left(\frac{\partial A_y}{\partial x} - \frac{\partial A_x}{\partial y}\right)\underset{\sim}{k} \tag{A.38}$$

$$\underset{\sim}{\nabla} \cdot (\underset{\sim}{\nabla} \times \underset{\sim}{A}) = 0 \tag{A.39}$$

$$\underset{\sim}{\nabla} \times (\underset{\sim}{\nabla} \times \underset{\sim}{A}) = \underset{\sim}{\nabla}(\underset{\sim}{\nabla} \cdot \underset{\sim}{A}) - \nabla^2 \underset{\sim}{A} \tag{A.40}$$

The scalar $\underset{\sim}{\nabla} \cdot \underset{\sim}{A}$ is called the divergence (sometimes denoted by div $\underset{\sim}{A}$), the vector $\underset{\sim}{\nabla} \times \underset{\sim}{A}$ the rotation of $\underset{\sim}{A}$ (rot $\underset{\sim}{A}$), and the scalar ∇^2 the Laplacian operator.

If Eq. (A.28) is squared and the relations given by Eq. (A.23) are used, the following relations can be derived:

$$\ell_{xx'}^2 + \ell_{yx'}^2 + \ell_{zx'}^2 = 1, \text{etc.} \tag{A.41a}$$

$$\ell_{xx'} \ell_{xy'} + \ell_{yx'} \ell_{yy'} + \ell_{zx'} \ell_{zy'} = 0, \text{etc.} \tag{A.41b}$$

In developing the theory of light scattering it becomes necessary to evaluate orientational averages of various combinations of the direction cosines with equal probability. For even and odd powers of $\ell_{xx'}$ we find that

$$\langle \ell_{xx'}^{2p} \rangle = \int_0^\pi (\cos\theta_{xx'})^{2p} \sin\theta_{xx'}\, d\theta_{xx'} \Big/ \int_0^\pi \sin\theta_{xx'}\, d\theta_{xx'} = \frac{1}{2p+1} \tag{A.42a}$$

$$\langle \ell_{xx'}^{2p+1} \rangle = 0 \tag{A.42b}$$

If we square Eq. (A.41a) and average with equal orientational probability, we get

$$\begin{aligned} &\sum_{m=x,y,z} \langle \ell_{mx'}^4 \rangle + \sum_{m \neq n}\sum \langle \ell_{mx'}^2 \ell_{nx'}^2 \rangle \\ &= 3/5 + 6 \langle \ell_{xx'}^2 \ell_{yx'}^2 \rangle = 1 \end{aligned} \tag{A.43}$$

where we have used Eq. (A.42a). This gives

$$\langle \ell_{xx'}^2 \ell_{yx'}^2 \rangle = \langle \ell_{xx'}^2 \ell_{xy'}^2 \rangle = 1/15 \tag{A.44}$$

It can be shown by similar calculations that

$$\langle \ell_{xy'} \ell_{yy'} \ell_{xz'} \ell_{yz'} \rangle = -1/30 \tag{A.45}$$

$$\langle \ell_{xy'} \ell_{xz'} \rangle = \langle \ell_{yx'} \ell_{zx'} \rangle = 0 \tag{A.46}$$

A.4. Dyadics and Orthogonal Transformations

A dyadic is synonymous with a tensor of the second rank. All tensors that appear in this book are dyadics.

In a Cartesian coordinate system (x, y, z), a dyadic is defined as a set of nine quantities a_{mn} $(m, n = x, y, z)$ which transform according to the rule

$$a_{m'n'} = \Sigma_m \Sigma_n \ell_{mm'} \ell_{nn'} a_{mn}, \qquad (m', n' = x', y', z') \tag{A.47}$$

when the coordinate system to which it refers undergoes a transformation from (x, y, z) to (x', y', z'). Each a_{mn} is called the (m, n) element of the dyadic. A set of nine products, such as $A_x B_y$, obtained from the components of two vectors $\underset{\sim}{A}$ and $\underset{\sim}{B}$ is a typical example of a dyadic, and is designated as $\underset{\sim}{A}\underset{\sim}{B}$. As a vector $\underset{\sim}{A}$ having three components (A_x, A_y, A_z) is written as $\underset{\sim}{i}A_x + \underset{\sim}{j}A_y + \underset{\sim}{k}A_z$, a dyadic $\underset{\sim}{a}$ consisting of a_{mn} $(m, n = x, y, z)$ is represented by

$$\begin{aligned} \underset{\sim}{a} = {} & a_{xx}\underset{\sim}{i}\underset{\sim}{i} + a_{xy}\underset{\sim}{i}\underset{\sim}{j} + a_{xz}\underset{\sim}{i}\underset{\sim}{k} \\ & + a_{yx}\underset{\sim}{j}\underset{\sim}{i} + a_{yy}\underset{\sim}{j}\underset{\sim}{j} + a_{yz}\underset{\sim}{j}\underset{\sim}{k} \\ & + a_{zx}\underset{\sim}{k}\underset{\sim}{i} + a_{zy}\underset{\sim}{k}\underset{\sim}{j} + a_{zz}\underset{\sim}{k}\underset{\sim}{k} \end{aligned} \tag{A.48}$$

Here the symbols $\underset{\sim}{i}\underset{\sim}{i}$, $\underset{\sim}{j}\underset{\sim}{k}$, $\underset{\sim}{k}\underset{\sim}{k}$, etc., are called unit dyads.

The single-dot product of a dyadic $\underset{\sim}{a}$ and vector $\underset{\sim}{A}$, written as $\underset{\sim}{a} \cdot \underset{\sim}{A}$, and that of two dyadics $\underset{\sim}{a}$ and $\underset{\sim}{b}$, written as $\underset{\sim}{a} \cdot \underset{\sim}{b}$, are defined by introducing the following multiplication rules

$$(\underset{\sim}{i}\underset{\sim}{j}) \cdot \underset{\sim}{j} = \underset{\sim}{i}(\underset{\sim}{j} \cdot \underset{\sim}{j}) = \underset{\sim}{i}, \quad (\underset{\sim}{i}\underset{\sim}{j}) \cdot \underset{\sim}{i} = (\underset{\sim}{i}\underset{\sim}{j}) \cdot \underset{\sim}{k} = 0, \text{etc.} \tag{A.49}$$

$$\underset{\sim}{i} \cdot (\underset{\sim}{i}\underset{\sim}{j}) = (\underset{\sim}{i} \cdot \underset{\sim}{i})\underset{\sim}{j} = \underset{\sim}{j}, \quad \underset{\sim}{j} \cdot (\underset{\sim}{i}\underset{\sim}{j}) = \underset{\sim}{k} \cdot (\underset{\sim}{i}\underset{\sim}{j}) = 0, \text{etc.} \tag{A.50}$$

and

$$(\underset{\sim}{i}\underset{\sim}{j}) \cdot (\underset{\sim}{j}\underset{\sim}{k}) = \underset{\sim}{i}(\underset{\sim}{j} \cdot \underset{\sim}{j})\underset{\sim}{k} = \underset{\sim}{i}\underset{\sim}{k}, \quad (\underset{\sim}{j}\underset{\sim}{k}) \cdot (\underset{\sim}{i}\underset{\sim}{j}) = \underset{\sim}{j}(\underset{\sim}{k} \cdot \underset{\sim}{i})\underset{\sim}{j} = 0, \text{etc.} \tag{A.51}$$

where a dot between a pair of unit vectors in parentheses indicates the scalar product of the two vectors, and Eqs. (A.23) have been used. By these rules we get

$$\begin{aligned} \underset{\sim}{a} \cdot \underset{\sim}{A} = {} & (a_{xx}A_x + a_{xy}A_y + a_{xz}A_z)\underset{\sim}{i} + (a_{yx}A_x + a_{yy}A_y + a_{yz}A_z)\underset{\sim}{j} \\ & + (a_{zx}A_x + a_{zy}A_y + a_{zz}A_z)\underset{\sim}{k} \end{aligned} \tag{A.52}$$

Thus the single-dot product of a dyadic and a vector gives another vector. It can be easily shown that $\underset{\sim}{a} \cdot \underset{\sim}{A} \neq \underset{\sim}{A} \cdot \underset{\sim}{a}$; i.e., the order of multiplication in the single-dot product of a dyadic and a vector is not interchangeable. The single-dot product of dyadics $\underset{\sim}{a}$ and $\underset{\sim}{b}$ gives another dyadic, and its (m, n) element $(\underset{\sim}{a} \cdot \underset{\sim}{b})_{mn}$ is given by

$$(\underset{\sim}{a} \cdot \underset{\sim}{b})_{mn} = a_{mx} b_{xn} + a_{my} b_{yn} + a_{mz} b_{zn} \tag{A.53}$$

where $m, n = x, y, z$.

The double-dot product of dyadics $\underset{\sim}{a}$ and $\underset{\sim}{b}$, $\underset{\sim}{a} : \underset{\sim}{b}$, is defined as the sum of $(\underset{\sim}{a} \cdot \underset{\sim}{b})_{mm}$, i.e., the diagonal elements of a dyadic $(\underset{\sim}{a} \cdot \underset{\sim}{b})$. Thus we have

$$\begin{aligned} \underset{\sim}{a} : \underset{\sim}{b} = \Sigma_m \Sigma_n a_{mn} b_{nm} &= (a_{xx} b_{xx} + a_{xy} b_{yx} + a_{xz} b_{zx}) \\ &+ (a_{yx} b_{xy} + a_{yy} b_{yy} + a_{yz} b_{zy}) \\ &+ (a_{zx} b_{xz} + a_{zy} b_{yz} + a_{zz} b_{zz}) \end{aligned} \tag{A.54}$$

In terms of this scalar quantity the Taylor series expansion of $f(\underset{\sim}{R} + \underset{\sim}{r})$ about $\underset{\sim}{R}$ can be written in compact form as

$$f(\underset{\sim}{R} + \underset{\sim}{r}) = f(\underset{\sim}{R}) + \underset{\sim}{r} \cdot \underset{\sim}{\nabla} f + (1/2!)\, \underset{\sim}{r}\, \underset{\sim}{r} : \underset{\sim}{\nabla}\, \underset{\sim}{\nabla} f + \ldots \tag{A.55}$$

We note that the second term on the right-hand side is a scalar product of two vectors $\underset{\sim}{r}$ and $\underset{\sim}{\nabla} f$.

Generalizing the analysis described above, we may define a dyadic $\underset{\sim}{a}$ in an arbitrary r-dimensional space by

$$\underset{\sim}{a} = \sum_{i=1}^{r} \sum_{j=1}^{r} a_{ij}\, \underset{\sim}{e}_i\, \underset{\sim}{e}_j \tag{A.56}$$

where $\underset{\sim}{e}_i$ is the unit vector in the direction of the i-th axis in the r-dimensional Cartesian coordinate system, so that

$$\underset{\sim}{e}_i \cdot \underset{\sim}{e}_j = \delta_{ij} \tag{A.57}$$

δ_{ij} being Kronecker's delta, which is unity for $i = j$ and zero otherwise. Defining rules similar to Eqs. (A.49) and (A.50) for the single-dot product of a unit dyad $\underset{\sim}{e}_i\, \underset{\sim}{e}_j$ and a unit vector $\underset{\sim}{e}_k$, we express the element a_{ij} as follows:

$$a_{ij} = \underset{\sim}{e}_i \cdot \underset{\sim}{a} \cdot \underset{\sim}{e}_j \tag{A.58}$$

When a dyadic $\underset{\sim}{a}$ is symmetric, i.e., $a_{ij} = a_{ji}$ for all i and j, we can transform it to a diagonal form, i.e., one which consists of unit dyads formed of the same unit vectors, by rotating the coordinate axes to another set of orthogonal axes. Denoting the unit vector of the i-th axis in this new coordinate system by $\underset{\sim}{e}_i{}'$, we have

$$\underset{\sim}{a} = \sum_{i=1}^{r} \lambda_i \underset{\sim}{e}_i{}' \underset{\sim}{e}_i{}' \tag{A.59}$$

and

$$\underset{\sim}{e}_i{}' = \sum_{j=1}^{r} \ell_{ji} \underset{\sim}{e}_j \quad \text{or} \quad \underset{\sim}{e}_i = \sum_{j=1}^{r} \ell_{ij} \underset{\sim}{e}_j{}' \tag{A.60}$$

where ℓ_{ij} is the "direction cosine" of the i-th axis in the unprimed coordinate system relative to the j-th axis in the primed coordinate system. The quantities λ_i are called the eigenvalues of $\underset{\sim}{a}$, and the vectors $\underset{\sim}{e}_i{}'$ the eigenvectors. The direction cosines satisfy the relations

$$\sum_{k=1}^{r} \ell_{ik} \ell_{jk} = \sum_{k=1}^{r} \ell_{ki} \ell_{kj} = \delta_{ij}, \qquad (i, j = 1, 2, \ldots, r) \tag{A.61}$$

which are the extensions of Eqs. (A.41a) and (A.41b). The rotation of the coordinate system which makes $\underset{\sim}{a}$ diagonal is called the **orthogonal transformation**.

Multiplication of Eq. (A.59) by $\underset{\sim}{e}_i{}'$ yields

$$\underset{\sim}{a} \cdot \underset{\sim}{e}_i{}' = \lambda_i \underset{\sim}{e}_i{}', \quad (i = 1, 2, \ldots, r) \tag{A.62}$$

Substituting Eq. (A.56) into this and transforming $\underset{\sim}{e}_i$ to $\underset{\sim}{e}_i{}'$ by Eq. (A.60), we obtain the following set of linear homogeneous equations for $\ell_{1i}, \ell_{2i}, \ldots, \ell_{ri}$:

$$\begin{aligned} (a_{11} - \lambda_i) \ell_{1i} + a_{12} \ell_{2i} + \ldots + a_{1r} \ell_{ri} &= 0 \\ a_{21} \ell_{1i} + (a_{22} - \lambda_i) \ell_{2i} + \ldots + a_{2r} \ell_{ri} &= 0 \\ \cdots\cdots\cdots\cdots\cdots\cdots\cdots\cdots & \\ a_{r1} \ell_{1i} + a_{r2} \ell_{2i} + \ldots + (a_{rr} - \lambda_i) \ell_{ri} &= 0 \end{aligned} \tag{A.63}$$

For this set to have nontrivial roots it is necessary that the determinant formed by the coefficients of $\ell_{1i}, \ell_{2i}, \ldots, \ell_{ri}$ vanish:

$$\begin{vmatrix} a_{11} - \lambda & a_{12} & \cdots & a_{1r} \\ a_{21} & a_{22} - \lambda & \cdots & a_{2r} \\ \cdots & \cdots & \cdots & \\ a_{r1} & a_{r2} & \cdots & a_{rr} - \lambda \end{vmatrix} = 0 \tag{A.64}$$

This equation for λ, known as the secular equation for the dyadic $\underset{\sim}{a}$, allows the eigenvalues $\lambda_1, \lambda_2, \ldots, \lambda_r$ to be evaluated. The well-known relation between roots and coefficients in algebraic equations gives

$$\lambda_1 \lambda_2 \ldots \lambda_r = |a| \tag{A.65}$$

where $|a|$ is the determinant of elements a_{ij}. For each λ_i we can determine ℓ_{ji} (and hence $\underset{\sim}{e}_i{}'$) from Eqs. (A.63) together with Eqs. (A.61).

A vector $\underset{\sim}{x}$ in an r-dimensional space may be represented by

$$\underset{\sim}{x} = \sum_{i=1}^{r} x_i \underset{\sim}{e}_i = \sum_{i=1}^{r} x_i{}' \underset{\sim}{e}_i{}' \tag{A.66}$$

Applying Eqs. (A.49) and (A.50), we find that

$$\Psi \equiv \underset{\sim}{x} \cdot \underset{\sim}{a} \cdot \underset{\sim}{x} = \sum_{i=1}^{r} a_{ij} x_i x_j \tag{A.67}$$

This sum is called a quadratic form in x_i. If Eq. (A.59) is used for $\underset{\sim}{a}$, Ψ can also be represented by

$$\Psi = \sum_{i=1}^{r} \lambda_i (x_i{}')^2 \tag{A.68}$$

The multivariate Gaussian distributions which appear in § 3.1.3 have the form

$$w(x_1, x_2, \ldots, x_r) = C \exp\left[-(\gamma/2)\Psi\right], \quad (\gamma > 0) \tag{A.69}$$

with Ψ of the form of Eq. (A.67). Since the Jacobian for the orthogonal transformation is unity, it follows that

$$\begin{aligned} J_0 &\equiv \int_{-\infty}^{\infty} \ldots \int_{-\infty}^{\infty} \exp(-\gamma\Psi/2)\, dx_1\, dx_2 \ldots dx_r \\ &= (2\pi/\gamma)^{r/2} \prod_{i=1}^{r} (1/\lambda_i) = (2\pi/\gamma)^{r/2} / |a| \end{aligned} \tag{A.70}$$

and

$$\begin{aligned} J_2 &\equiv \int_{-\infty}^{\infty} \ldots \int_{-\infty}^{\infty} x_i x_j \exp(-\gamma\Psi/2)\, dx_1\, dx_2 \ldots dx_r \\ &= \int_{-\infty}^{\infty} \ldots \int_{-\infty}^{\infty} \left(\sum_{k=1}^{r} \ell_{ik} \ell_{jk} x_k{}'^2 \right) \times \end{aligned}$$

$$\exp\left(-\frac{\gamma}{2}\sum_{i=1}^{r}\lambda_i x_i'^2\right) dx_1' dx_2' \ldots dx_r'$$

$$= (2\pi/\gamma)^{r/2} (\gamma |a|)^{-1} \sum_{k=1}^{r} (\ell_{ik}\ell_{jk})/\lambda_k \tag{A.71}$$

Thus we find that

$$\langle x_i x_j \rangle = J_2/J_0 = \gamma^{-1} \sum_{k=1}^{r} (\ell_{ik}\ell_{jk})/\lambda_k \tag{A.72}$$

This sum can be evaluated in the following way. We define a dyadic $\underset{\sim}{I}$, called the unit dyadic or the idem factor, by

$$\underset{\sim}{I} = \sum_{i=1}^{r} \underset{\sim}{e}_i \underset{\sim}{e}_i \tag{A.73}$$

and the reciprocal dyadic $\underset{\sim}{a}^{-1}$ by

$$\underset{\sim}{a}^{-1} \cdot \underset{\sim}{a} = \underset{\sim}{a} \cdot \underset{\sim}{a}^{-1} = \underset{\sim}{I} \tag{A.74}$$

The (i, j) element of $\underset{\sim}{a}^{-1}$ is given by

$$(\underset{\sim}{a}^{-1})_{ij} = \Delta_{ji}/|a| \tag{A.75}$$

provided that $|a| \neq 0$. Here Δ_{ji} is the cofactor of the element (a_{ji}) in the determinant $|a|$, and satisfies the relations

$$\sum_{i=1}^{r} a_{ij}\Delta_{ik} = \sum_{i=1}^{r} a_{ji}\Delta_{ki} = |a|\,\delta_{jk} \tag{A.76}$$

By using Eq. (A.58) we can express $(\underset{\sim}{a}^{-1})_{ij}$ as $\underset{\sim}{e}_i \cdot \underset{\sim}{a}^{-1} \cdot \underset{\sim}{e}_j$. Substituting Eq. (A.60) for $\underset{\sim}{e}_i$ and $\underset{\sim}{e}_j$ and applying Eq.(A.62), or $\underset{\sim}{a}^{-1} \cdot \underset{\sim}{e}_k' = \lambda_k^{-1}\underset{\sim}{e}_k'$, we can carry out the following operation:

$$\begin{aligned}(\underset{\sim}{a}^{-1})_{ij} &= (\Sigma_h \ell_{ih}\underset{\sim}{e}_h') \cdot \underset{\sim}{a}^{-1} \cdot (\Sigma_k \ell_{jk}\underset{\sim}{e}_k') \\ &= \Sigma_k [\Sigma_h \ell_{ih}\underset{\sim}{e}_h' \cdot (\ell_{jk}/\lambda_k)\underset{\sim}{e}_k'] = \Sigma_k (\ell_{ik}\ell_{jk})/\lambda_k\end{aligned} \tag{A.77}$$

From Eqs. (A.75) and (A.77) we get the desired expression for the sum considered, and $\langle x_i x_j \rangle$ becomes

$$\langle x_i x_j \rangle = \gamma^{-1}\Delta_{ji}/|a| \tag{A.78}$$

Finally, we touch upon matrices. Let $\underset{\sim}{a}$ be a square matrix of rank r

$$\underset{\sim}{a} = \begin{bmatrix} a_{11} & a_{12} & \dots & a_{1r} \\ a_{21} & a_{22} & \dots & a_{2r} \\ \dots & \dots & \dots & \dots \\ a_{r1} & a_{r2} & \dots & a_{rr} \end{bmatrix} \tag{A.79}$$

and $\underset{\sim}{x}$ be a column vector

$$\underset{\sim}{x} = \begin{bmatrix} x_1 \\ x_2 \\ \cdot\cdot \\ \cdot\cdot \\ \cdot\cdot \\ x_r \end{bmatrix} \tag{A.80}$$

Then the quadratic for Ψ defined by Eq. (A.67) can be represented by

$$\Psi = \underset{\sim}{x}^t \underset{\sim}{a} \underset{\sim}{x} \tag{A.81}$$

where $\underset{\sim}{x}^t$ denotes the transpose of $\underset{\sim}{x}$, i.e., a row vector $\underset{\sim}{x}^t = [x_1 \; x_2 \; \dots x_r]$. The basic rules of matrix algebra are

$$\begin{aligned} (\underset{\sim}{a} + \underset{\sim}{b})_{ij} &= a_{ij} + b_{ij} \\ (c\underset{\sim}{a})_{ij} &= c\underset{\sim}{a}_{ij} \\ (\underset{\sim}{a}\underset{\sim}{b})_{ij} &= \Sigma_k \, a_{ik} \, b_{kj} \end{aligned} \tag{A.82}$$

The inverse of matrix $\underset{\sim}{a}$ is written $\underset{\sim}{a}^{-1}$ and defined by

$$\underset{\sim}{a}^{-1} \underset{\sim}{a} = \underset{\sim}{a} \underset{\sim}{a}^{-1} = \underset{\sim}{E} \tag{A.83}$$

where $\underset{\sim}{E}$ denotes a unit matrix as in Eq. (1.3.54). The (i, j) element of $\underset{\sim}{a}^{-1}$ is represented by Eq. (A.75). For further details concerning dyadics, tensors, and matrices the reader is referred to appropiate textbooks.

A.5. Electric Field of Scattered Light

In this appendix, Eq. (3.2.5) with $\underset{\sim}{p}^o$ given by Eq. (3.2.52) for the electric field of scattered light is derived by the treatment of Landau and Lifshitz [1].

We consider a nonconducting, nonmagnetic medium, and denote the electric field, the magnetic field, and the electric flux density of light scattered by it by $\underset{\sim}{E}$, $\underset{\sim}{H}$, and $\underset{\sim}{D}$, respectively. These quantities are governed by the Maxwell equations

$$\underset{\sim}{\nabla} \times \underset{\sim}{E} = -\mu_0 (\partial \underset{\sim}{H}/\partial t) \tag{A.84}$$

$$\underset{\sim}{\nabla} \times \underset{\sim}{H} = \partial \underset{\sim}{D}/\partial t \tag{A.85}$$

$$\underset{\sim}{\nabla} \cdot \underset{\sim}{H} = 0 \tag{A.86}$$

and

$$\underset{\sim}{\nabla} \cdot \underset{\sim}{D} = 0 \tag{A.87}$$

where μ_0 is the magnetic permeability of the medium.

If the electric field and electric flux density of the incident light are denoted by $\underset{\sim}{E}_0$ and $\underset{\sim}{D}_0$, the material equation which relates the total electric flux density $\underset{\sim}{D}_0 + \underset{\sim}{D}$ to the total electric field $\underset{\sim}{E}_0 + \underset{\sim}{E}$ may be written as

$$\underset{\sim}{D}_0 + \underset{\sim}{D} = (\epsilon \underset{\sim}{I} + \Delta \underset{\sim}{\epsilon}) \cdot (\underset{\sim}{E}_0 + \underset{\sim}{E}) = \epsilon(\underset{\sim}{E}_0 + \underset{\sim}{E}) + \Delta \underset{\sim}{\epsilon} \cdot \underset{\sim}{E}_0 \tag{A.88}$$

Here ϵ is the average permittivity of the medium, $\Delta \underset{\sim}{\epsilon}$ is the fluctuation of local permittivity, $\underset{\sim}{I}$ is the unit dyadic, and the single dot has the meaning given by Eq. (A.52). Note that as finally expressed in Eq. (A.88), the term $\Delta \underset{\sim}{\epsilon} \cdot \underset{\sim}{E}$ has been dropped, since $\underset{\sim}{E}$ is much smaller in amplitude than $\underset{\sim}{E}_0$. With the relation $\underset{\sim}{D}_0 = \epsilon \underset{\sim}{E}_0$, Eq. (A.88) reduces to

$$\underset{\sim}{D} = \epsilon \underset{\sim}{E} + \Delta \underset{\sim}{\epsilon} \cdot \underset{\sim}{E}_0 \tag{A.89}$$

Elimination of $\underset{\sim}{H}$ from Eqs. (A.84) and (A.85) yields

$$\underset{\sim}{\nabla} \times \underset{\sim}{\nabla} \times \underset{\sim}{E} = -\mu_0 (\partial^2 \underset{\sim}{D}/\partial t^2) \tag{A.90}$$

If this is combined with Eq. (A.89), and Eqs. (A.40) and (A.87) are used, a wave equation for $\underset{\sim}{D}$ is obtained:

$$\nabla^2 \underset{\sim}{D} - c^{-2} (\partial^2 \underset{\sim}{D}/\partial t^2) = -\underset{\sim}{\nabla} \times \underset{\sim}{\nabla} \times (\Delta \underset{\sim}{\epsilon} \cdot \underset{\sim}{E}_0) \tag{A.91}$$

where

$$c^2 = (\epsilon \mu_0)^{-1} \tag{A.92}$$

The quantity c represents the velocity of the wave propagating in the medium.

Equation (A.91) may be solved by introducing the Hertz vector potential $\underset{\sim}{\Pi}$ defined as

$$\underset{\sim}{D} = \underset{\sim}{\nabla} \times \underset{\sim}{\nabla} \times \underset{\sim}{\Pi} \tag{A.93}$$

In terms of $\underset{\sim}{\Pi}$, Eq. (A.91) can be written

$$\nabla^2 \underset{\sim}{\Pi} - c^{-2} (\partial^2 \underset{\sim}{\Pi}/\partial t^2) = -(\Delta \underset{\sim}{\epsilon} \cdot \underset{\sim}{E}_0) \tag{A.94}$$

We now consider the geometrical situation in Fig. A.1 and suppose that the light scattered from a small region of volume V in a solution is observed at point P. The Hertz vector potential $\underset{\sim}{\Pi}$ at P is the sum of contributions $\underset{\sim}{\Pi}^{\alpha}$ due to light scattered from volume element $\Delta\nu^{\alpha}$ at a position $\underset{\sim}{r} = \underset{\sim}{r}^{\alpha}$ inside V. Assuming that $\Delta\underset{\sim}{\epsilon} \cdot \underset{\sim}{E}_0$ is uniform within each volume element and assigning its sum over the α-th volume element, $(\Delta\underset{\sim}{\epsilon} \cdot \underset{\sim}{E}_0)\,\Delta\nu^{\alpha}$, to point $\underset{\sim}{r}^{\alpha}$, we can derive from Eq. (A.94)

$$\nabla^2 \underset{\sim}{\Pi}^{\alpha} - c^{-2}\,(\partial^2 \underset{\sim}{\Pi}^{\alpha}/\partial t^2) = -\,\underset{\sim}{p}^{\alpha}(t)\,\delta\,(\underset{\sim}{r} - \underset{\sim}{r}^{\alpha}) \tag{A.95}$$

with

$$\underset{\sim}{p}^{\alpha}(t) \equiv [\Delta\underset{\sim}{\epsilon}\,(\underset{\sim}{r}^{\alpha}, t) \cdot \underset{\sim}{E}_0\,(\underset{\sim}{r}^{\alpha}, t)]\,\Delta\nu^{\alpha} \tag{A.96}$$

where $\delta\,(\underset{\sim}{r} - \underset{\sim}{r}^{\alpha})$ is the three-dimensional delta function. The formal solution to Eq. (A.95) for a point source is given by

$$\underset{\sim}{\Pi}^{\alpha}(\underset{\sim}{r}, t) = \underset{\sim}{p}^{\alpha}(t')/4\pi\,|\underset{\sim}{r} - \underset{\sim}{r}^{\alpha}| \tag{A.97}$$

where

$$t' = t - c^{-1}\,|\underset{\sim}{r} - \underset{\sim}{r}^{\alpha}| \tag{A.98}$$

These equations show that the state of the α-th volume element at time t' represented by $\underset{\sim}{p}^{\alpha}(t')$ affects the Hertz vector potential at a position $\underset{\sim}{r}$ after a time interval $\Delta t \equiv t - t' = c^{-1}\,|\underset{\sim}{r} - \underset{\sim}{r}^{\alpha}|$. For this reason, Δt is called the retardation time.

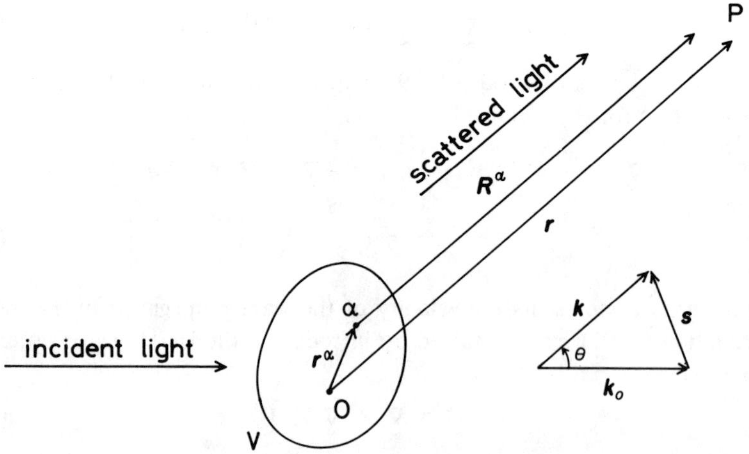

Fig. A.1. Relation between incident and scattered light. α, scatterer; P, point of observation (detector); $\underset{\sim}{k}_0$ and $\underset{\sim}{k}$, wave vectors for incident and scattered beams of light; $\underset{\sim}{s}$, scattering vector.

Summing $\underset{\sim}{\Pi}^{\alpha}$ over all volume elements in V and going to the continuous limit, we obtain

$$\underset{\sim}{\Pi}(\underset{\sim}{r}, t) = \frac{1}{4\pi} \int_V \frac{\Delta\underset{\sim}{\epsilon}(\underset{\sim}{r}^{\alpha}, t') \cdot \underset{\sim}{E}_0(\underset{\sim}{r}^{\alpha}, t')}{|\underset{\sim}{r} - \underset{\sim}{r}^{\alpha}|} d\underset{\sim}{r}^{\alpha} \tag{A.99}$$

If Eqs. (A.89) and (A.93) are equated with $\underset{\sim}{E}_0 = 0$ at point P and Eq. (A.99) is substituted, it is seen that

$$\underset{\sim}{E}(\underset{\sim}{r}, t) = \frac{1}{4\pi\epsilon} \underset{\sim}{\nabla} \times \underset{\sim}{\nabla} \times \left[\int_V \frac{\Delta\underset{\sim}{\epsilon}(\underset{\sim}{r}^{\alpha}, t') \cdot \underset{\sim}{E}_0(\underset{\sim}{r}^{\alpha}, t')}{|\underset{\sim}{r} - \underset{\sim}{r}^{\alpha}|} d\underset{\sim}{r}^{\alpha} \right] \tag{A.100}$$

We denote the unit vectors in the directions of the incident light and the scattered light by $\underset{\sim}{e}_0$ and $\underset{\sim}{e}$, respectively; and define the wave vectors $\underset{\sim}{k}_0$ and $\underset{\sim}{k}$ by

$$\underset{\sim}{k}_0 \equiv (2\pi/\lambda)\underset{\sim}{e}_0, \quad \underset{\sim}{k} \equiv (2\pi/\lambda)\underset{\sim}{e} \tag{A.101}$$

and the scattering vector $\underset{\sim}{s}$ by

$$\underset{\sim}{s} \equiv \underset{\sim}{k} - \underset{\sim}{k}_0 \quad [\text{hence } s \equiv |\underset{\sim}{s}| = (4\pi/\lambda)\sin(\theta/2)] \tag{A.102}$$

where λ is the wavelength of the incident light in the solution and θ is the scattering angle (see Fig. A.1). Since the point of observation P is taken far from the scattering volume, the following approximation is valid:

$$|\underset{\sim}{r} - \underset{\sim}{r}^{\alpha}| = r - \underset{\sim}{r}^{\alpha} \cdot \underset{\sim}{e} = r - (\lambda/2\pi)\underset{\sim}{r}^{\alpha} \cdot \underset{\sim}{k} \tag{A.103}$$

Hence

$$t' = t - c^{-1} r + \omega^{-1} \underset{\sim}{r}^{\alpha} \cdot \underset{\sim}{k} \tag{A.104}$$

where ω is the angular frequency defined by

$$\omega \equiv 2\pi c/\lambda = kc \tag{A.105}$$

Then the incident electric field $\underset{\sim}{E}_0(\underset{\sim}{r}^{\alpha}, t')$ in Eq. (A.100) can be expressed as

$$\begin{aligned} \underset{\sim}{E}_0(\underset{\sim}{r}^{\alpha}, t') &= \underset{\sim}{E}_0{}^{o} \exp(i\omega t' - i\underset{\sim}{k}_0 \cdot \underset{\sim}{r}^{\alpha}) \\ &= \underset{\sim}{E}_0{}^{o} \exp(i\underset{\sim}{s} \cdot \underset{\sim}{r}^{\alpha}) \exp(i\omega t - ikr) \end{aligned} \tag{A.106}$$

where $i = (-1)^{1/2}$ and $\underset{\sim}{E}_0{}^{o}$ is the amplitude of $\underset{\sim}{E}_0$.

Substituting Eq. (A.106) into Eq. (A.100) and neglecting the time dependence of $\Delta\underset{\sim}{\epsilon}$ (this dependence leads to a frequency spectrum of scattered light, a matter outside the scope of the present treatise), we obtain

$$\underset{\sim}{E}(\underset{\sim}{r}, t) = \frac{1}{4\pi\epsilon} \underset{\sim}{\nabla} \times \underset{\sim}{\nabla} \times \left[\exp(i\omega t - ikr) \times \int_V \frac{\Delta \underset{\sim}{\epsilon}^{\alpha} \cdot \underset{\sim}{E}_0{}^{o}}{|\underset{\sim}{r} - \underset{\sim}{r}^{\alpha}|} \exp(i\underset{\sim}{s} \cdot \underset{\sim}{r}^{\alpha}) \cdot d\underset{\sim}{r}^{\alpha} \right] \tag{A.107}$$

It should be noted here that the gradient operator $\underset{\sim}{\nabla}$ acts on both exp (ikr) and $|\underset{\sim}{r} - \underset{\sim}{r}^{\alpha}|^{-1}$. Performing the differentiation, expanding the result in powers of r^{-1}, and neglecting terms higher than order r^{-1}, we arrive at the desired expression for the electric field of the scattered light

$$\underset{\sim}{E}(\underset{\sim}{r}, t) = -\frac{1}{4\pi\epsilon_0} \left(\frac{2\pi}{\lambda_0}\right)^2 \exp(i\omega t - ikr) \frac{\underset{\sim}{r} \times (\underset{\sim}{r} \times \underset{\sim}{p}^{o})}{r^3} \tag{A.108}$$

where

$$\underset{\sim}{p}^{o} \equiv \int_V (\Delta \underset{\sim}{\epsilon}^{\alpha} \cdot \underset{\sim}{E}_0{}^{o}) \exp(i\underset{\sim}{s} \cdot \underset{\sim}{r}^{\alpha})\, d\underset{\sim}{r}^{\alpha} \tag{A.109}$$

ϵ_0 is the permittivity of vacuum, and λ_0 is the wavelength of the incident light in vacuum. We note that in deriving Eq. (A.108) we have used the relations

$$\lambda_0/\lambda = c_0/c = (\epsilon/\epsilon_0)^{1/2} = \tilde{n} \tag{A.110}$$

which are derivable from Eqs. (A.92) and (A.105). Here c_0 denotes the velocity of light in vacuum and $\tilde{n}$ is the refractive index of the dielectric medium under consideration.

It often becomes necessary to treat a discrete distribution of point scatterers, in which the source function $\Delta \underset{\sim}{\epsilon} \cdot \underset{\sim}{E}_0$ cannot be regarded as a continuous function of position. An ideal gas is an example. If the amplitude of the oscillating dipole moment of a scatterer i is denoted by $\underset{\sim}{p}_i{}^{o}$ and its instantaneous position by $\underset{\sim}{r}_i$, the source function $\Delta \underset{\sim}{\epsilon}^{\alpha} \cdot \underset{\sim}{E}_0{}^{o}$ in Eq. (A.109) may be expressed by

$$\Delta \underset{\sim}{\epsilon}^{\alpha} \cdot \underset{\sim}{E}_0{}^{o} = \sum_{i=1}^{N_t} \underset{\sim}{p}_i{}^{o}\, \delta(\underset{\sim}{r}_i - \underset{\sim}{r}^{\alpha}) \tag{A.111}$$

where the summation extends over all N_t scatterers contained in the medium. Substitution of Eq. (A.111) into Eq. (A.109) gives

$$\underset{\sim}{p}^{o} = \sum_{i=1}^{N_t} \underset{\sim}{p}_i{}^{o} \exp(i\underset{\sim}{s} \cdot \underset{\sim}{r}_i)\, m(\underset{\sim}{r}_i) \tag{A.112}$$

where $m(\underset{\sim}{r}_i)$ is the step function defined by Eq. (3.1.84), which is unity if the scatterer i is inside the scattering volume V and zero otherwise. We may express $\underset{\sim}{p}_i^o$, the amplitude of $\underset{\sim}{p}_i$, in the form

$$\underset{\sim}{p}_i^o = \underset{\sim}{\alpha}_i \cdot \underset{\sim}{E}_0^o \tag{A.113}$$

For an ideal gas the quantity $\underset{\sim}{\alpha}_i$ can be identified with the polarizability tensor of the scatterer i. However, for condensed media, this quantity is only an effective excess polarizability, and its molecular theoretical interpretation may be an exceedingly difficult task.

Reference

1. L. D. Landau and E. M. Lifshitz, "Electrodynamics of Continuous Media," Addison-Wesley, Reading, MA, 1960.

Index

Index